D0917413

THE ALKALOIDS

Chemistry and Physiology

Volume XX

THE ALKALOIDS

Chemistry and Physiology

Founding Editor

R. H. F. Manske

Edited by

R. G. A. Rodrigo

Department of Chemistry

Wilfrid Laurier University

Waterloo, Ontario, Canada

VOLUME XX

1981

ACADEMIC PRESS

A Subsidiary of Harcourt Brace Jovanovich, Publishers

NEW YORK • LONDON • TORONTO • SYDNEY • SAN FRANCISCO

ACADEMIC PRESS, INC.
111 Fifth Avenue, New York, New York 10003

United Kingdom Edition published by
ACADEMIC PRESS, INC. (LONDON) LTD.
24/28 Oval Road, London NW1 7DX

Library of Congress Cataloging in Publication Data

Manske, Richard Helmuth Fred, Date.
The alkaloids; chemistry and physiology.

Vols. 8-16 edited by R. H. F. Manske; vols. 17- edited by R. H. F. Manske, R. G. A. Rodrigo.
1. Alkaloids. 2. Alkaloids--Physiological effect.
I. Holmes, Henry Lavergne, joint author. II. Title.
QD421.M3 547.7'2 50-5522
ISBN 0-12-469520-5 (v. 20) AACR1

PRINTED IN THE UNITED STATES OF AMERICA

81 82 83 84 9 8 7 6 5 4 3 2 1

CONTENTS

Chapter 1. Bisindole Alkaloids

Geoffrey A. Cordell and J. Edwin Saxton

Chapter 2. The Eburnamine–Vincamine Alkaloids

Werner Döpke

LIST OF CONTRIBUTORS

Numbers in parentheses indicate the pages on which the authors' contributions begin.

GEOFFREY A. CORDELL (1), College of Pharmacy, University of Illinois, Chicago, Illinois 60612

WERNER DÖPKE (297), Sektion Chemie, Humboldt Universität, Berlin, German Democratic Republic

J. EDWIN SAXTON (1), Department of Organic Chemistry, The University, Leeds LS29JT, England

PREFACE

This volume is largely devoted to a review of the bisindole type and includes all alkaloids containing two tryptophan-derived nuclei. Various aspects of the chemistry of such indole alkaloids have been covered in several earlier volumes up to and including Volume XI of this series. All structural types are included in the present review and the remarkable advances of recent years in our knowledge of these very complex compounds are discussed and evaluated. A smaller chapter reviews the eburnamine–vincamine group of indole alkaloids which has not received any attention in this series since Volume XI.

The editor wishes to thank the authors for their cooperation.

R. G. A. Rodrigo

CONTENTS OF PREVIOUS VOLUMES

Contents of Volume I

Contents of Volume II

Contents of Volume III

Contents of Volume IV

Contents of Volume V

Contents of Volume VI

Contents of Volume VII

Contents of Volume VIII

Contents of Volume IX

Contents of Volume X

Contents of Volume XI

Contents of Volume XII

Contents of Volume XIII

Contents of Volume XIV

Contents of Volume XV

Contents of Volume XVI

Contents of Volume XVII

Contents of Volume XVIII

Contents of Volume XIX

——CHAPTER 1——

BISINDOLE ALKALOIDS

GEOFFREY A. CORDELL

College of Pharmacy,
University of Illinois,
Chicago, Illinois

AND

J. EDWIN SAXTON

Department of Chemistry
The University,
Leeds, England

THE ALKALOIDS, VOL. XX

ISBN 0-12-469520-5

I. Introduction

In previous volumes of this series there has been some discussion of various bisindole alkaloids (*1–12*). However, this coverage was somewhat sporadic and diffuse because it was part of the broader discussion of alkaloids of a particular genus or of a limited structure type. The remarkable progress made in this area during the past few years indicated that a more unified approach to these alkaloids would be appropriate. Gorman *et al.* have used a similar system in a previous review of this field (*13*).

Fundamentally, this chapter deals with those alkaloids which contain two nuclei derived from tryptophan, and is organized approximately along the lines of a progressing biosynthetic pathway. Because of the divergence from previous organizational alignments, it appears necessary to mention the prior notations of some of the alkaloids and also to indicate the volumes in this series in which certain of these alkaloids have been described earlier.

At the outset it should be said, as Gorman *et al.* have pointed out, that very few of these alkaloids are "dimeric" in the strict sense of the word, i.e., comprised of two identical units. Rather, it is more common that the two (or more) units differ; indeed, these differences are usually not minor but typically involve quite dissimilar skeleta. The problems associated with the structure elucidation of natural products of this complexity are manifold, and this area was almost unexplored before the structures of vincaleukoblastine (*14*) and voacamine (*15*) were solved. Since then, many new bisindole alkaloids of divergent structure have been isolated, and their characterization has been greatly facilitated by the critical application of spectroscopic techniques. Particularly in the recent past, ^{13}C-NMR spectroscopy has contributed substantially to the solution of structural problems in this area.

In the organization of this chapter the simplest alkaloids derived from tryptophan are considered first. These are followed by a group formed from a monoterpenoid indole alkaloid and tryptophan, and finally by the groups formed by the union of two monoterpenoid indole alkaloid types arranged biogenetically. Besides annual reviews of this field (*13*, *16–21*), there have been several other reviews of various of the alkaloid groups discussed here (*15*, *22–30*).

II. Tryptamine–Tryptamine Type

A. STAUROSPORINE (1)

Staurosporine, $[\alpha]_D$ +35.0°, was obtained from the fungus *Streptomyces staurosporeus* (*31*), and preliminary data indicated that the alkaloid probably contained two indole nuclei. Structure **1** was deduced by X-ray crystallographic analysis (*32*) of the methanol solvate which crystallized in the monoclinic space group C2. The final *R* value from analysis of 2352 independent reflections was 4.7%. No biosynthetic studies have been conducted, but one can imagine a derivation from indole-3-acetic acid and tryptamine. The oxidative coupling of two C-2 carbons on indole nuclei is a unique feature of the structure. The alkaloid showed hypotensive activity and antimicrobial activity against both fungi and yeasts (*31*).

With the structure defined, a brief discussion of the spectral data is warranted. Staurosporine displayed principal absorption maxima at 243

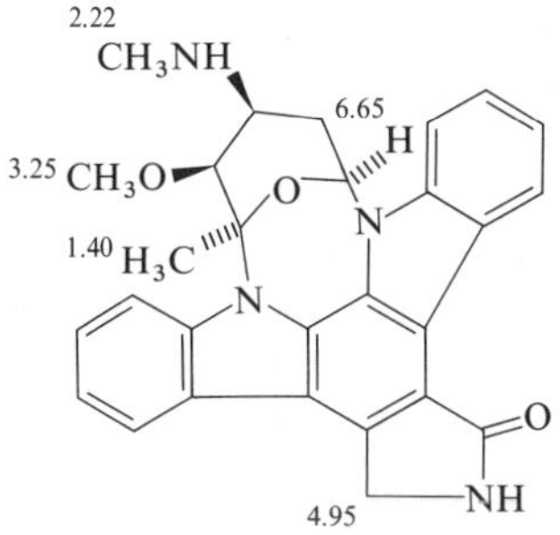

NH at 9.25
and 9.30 ppm

1 Staurosporine

and 292 nm, but the chromophore extended to 322, 335, 356, and 372 nm. A carbonyl band in the IR spectrum at 1675 cm^{-1} indicative of an α,β-unsaturated ketone intimates that there is little orbital overlap between the nitrogen lone pair and the carbonyl group. Some of the principal ^{1}H-NMR assignments are shown on structure **1** (*31*).

B. Trichotomine and Related Alkaloids

There are very few *blue* alkaloids. The fruits of *Clerodendron trichotomum* Thumb. (Verbenaceae) are blue, and in 1946 (*33*) and 1950 (*34*), efforts were made to isolate the pigment responsible for this coloration. These efforts were unsuccessful, and it was not until 1974 that the pigment was obtained in pure form by Hirata and co-workers (*35, 36*).

Acetone extracts of the fresh blue fruits of *C. trichotomum* were chromatographed on Sephadex LH-20 to afford trichotomine in $1.5 \times 10^{-3}\%$ yield. No melting point was observed below 300°, and the analytical data suggested a molecular formula $C_{15}H_{10}N_2O_3$ or some multiple thereof. The IR spectrum indicated carboxylic acid absorption (1712 cm^{-1}) and undefined unsaturation at 1650 and 1600 cm^{-1}.

The UV spectrum was quite extraordinary, showing λ_{max} 242, 337, 353, 618, and 660 nm; the last two absorptions had ε values of 67,000 and 69,000, respectively. In the NMR spectrum eight aromatic protons were observed. From the nature of the coupling constants, the aromatic nuclei were deduced to be 1,2-disubstituted. A two-proton singlet was observed at 7.22 ppm and attributed to vinyl protons deshielded by location in the center of a conjugated system. Two equivalent exchangeable protons were observed at 4.05 ppm, and the remainder of the spectrum indicated a —CH(X)CH_2— system. The X group must be highly deshielding group, for the methine proton appeared as a doublet of doublets ($J = 2.0$, 7.0 Hz) at 5.20 ppm. The associated methylene protons were also observed as two doublets of doublets at 3.42 ($J = 7.0$, 17.5 Hz) and 3.78 ppm ($J = 2.0$, 17.5 Hz). The simplicity of the spectrum suggested that the molecule was symmetrical, and because the IR spectrum indicated the presence of a carboxylic acid, it was thought that this might be the deshielding group X.

Treatment of trichotomine with diazomethane gave a methyl ester which by osmometry was shown to have a molecular weight of 569 ± 10. The IR spectrum now showed a carbonyl absorption at 1745 cm^{-1} and the NMR spectrum a singlet at 3.64 ppm for a carbomethoxy group. The remainder of the spectrum was quite similar to that of the parent compound, except for the exchangeable protons which now appeared at 10.97 ppm.

The simplicity of the NMR spectra of trichotomine and its methyl ester together with the molecular weight determination suggested that trichotomine was a symmetrical, dimeric compound. In addition, the strongly deshielding group X must now be regarded as a carboxylic acid in trichotomine.

The exchangeable protons were removed on acetylation of trichotomine with Ac_2O–py overnight to afford a mixture of mono- and diacetylated compounds. From the position of the acetate singlet (2.70 ppm), the exchangeable proton must be an NH group.

The nature of the nucleus of trichotomine was deduced when catalytic hydrogenation (PtO_2 in EtOAc–MeOH) of trichotomine methyl ester gave a hexahydro derivative having a UV spectrum very similar to that of a 2,3-disubstituted indole. A five-membered lactam was observed in the IR spectrum at 1689 cm^{-1}. This group was not observed in the parent compound, but a band at 1672 cm^{-1} was no longer present. On this basis it was concluded that the lactam in trichotomine itself must be unsaturated in some way.

The NMR spectrum of the hexahydro derivative indicated the presence of the two exchangeable NH protons at 7.70 ppm, the four adjacent aromatic protons, a methylene–methine system, and the carbomethoxy groups. The six newly introduced protons were observed as a methine doublet of doublets ($J = 7.6$, 9.4 Hz) at 4.92 ppm, a methine doublet of doublets ($J = 7.0$, 7.6 Hz) at 3.33 ppm, and two protons of a methylene group as multiplets at 2.31 and 1.42 ppm. The interrelationships of these groups were confirmed by double-resonance studies.

At this point it was suggested that unit **2** was present in the hexahydroderivative. As only one of these protons is present in trichotomine itself, it follows that this unit is present as either unit **3** or unit **4** in the parent compound. Biogenetically, a tryptophan derivation appeared likely; therefore, the lactam nitrogen must be derived from the amine nitrogen of the amino acid. The symmetrical hexahydro derivative should therefore have structure **5**. It follows that trichotomine must be **6**, in which the stereochemical nature of the central double bond remained to be determined (*35*, *36*).

The close proximity of the vinyl proton to the indole nitrogen was confirmed when *N,N'*-dimethyltrichotomine methyl ester (**7**) gave an observable (15%) NOE effect between the N_a-methyl and the vinyl protons (*35, 36*)

The structure of trichotomine was confirmed by X-ray analysis (*37*) of *N,N'*-di-(*p*-bromobenzoyl)trichotomine dimethyl ester (**8**), mp 272°–3°, obtained as deep blue needles. This result indicated that the stereochemistry of the central ring junction was trans. More surprisingly, it was found that there was a dihedral angle of 38.6° between the planes of the aromatic nuclei (*37*). Trichotomine therefore has the structure and configuration shown in **9**.

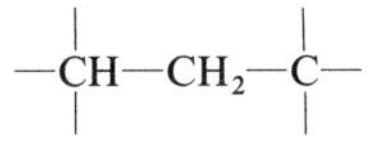

—CH—CH$_2$—C—

2 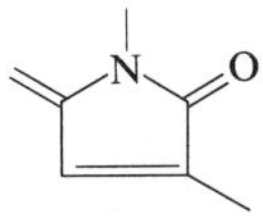 **3** 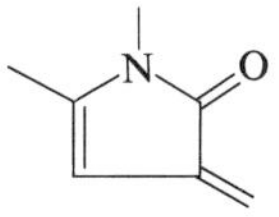 **4**

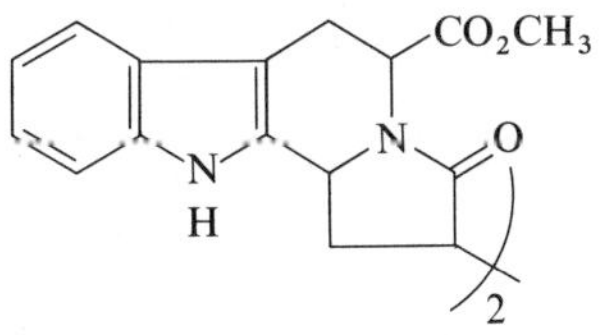

5

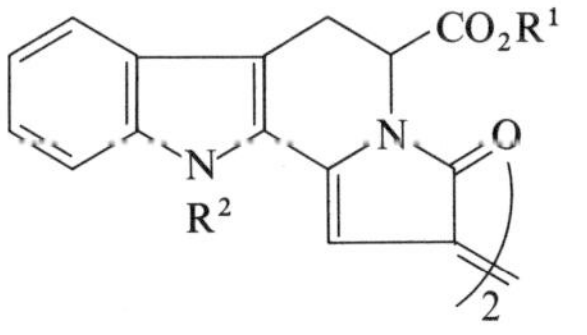

6 $R^1 = R^1 = H$
7 $R^1 = R^2 = CH_3$
8 $R^1 = CH_3$, $R^2 = COC_6H_4p$-Br

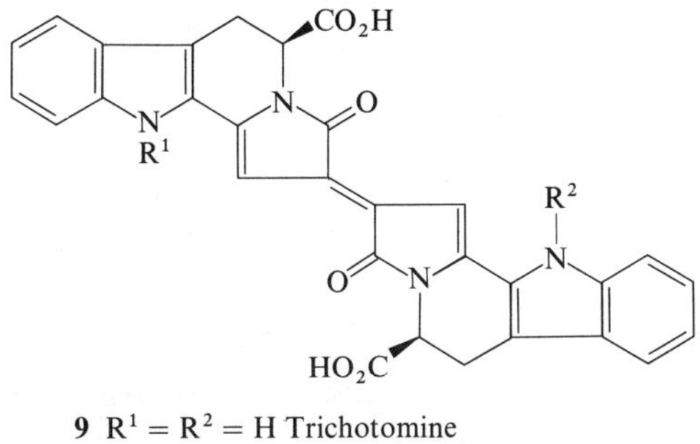

9 $R^1 = R^2 = H$ Trichotomine
17 $R^1 = \beta$-D-glucopyranosyl, $R^2 = H$ Trichotomine G_1
18 $R^1 = R^2 = \beta$-D-glucopyranosyl

The structure assigned to trichotomine has also been confirmed by synthesis (Scheme 1) (*38, 39*). A crucial intermediate is the *N*-tryptophanyl succinimide (**10**) prepared when L-tryptophanyl methyl ester (**11**) was condensed with succinic anhydride and the amide acid **12**, converted to the corresponding acid chloride with thionyl chloride, and cyclized at 200°–230°

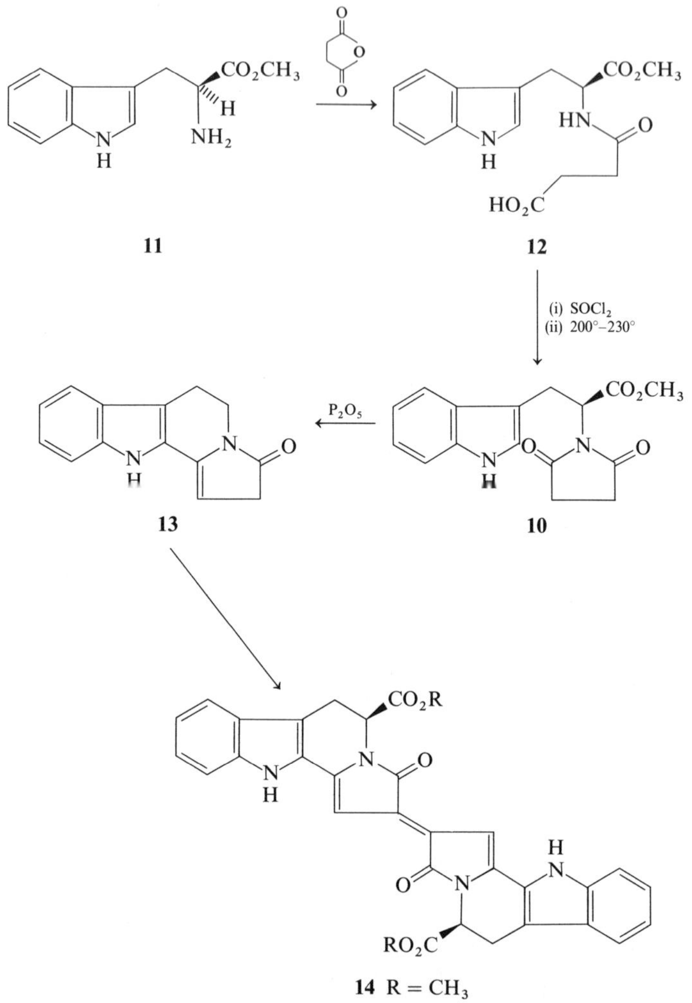

11

12

(i) $SOCl_2$
(ii) 200°–230°

P_2O_5

13

10

14 R = CH_3
9 R = H

SCHEME 1. *Iwadare and co-workers' synthesis of trichotomine* (**9**) (*38, 39*).

under reduced pressure to give **10**. Bischler–Napieralski cyclization of **10** with phosphorus pentoxide gave **13** in low yield. When this amide was heated at 80°–90° in air, trichotomine dimethyl ester (**14**) was produced, which was hydrolyzed with alkali (4% potassium hydroxide) to give trichotomine (**9**) (*38, 39*).

Kapadia and Rao (*40*) have reported a synthesis of trichotomine (**9**) which is much more convenient than that of Iwadare and co-workers (*38, 39*). The idea was to use a more biomimetic approach in which an α-keto acid is used as a starting material. Previous work had established that in other alkaloid series, such keto acids are not only biosynthetic intermediates (*41*, *42*) but can be used for *in vitro* synthesis (*43*).

When an aqueous solution of L-tryptophan (**15**) and α-ketoglutaric acid (**16**) in water was stirred at room temperature (rt) for 5 days, a blue–green

CO_2H H NH_2 N H + HO O O O OH (i) rt, 5 days (ii) CH_2N_2 → **14**

15 **16**

precipitate was produced. This was filtered, dissolved in methanol, and treated with diazomethane to give in low yield trichotomine dimethyl ester (**14**) (*40*), identical to a sample prepared by the procedure of Iwadare *et al.* (*38*, *39*).

Kapadia and Rao also reported (*40*) an improvement in the Iwadare procedure whereby the imide ester **10** was obtained by direct treatment of L-tryptophan methyl ester (**11**) with succinimide.

Trichotomine has shown bronochodilating, hypotensive, and sedative activities (*39*).

Besides trichotomine (**9**), two other blue pigments were obtained from the fruits of *C. trichotomum* (*36*): they are trichotomine G_1 and a mixture of anomers of *N,N′*-di(glucopyranosyl)trichotomine.

Trichotomine G_1, like trichotomine, had no melting point below 300°, was blue (λ_{max} 247, 336, 350, 617, and 658 nm), and showed very strong absorption in the IR spectrum for a hydroxyl group, as well as absorptions for a carboxylic acid residue (ν_{max} 1760 cm^{-1}).

The NMR spectrum of trichotomine G_1 taken in acetone–D_2O indicated a complex pattern of eight aromatic protons in the range 6.80–7.80 ppm, two methine protons as a singlet at 7.20 ppm, and an anomeric proton as a doublet ($J = 7.2$ Hz) at 5.70 ppm. Absorptions in the region 3.50–4.00 ppm also indicated the presence of one sugar unit. This was confirmed when trichotomine G_1 gave a tetraacetate derivative (mp 176°–178°, $C_{44}H_{38}N_4O_{15}$), three of the sugar methine protons of which had shifted downfield to the region 4.95–5.40 ppm, and which showed only a single exchangeable NH at 9.34 ppm.

Hydrolysis of trichotomine G_1 with methanolic HCl afforded trichotomine dimethyl ester (**14**) and D-glucose, and from the coupling constant of the anomeric proton a β-glycosidic bond must be present in the parent compound. Trichotomine G_1 is therefore *N*-β-D-glucopyranosyltrichotomine (**17**) (*36*).

From the NMR spectrum, a second polar pigment was also thought to contain a sugar unit, and indeed, hydrolysis with 3 *N* methanolic hydrochloric acid gave trichotomine dimethyl ester (**14**) and D-glucose. Acetylation gave an octaacetate derivative which in the region of the anomeric protons showed two signals, a doublet ($J = 3$ Hz) at 6.32 ppm integrating for two-thirds of a proton, and a doublet ($J = 7.2$ Hz) at 5.72 ppm integrating for one and one-third protons. This indicates that the isolate is a mixture of anomers of *N,N'*-di-β-(D-glucopyranosyl)trichotomine (**18**) (*36*). These compounds were the first *N*-D-glucopyranosylindole derivatives to be obtained from natural sources.

C. Calycanthine, Chimonanthine, and Related Alkaloids

There are two well-known dimeric indole alkaloids derived from a simple tryptamine unit, namely, calycanthine (**19**) and chimonanthine (**20**). The isolation and structure elucidation of these alkaloids have been discussed in previous volumes of this series (*1*, *6*).

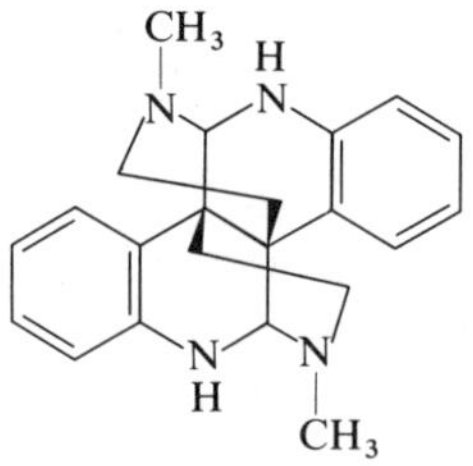

19 Calycanthine

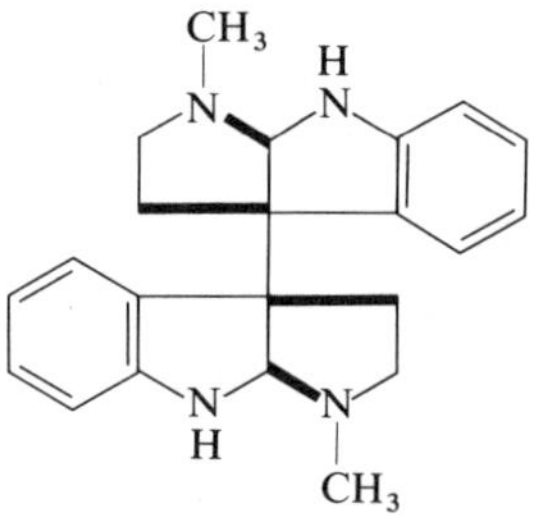

20 (−)-Chimonanthine

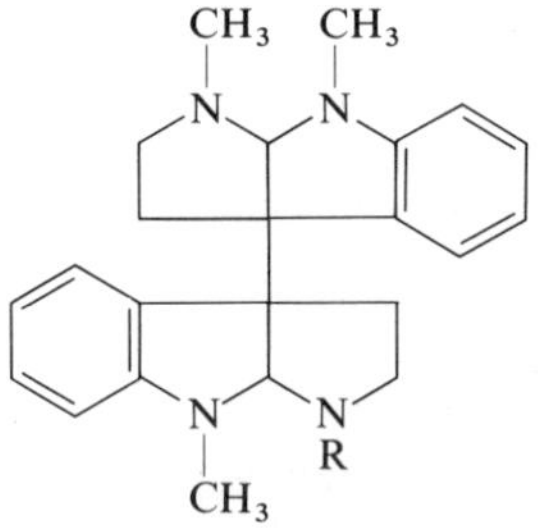

21 R = CH_3 Calycanthidine
22 R = H Folicanthine

Each of these compounds, together with the two other members of this group, calycanthidine (**21**) and folicanthine (**22**), has been isolated from plants in the family Calycanthaceae. This plant family is represented only by the genera *Calycanthus* and *Chimonanthus*.

Calycanthine has recently been isolated from two quite different plant families, the Celastraceae and the Rubiaceae. Thus, Culvenor and co-workers (*44*) reported the isolation of calycanthine from *Bhesa archboldiana* (Merr. Perry) Ding Hou, a plant of the Celastraceae, and Woo-Ming and Stuart (*45*) obtained calycanthine from the stems and leaves of *Palicourea alpina* (Sw.) DC of the Rubiaceae. In the latter case, not all samples contained calycanthine, and other indole alkaloids were also noted (*46*, *47*). On the other hand, in the former case co-occurrence was noted with the pyrrolizidine alkaloid 9-angeloylretronecine-*N*-oxide (**23**).

In the previous discussion of this alkaloid group (*6*) brief mention was made of the X-ray structure determination of chimonanthine. This work has now been published in full (*48*).

The dihydrobromide salt of chimonanthine was used for the analysis, the crystals belonging to the tetragonal space group $P4_12_12$ with eight molecules in a unit cell of dimensions $a = b = 13.95$, $c = 26.67$ Å. The final value of R was 14.9% over 2093 independent reflections. These results indicated structure **24** for chimonanthine itself, in which some of the key interatomic distances are also shown. In the C and C′ rings the average valence angle is

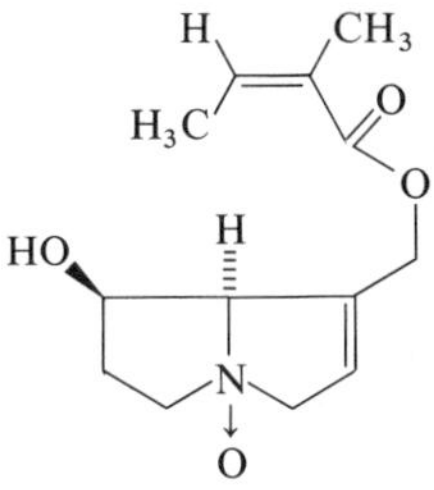

23

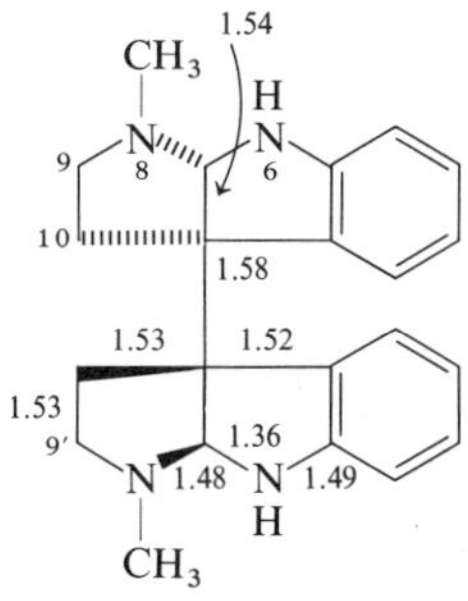

24 Chimonanthine (meso); interatomic distances in Å

only 104°, considerably less than for a true tetrahedral atom. These rings adopt an envelope conformation in which C-9 and C-9′ constitute the flaps of the envelope.

1. Chemistry

Although the degradation of both calycanthine and chimonanthine has been described in Volume VIII of this series, Takamizawa and co-workers (*49*) have reported some additional degradative experiments which are of some interest.

Treatment of calycanthine with benzoyl chloride, as reported by Manske (*50*), gave a neutral benzoylation product which failed to crystallize. Potassium permanganate oxidation gave *N*-benzoyl-*N*-methyltryptamine (**25**).

The Japanese workers suggested structure **26** for the intermediate prior to oxidation, although no data were presented to support this suggestion (*49*).

Reaction of calycanthine with Grignard reagent followed by acylation with methyl benzoate gave a product believed to be benzoylcalycanthine (**27**). Oxidation with potassium permanganate then gave calycanine (**28**) by elimination of the ethanamine bridges. Again, no evidence was presented for the structure of the intermediate prior to oxidation (*49*).

The Japanese workers also investigated the acid hydrolysis of calycanthine. Treatment with dilute HCl at 100° gave an oil formulated as **29** because it underwent diazo coupling. No spectral data were presented. Vacuum

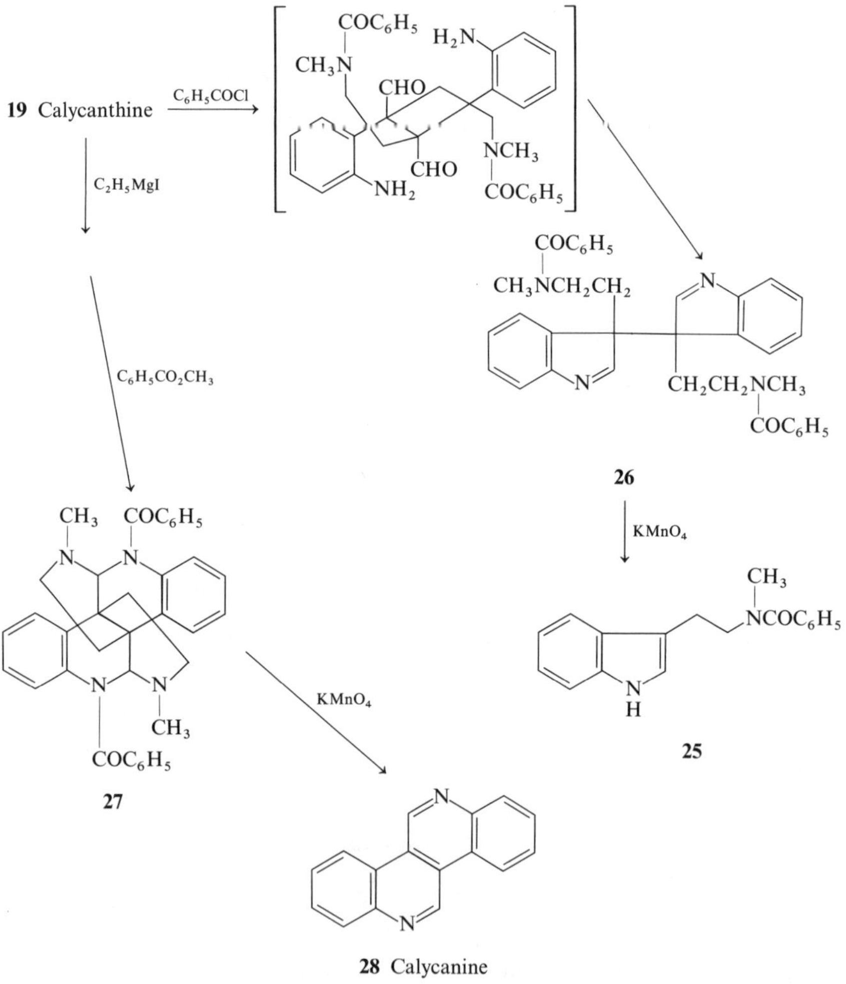

19 Calycanthine $\xrightarrow[100°, 72\ hr]{HCl}$ **29** $\xrightarrow[(ii)\ \Delta]{(i)\ C_6H_5COCl}$ **31**

19 Calycanthine $\xrightarrow{HNO_2,\ 5°}$ **32** $\rightarrow$ **33**

29 $\xrightarrow{\Delta}$ **30**

29 R = $—CH_2CH_2NHCH_3$

30

31

32

33 R = $—CH_2CH_2NHCH_3$

pyrolysis of the base gave *N*-methyltryptamine (**30**), and pyrolysis of the *N*-benzoyl derivative of **29** afforded **31**. The structure of the latter compound was based solely on the molecular formula and the IR spectrum. No yield was given (*49*).

Hydrolysis of calycanthine with either dilute HCl at 180° for 1 hr or dilute acetic acid at 120° for 70 hr also gave *N*-methyltryptamine (**30**).

In 1905, Gordin reported (*51*) that calycanthine afforded a dinitroso derivative. The Japanese group (*49*) found that treatment of **19** with nitrous acid at 5° gave a yellow precipitate, suggested on the basis of microanalysis to be the dinitrosamine **32**. The compound was soluble in 4 *N* HCl solution and after 18 hr had decomposed to a product proposed to be the bis-diazonium species **33**. No spectral data were presented which might give evidence for this structure (*49*).

This potentially interesting chemistry of calycanthine is unfortunately marred by lack of adequate data in support of any of the above complex structures and requires complete verification by modern spectroscopic techniques.

In 1966 Mason and Vane reported the CD spectrum of calycanthine in the range 28–45 kK (*52*). More recently, Mason and co-workers (*53*) examined the spectrum in the range 28–53 kK and found that, although the first two CD bands correspond closely with those calculated theoretically by the point-dipole exciton method, the third, higher-energy band, did not. Improved data, however, were obtained using the π-electron dipole-velocity procedure based on the known chirality of the dimeric π system.

2. Synthesis

The Hendrickson synthesis (*54*) of meso and racemic chimonanthine from the oxindole **34** by oxidation with iodine in THF has been discussed in a previous volume of this series (*6*).

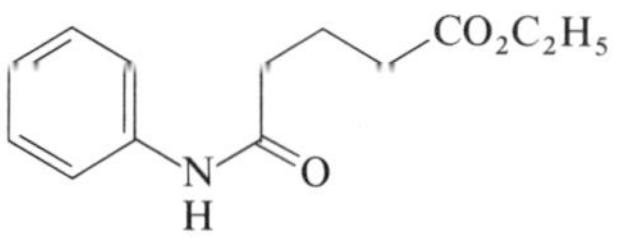

$CO_2C_2H_5$

N O

H

34

Subsequently, Scott and co-workers (*55*) described their synthetic approach to this problem. With the knowledge that indolylmagnesium halides are ionic and that ferric chloride can generate diphenyl from phenyl Grignard, a similar reaction was attempted with a tryptamine derivative.

Thus, the Grignard derivative of *N*-methyltryptamine (**30**) in ether was treated with anhydrous ferric chloride at room temperature after mild hydrolysis with ammonium chloride. Besides unreacted starting material, four products were obtained: racemic chimonanthine (**35**), *meso*-chimonanthine (**36**), *meso*-dehydrocalycanthine (**37**), and the meso compound **38**. The yields were 19%, 7%, 3.5%, and 3.5%, respectively.

The structure of *meso*-chimonanthine (**36**) was suggested on the basis of Hendrickson's work (*54*) and was fortuitously confirmed by an almost simultaneous isolation from an alkaloid fraction of *Calycanthus floridus*. This alkaloid, mp 198°–202°, $[\alpha]_D$ 0°, showed identical mass and UV spectral data to racemic chimonanthine. The main spectral difference was in the ^{1}H-NMR spectrum, where the *N*-methyl resonance appeared at 2.37 ppm vs. 2.30 ppm in chimonanthine (*55*). When the solvent used for the coupling reaction was changed from ether to THF, little dimeric material was produced, presumably because of better radical scavenging.

Heating racemic chimonanthine (**35**) in dilute acetic acid at 90° for 30 hours afforded racemic calycanthine in 70% yield. *meso*-Chimonanthine (**36**) afforded *meso*-calycanthine under similar conditions (*55*).

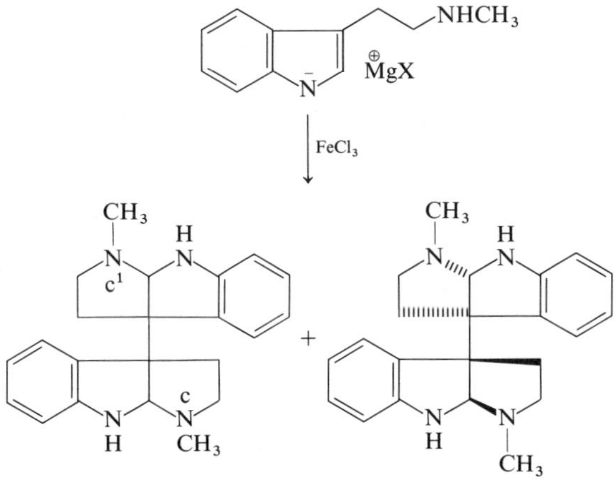

$FeCl_3$

35 (±)-Chimonanthine **36** *meso*-Chimonanthine

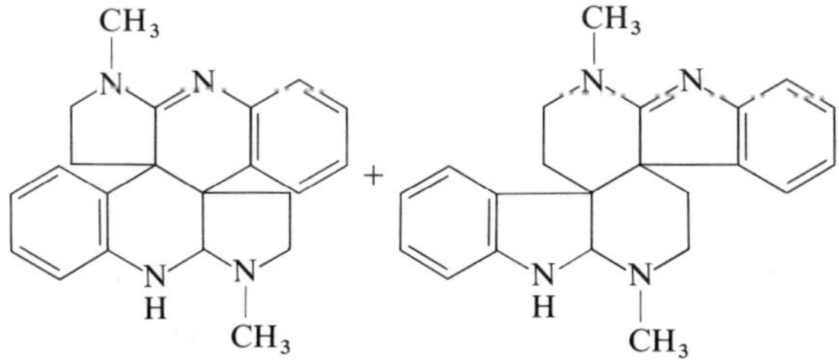

37 *meso*-Dehydrocalycanthine **38**

3. Biosynthesis

Two groups have discussed some preliminary experiments on the biosynthesis of the tryptamine dimers (*56*, *57*), and their results support the biogenetic speculation of Robinson and Teuber (*58*) that these compounds were derived by β,β'-oxidative dimerization of two tryptamine units.

In 1965 Schütte and Maier (*56*) described a series of tracer experiments using (±)-[2-^{14}C]tryptophan as a precursor of the alkaloids of *Calycanthus floridus*. Specific incorporation into calycanthine (**19**) was in the range 0.61–2.0%, but the location of the label was not determined. Several other alkaloids were also labeled but were not identified owing to lack of comparison samples.

These results were confirmed and extended by O'Donovan and Keogh (*57*), who showed that (±)-[2-^{14}C]tryptophan labeled folicanthine (**22**) to the extent of 0.57%. The location of the label was determined by acid cleavage of **22** to 1-methyl-3-(2′-methylaminoethyl)indole (**39**), which could

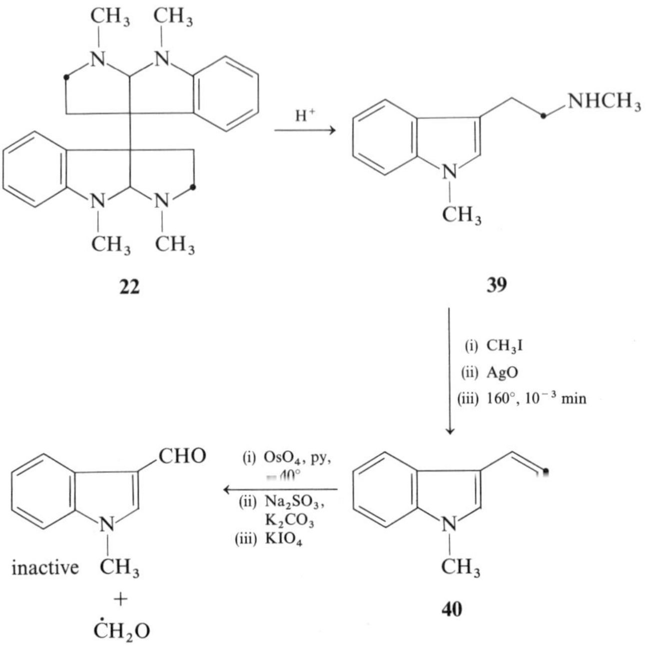

CH_3 CH_3 N N N N CH_3 CH_3 **22** $\xrightarrow{H^+}$ $NHCH_3$ N CH_3 **39**

(i) CH_3I
(ii) AgO
(iii) 160°, 10^{-3} min

CHO N inactive CH_3 + $\dot{C}H_2O$ $\xleftarrow{}$ (i) OsO_4, py, −40° (ii) Na_2SO_3, K_2CO_3 (iii) KIO_4 N CH_3 **40**

be degraded to 1-methyl-3-vinylindole (**40**), and the terminal methylene oxidatively removed as formaldehyde. The dimedone derivative of this formaldehyde contained all the activity.

No other precursor relationships have been established for these alkaloids.

D. The Chaetocins and Verticillins

Although they are not plant alkaloids, several fungal metabolites are known which bear a strong structural similarity to the chimonanthine-type compounds; these are the chaetocins and verticillins. Only the leading references for these compounds will be given and no detailed discussion of their structure elucidation presented.

In 1944 Waksman and Bugie obtained from cultures of *Chaetomium cochliodes* an antibacterial metabolite which they named chaetomin (*59*). They succeeded in extracting the main metabolite (*60*); and subsequently, Geiger (*61*) was able to show that the material had $[\alpha]_D$ +360° and a UV spectrum similar to that of 3-methylindole.

This problem remained dormant until 1972, when Safe and Taylor (*62*) reported the structure elucidation of chaetomin (**41**) obtained from *C. cochliodes* and *C. globosum*. Shortly prior to this, a group at Sandoz (*63*)

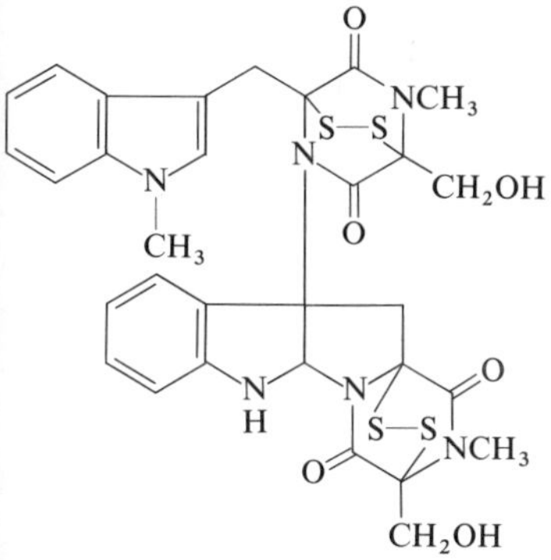

41 Chaetomin

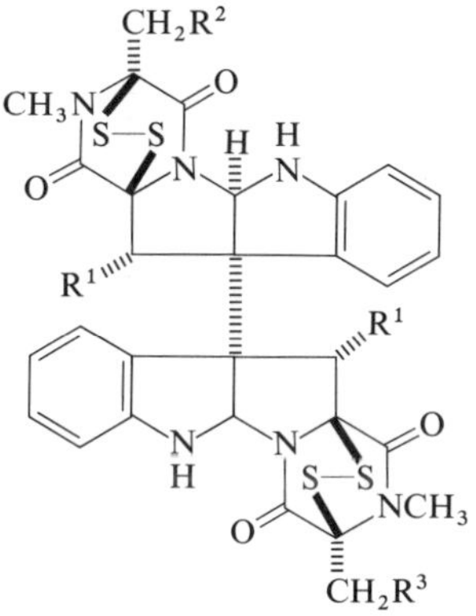

42 $R^1 = H, R^2 = R^3 = OH$ Chaetocin
43 $R^1 = R^2 = R^3 = OH$ 11α, 11′α-Dihydroxychaetocin
44 $R^1 = OH, R^2 = R^3 = H$ Verticillin A
45 $R^1 = R^3 = OH, R^2 = H$ Verticillin B

had described the structure elucidation of chaetocin (**42**), a metabolite of the fungus *C. minutum*. The same group (*64*) then obtained 11α,11′α-dihydroxychaetocin (**43**) from *Verticillium tenerum*.

At about this time, Japanese workers (*65*, *66*) isolated and determined the structure of three new antibiotics—verticillin A (**44**), verticillin B (**45**), and verticillin C—from a *Verticillium* species.

E. Hodgkinsine and a Related Isomer

Hodgkinsine, mp 128°, $[\alpha]_D$ +60°, an alkaloid of *Hodgkinsonia frutescens* F. Muell. (Rubiaceae), was first reported by Anet *et al.* in 1961 (*67*), and on the basis of a suggested molecular formula $C_{22}H_{26}N_4$, was regarded as being isomeric with calycanthine and chimonanthine (*67–69*). These speculations have been discussed elsewhere in this series (*6*).

Two groups (*70–72*) subsequently deduced that the previous molecular formula was incorrect and that hodgkinsine was actually a trimeric indole alkaloid of molecular formula $C_{33}H_{38}N_6$.

The Australian group (*70*, *71*), recognizing that a crystalline adduct was produced between hodgkinsine and benzene, prepared a similar adduct with bromobenzene. The bromine atom could not, however, be located in the crystal lattice, presumably because of lack of precise interstitial binding. However, work with the trimethiodide derivative did prove successful. Analysis of the data from 3198 reflections, after the reliability index had been reduced to 0.14, afforded structure **46** for hodgkinsine itself, in which

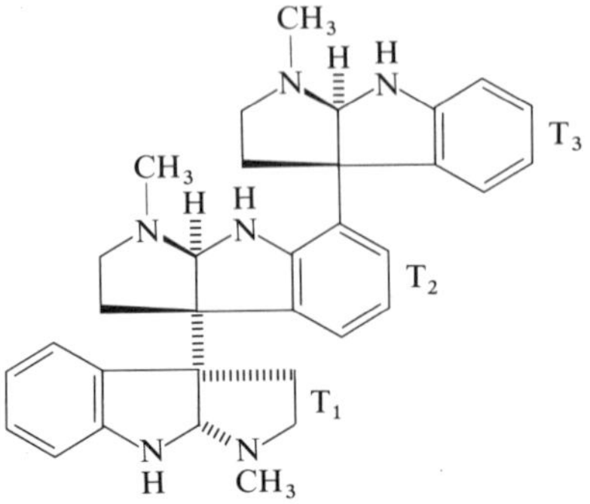

46 Hodgkinsine

each of the five-membered rings is cis-fused. The units T_1 and T_2 are of opposite configuration; therefore, their contributions to the CD spectrum of hodgkinsine are internally compensated. The only contributor to the CD spectrum is unit T_3 (*70, 71*).

The Manchester group established the structure of hodgkinsine by chemical degradation (*72*). On treatment with alkali, the trimethiodide of hodgkinsine afforded trisindolenine **47**, which could be reduced with sodium borohydride to give a tetrahydro derivative **48** in which the middle of the three units was not reduced. This compound fragmented to two products, *N,N*-dimethyltryptamine (**49**) and an indole–indoline **50** having a new type of bond in this series, one between an aliphatic carbon and the aromatic system. The nature of this linkage was determined by deuterium-labeling studies.

Thus in 8 *N* D_2SO_4–D_2O, **50** gave a hexadeuterio species, but the diindoline **51** only exchanged three aromatic hydrogens on acid-catalyzed deuteration. On boiling under strongly acidic conditions, *N,N*-dimethyltryptamine (**49**) and *N,N*-dimethyl-2,3-dihydrotryptamine (**52**) were produced.

It is well established that indolines only exchange hydrogen for deuterium at positions ortho and para to the nitrogen, whereas a 3-substituted indole will exchange all five remaining aromatic protons. Because the UV spectrum of **51** is that of a bisindoline in which the chromophores do not overlap, one of the positions either ortho or para to the newly introduced indoline in **51** must be substituted and be the point of attachment to position 3 of the second unit. Note that the formation of a trisindolenine indicates that carbons 3, 3′, and 3″ are all quaternary.

The two possible structures for the indole–indoline are, therefore, one in which C-3 and C-7′ are linked and a second one in which C-3 and C-5′ are joined. The possibility of a C-3–C-5′ linkage was eliminated by careful nitration of the diformyl derivative of **51**. Acid cleavage of the product

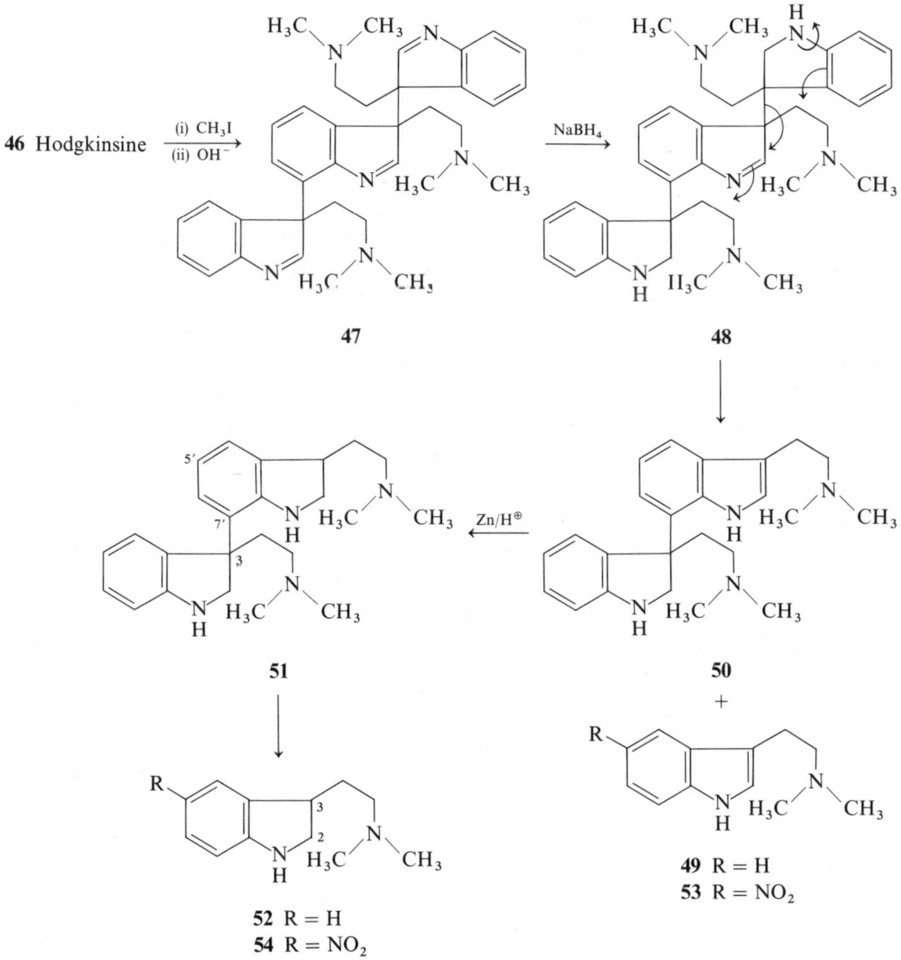

46 Hodgkinsine $\xrightarrow[\text{(ii) OH}^-]{\text{(i) CH}_3\text{I}}$ 47 $\xrightarrow{\text{NaBH}_4}$ 48 → 50 + 49 R = H; 53 R = NO_2

50 $\xrightarrow{\text{Zn/H}^\oplus}$ 51 → 52 R = H; 54 R = NO_2

afforded a mixture of 5-nitro-*N*,*N*-dimethyltryptamine (**53**) and 5-nitro-*N*,*N*-dimethyl-2,3-dihydrotryptamine (**54**). The corresponding 7-nitro-substituted derivatives of **49** or **52** were not detected. The linkage is, therefore, between positions 3 and 7′. On this basis the tris–indolenine derivative was assigned structure **48** and hodgkinsine structure **46** (without stereochemistry) (*72*).

A minor base obtained by an Australian group from *Psychotria beccarioides* Wernh. (Rubiaceae) (*73*) displayed similar IR and mass spectra to hodgkinsine (**46**), but had an optical rotation $\{[\alpha]_D +24^\circ\}$ opposite to but almost half that of hodgkinsine. Some differences were noted in the *N*-methyl group aromatic regions of the NMR spectrum of the new base and hodgkinsine. No CD spectrum was reported for this base, although one might have been informative. On the evidence presented, it would appear to be a mixture of stereoisomers.

F. The Quadrigemines

The leaves of *Hodgkinsonia frutescens*, in addition to yielding hodgkinsine (**46**), has also afforded two minor tetrameric amorphous alkaloids, quadrigemine A (**55**) and quadrigemine B (**56**) (*74*).

The existence of such tetrameric bases was indicated initially from the mass spectra of fractions obtained after the removal of hodgkinsine. A combination of extensive countercurrent distribution, followed by preparative thin-layer chromatography (TLC), afforded quadrigemine A (0.001% yield) and quadrigemine B (0.01% yield). The bases are isomeric, each having the molecular formula $C_{44}H_{50}N_8$, and a molecular ion at *m/e* 690. The mass spectra of the two bases were quite similar and also resembled that of hodgkinsine in the region *m/e* 200–350. Very little information could be deduced from the ^{1}H-NMR spectrum of quadrigemine A, which showed a broad envelope in the region 7.1–6.4 ppm for the aromatic protons and a broad singlet at 2.36 ppm for the *N*-methyl groups. Protons in several regions of the spectrum exchanged on addition of D_2O.

A characteristic feature of the mass spectrum of quadrigemine A was that an ion corresponding to half the molecular weight was the base peak. This observation suggests that the four units are disposed in a symmetrical manner, and this fact has been confirmed chemically.

Reaction of quadrigemine A with methyl iodide gave an amorphous tetramethiodide which on elimination with base gave a tetraindolenine (**57**). The NMR spectrum of this compound showed four 6-proton singlets at 1.92, 1.96, 1.98 and 2.00 ppm for the *N*-methyl groups and four one-proton singlets at 8.08, 8.18, 8.44 and 8.68 ppm for the protons at the indolenine 2-position. Potassium borohydride reduction gave a single product in 73% yield identical in all its spectral properties with the indole–indoline **50** derived from hodgkinsine (**46**). However, the optical rotation of this base was only half that of compound **50**, and was identical (+32°) with the optical rotation of the parent quadrigemine A.

Because a pure optically active C-3 center having an *R* configuration has an optical rotation of +60°, it is clear that the chiral centers C-3 and C-3‴ are in an *R*:*S* ratio of about 3:1.

Parry and Smith interpreted these data in terms of quadrigemine A being a 1:1 mixture of the *RRSR* diastereoisomer **58** and of the diastereoisomer *RRSS*, or *RSRS*, or a mixture of these two meso diastereoisomers. Such an

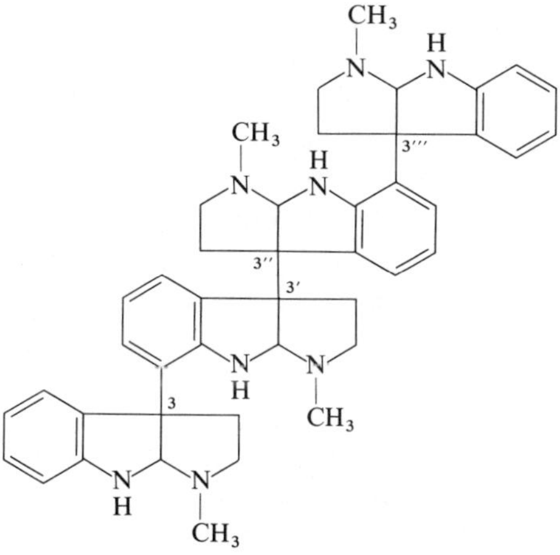

55 Quadrigemine A

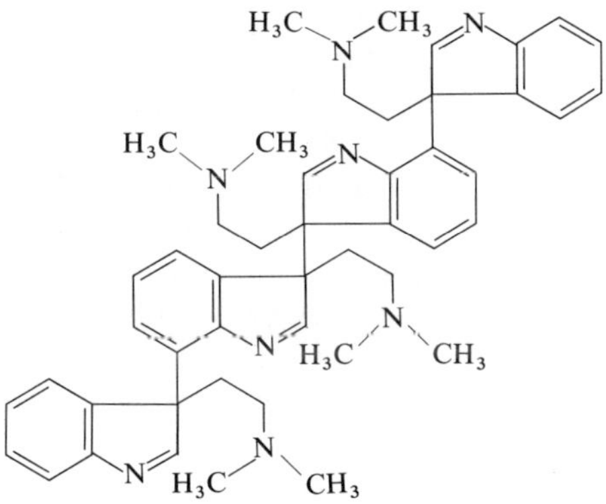

57

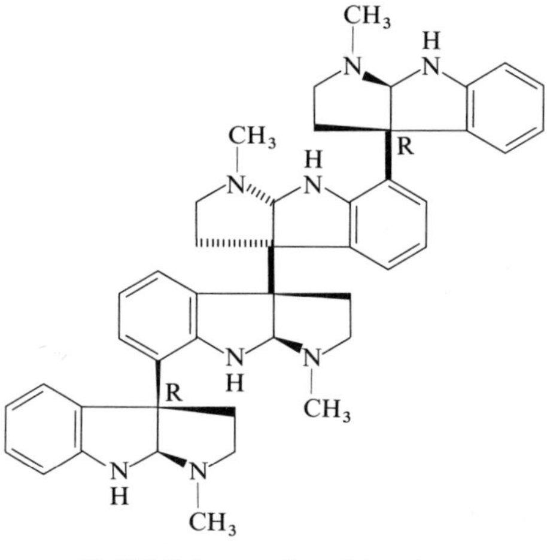

58 *RRSR* isomer of quadrigemine A

interpretation would also explain the complexity of the methine and *N*-methyl group regions in the spectrum of **55** (*74*).

The ^{1}H-NMR spectrum of quadrigemine B (**56**) was quite similar to that of hodgkinsine (**46**), except for an additional one-proton doublet ($J = 3$ Hz) at 4.9 ppm for one of the C-2 protons. In the IR spectrum, an additional band was noted at 3370 cm^{-1} for the extra NH group.

In the mass spectrum fragment ions were observed at *m/e* 172, 173, 516, and 517, suggesting the presence of a chimonanthine-type moiety at one end of the molecule.

Hofmann degradation of the tetramethiodide of quadrigemine B gave a tetraindolenine (**59**) showing singlets for the *N*-methyl groups at 2.04, 1.92, and 1.90 ppm (ratio 2:1:1) and singlets for the C-2 protons at 8.74, 8.64, 8.10, and 8.02 ppm. Reductive cleavage of the β,β'-bonds with potassium borohydride gave *N*,*N*-dimethyltryptamine (**49**) and trimer **60**, $C_{36}H_{48}N_6$, which was shown to be an indole–bisindoline. This compound exchanged seven aromatic hydrogens for deuterium in D_2SO_4–D_2O, indicating that two positions ortho or para to the indoline nitrogens are involved in additional carbon–carbon bonding.

Zinc–acid reduction of the bisindoline–indole gave an unstable triindoline **61**, which after complete N-formylation, nitration, and acid cleavage (3 *N* HCl at 95°) gave 5-nitro-*N*,*N*-dimethyltryptamine (**53**) and 5-nitro-*N*,*N*-dimethyl-2,3-dihydrotryptamine (**54**), among other products. No 7-nitro products were isolated. Thus, linkages to the aromatic nuclei are

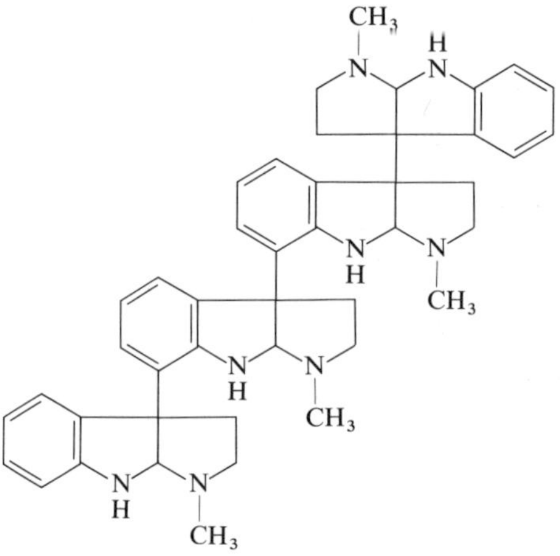

56 Quadrigemine B

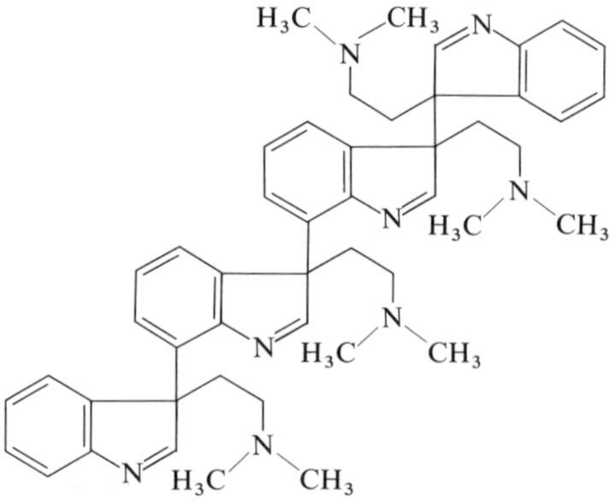

59

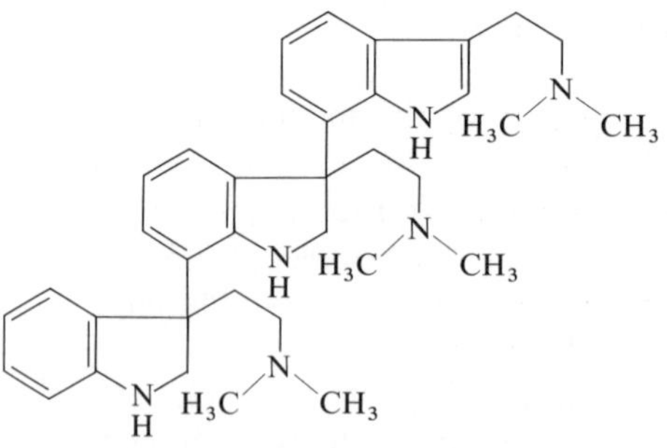

60

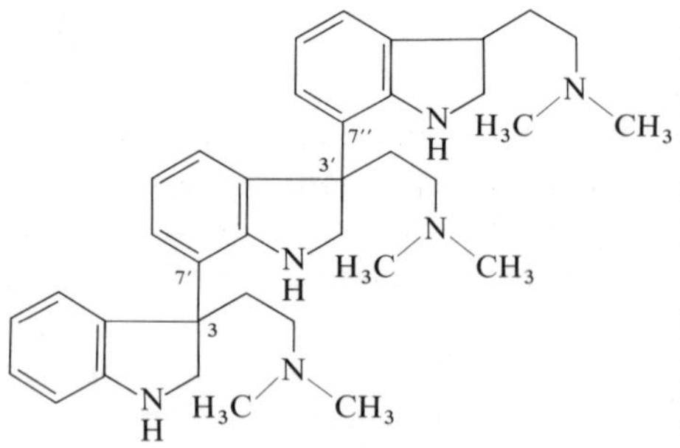

61

between carbons 3 and 7′ and 3′ and 7″. The triindoline should therefore have structure **61** and quadrigemine B structure **56**, without stereochemistry (*74*).

G. Psychotridine (62)

Recently, Hart and co-workers (*73*) reported the isolation of an alkaloid even more complex than the quadrigemines. This compound is psychotridine (**62**), the major alkaloid of *Psychotria beccarioides* Wernh., a plant in the family Rubiaceae, native to New Guinea.

Psychotridine, mp 180°–181°, $[\alpha]_D$ −38°, gave a molecular ion in the mass spectrum at *m/e* 862, in agreement with the molecular formula $C_{55}H_{62}N_{10}$. Major fragment ions were observed at *m/e* 516 ($C_{33}H_{36}N_6$) and 344 ($C_{22}H_{24}N_4$) (base peak), which were assigned structures **63** and **64** based on the mass spectra of the simple molecules. From a structural aspect, this would indicate a C-3″–C-3‴ linkage between the second and third units of an *N*-methytryptamine-derived pentamer. Notably, no $M^+ - 173$ ions were observed which would indicate a terminal *N*-methyltryptamine unit linking C-3 to C-3′ as in hodgkinsine (**46**). The crucial assumption here is that the nature of the linkages are the same as those in the hodgkinsine/quadrigemine compounds. On this basis, psychotridine was suggested to have structure **62**, and the additional chemical evidence supports this assignment.

Psychotridine formed an amorphous pentamethiodide which underwent elimination with dilute sodium hydroxide to afford a pentaindolenine **65**. The NMR spectrum of this base showed the presence of five 6-proton singlets at 1.94, 1.99, 2.02, 2.07, and 2.10 ppm for the dimethylamino groups and five 1-proton singlets at 8.10, 8.18, 8.68, 8.84, and 8.89 ppm for the C-2 methine protons. This evidence rules out the possibility of any N_a-methyl groups in psychotridine.

Further information on the structure of psychotridine came from an examination of the ^{13}C-NMR spectra of hodgkinsine (**46**) and psychotridine (**62**). Needless to say, the spectra were extremely complex and only quite general assignments could be made. Thus, signals at about 35 ppm could be assigned to *N*-methyl groups and those at about 52 ppm to the C-2 methylene groups adjacent to N_b. The fully substituted C-3a carbons appeared in the range 60–64 ppm. The methine carbon between the two nitrogen atoms appeared between 82 and 87 ppm. Unsubstituted C-7 aromatic carbon atoms were observed in the range 108–109 ppm, whereas the C-7a carbons exhibited the most deshielded resonances, at 149–152 ppm.

In the spectrum of hodgkinsine, the C-3a carbons resonated at 60.8, 63.2, and 63.9 ppm and the C-8a methine carbons at 82.4, 83.1, and 87.0 ppm.

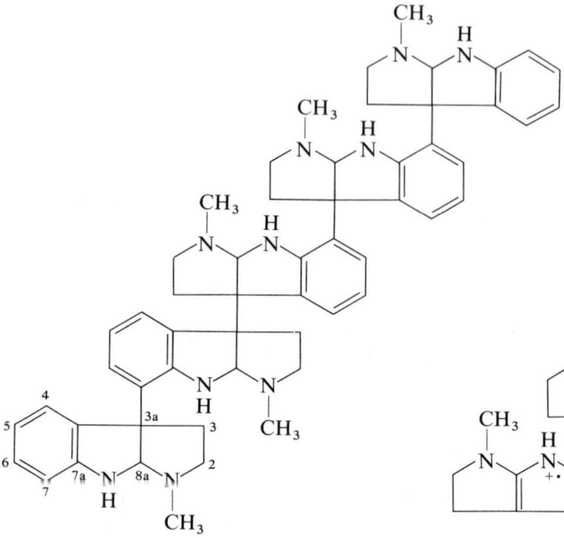

62 Psychotridine

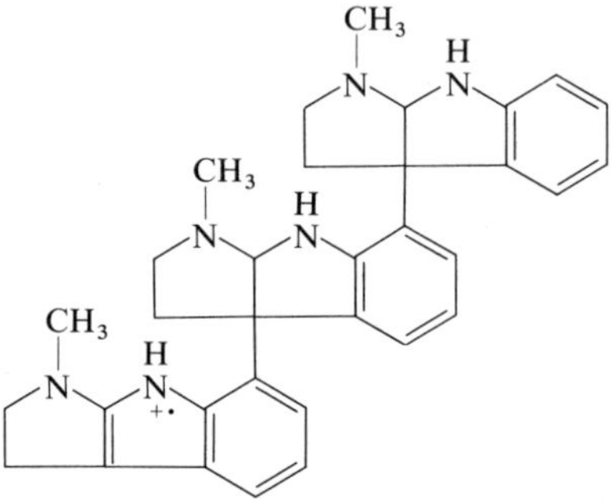

63 *m/e* 516

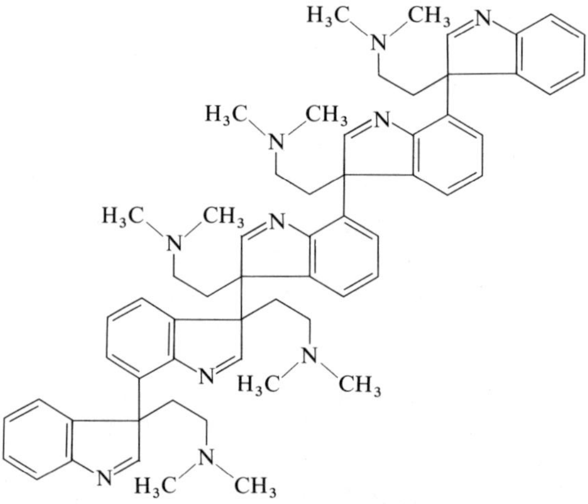

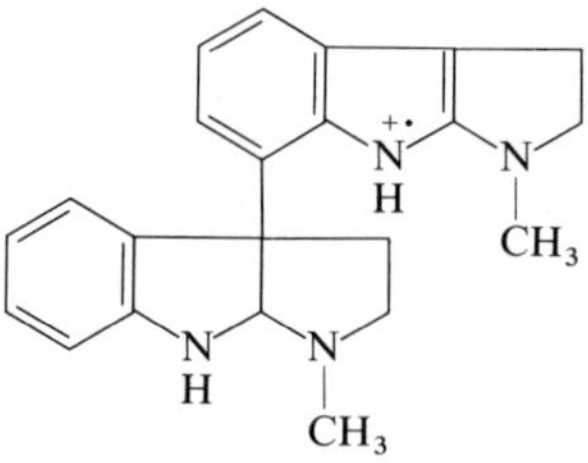

64 *m/e* 344

65

For psychotridine, the C-3a carbons appeared at 60.0, 60.6, 60.8, and 62.9 ppm, the last representing two carbon atoms. The C-8a carbons were observed at 82.2, 82.5, 85.4, 86.0, and 87.0 ppm, in agreement with the presence of five units derived from *N*-methyltryptamine in psychotridine. The C-7a resonances appeared at 150.7 and 148.9 ppm, the latter being more intense and apparently representing three carbons.

These data were interpreted by the Australian group to indicate that psychotridine has the gross structure **62** and is a single isomer (*73*).

III. Tryptamine–Tryptamine Type with an Additional Monoterpene Unit

Borreverine (68)

In 1973 Cavé and co-workers obtained the novel monomeric alkaloid borrerine (**66**) from *Borreria verticillata* (Rubiaceae) (*75*). More recently, the same group obtained (*76*) another alkaloid, borreverine, from the same source, and determined its structure by X-ray crystallography.

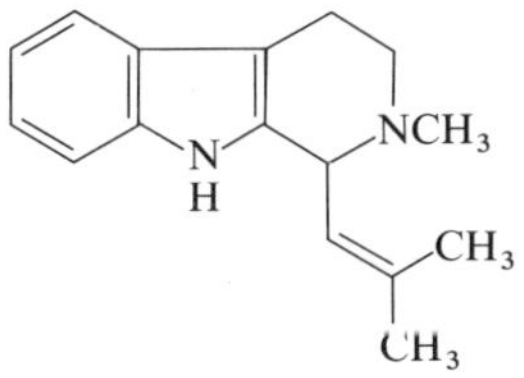

66 Borrerine

Borreverine, mp 193°, $[\alpha]_D$ 0°, showed a molecular ion which analyzed for $C_{32}H_{40}N_4$ and major fragment ions at *m/e* 437 (base peak), 250, 198, 196, 185, 182, and 172. The UV spectrum (λ_{max} 225, 285, and 295 nm) was typical for that of an indole–dihydroindole system. The ^{1}H-NMR spectrum indicated the presence of two quaternary methyl groups at 0.32 and 0.90 ppm, a vinyl methyl group (1.70 ppm), a single olefinic proton (5.60 ppm) and two *N*-methyl groups (2.53 and 2.56 ppm).

The alkaloid gave a diacetate derivative, mp 272°–274°, from which crystals suitable for X-ray analysis could be obtained. The crystals belonged to the monoclinic space group $P2_1/n$ and had unit cell dimensions of $a =$ 10.156, $b = 13.703$, and $c = 22.623$ Å. Least-squares refinements based on 2375 reflections gave an *R* factor of 0.10 and, by direct methods, structure **67** for diacetylborreverine.

Clearly, however, this is a bisindole alkaloid. From this and other spectroscopic data it was deduced that borreverine had undergone a facile rearrangement on acetylation. Eventually, highly unstable crystals of a methanol–water solvate were obtained. These, too, belonged to the monoclinic space group $P2_1/n$ and had unit cell dimensions of $a = 17.501$, $b =$ 10.495, and $c = 18.409$ Å. The structure was solved by least-squares refinements of 3038 reflections to give an *R* factor of 0.06, yielding **68** as the molecular arrangement for borreverine (*76*).

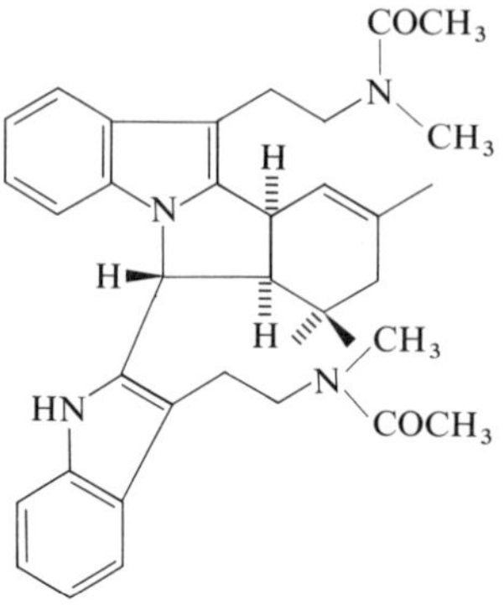

67

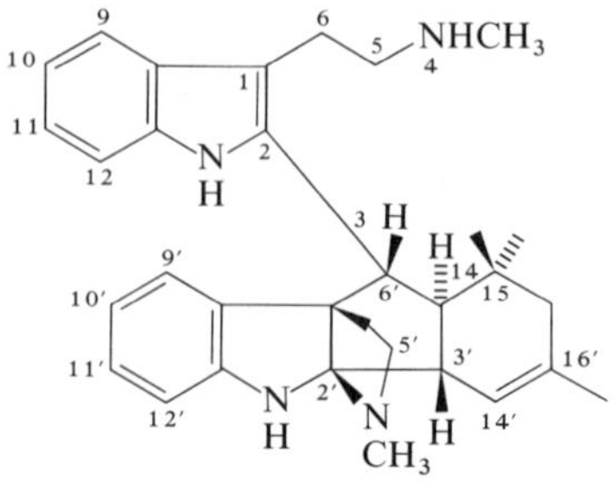

68 Borreverine

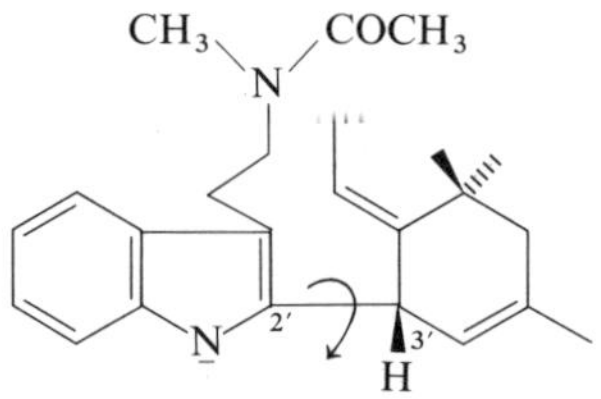

69

The structure of borreverine (**68**) may be formally derived by the joining of two 2-isoprenylated *N*-methyltryptamine residues. Rearrangement to diacetylborreverine (**67**) may occur by fragmentation to **69** with participation from the indoline nitrogen followed by rotation about the 2–12 bond and recyclization (*76*).

Bearing in mind the co-occurrence of borrerine and borreverine, an alternative biogenetic hypothesis was also considered, and it was this route which was conducted successfully *in vitro*. Although several acidic conditions proved successful, the optimum conditions involved heating borrerine (**66**) in a mixture of trifluoroacetic acid and benzene for 30 min to give an approximately 1 : 1 mixture of borreverine (**68**) and isoborreverine (**70**) (*77*). The proposed mechanism of this interesting dimerization is shown in Scheme 2.

The formation of isoborreverine in the reaction suggested that it too might be natural product. Indeed, closer examination of the additional fractions of *B. verticillata* indicated its presence (*78*), and further work demonstrated the presence of both alkaloids in *Flindersia fournieri* Panch. and Sieb. (Flindersiaceae) (*79*).

Isoborreverine (**70**), $[\alpha]_D$ 0°, displaying an indolic UV spectrum, gave a base peak at *m/e* 394 (loss of two $—CH_2NHCH_3$ units). The ^{1}H-NMR

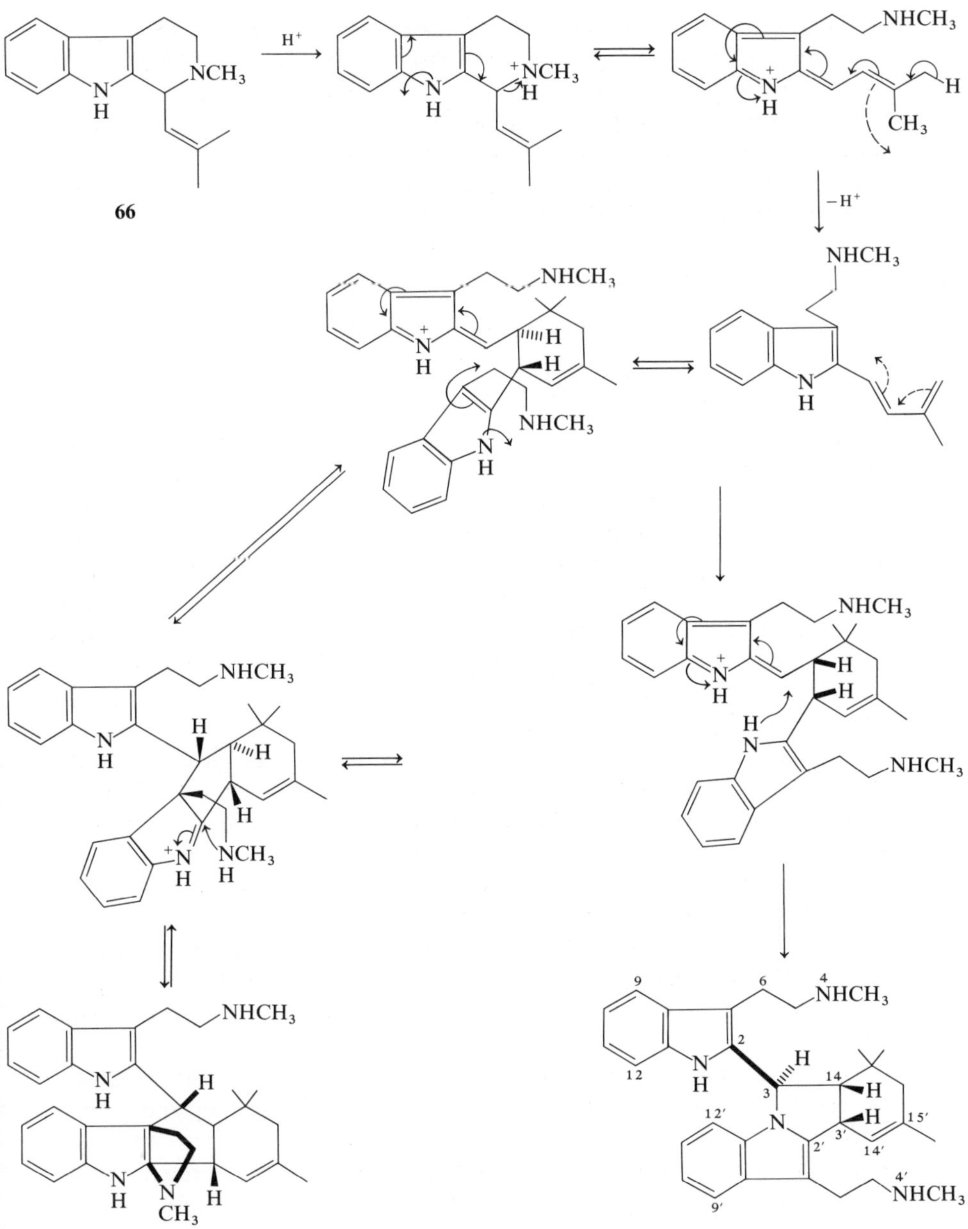

66

68

70

SCHEME 2. *Mechanism of formation of borreverine* (**68**) *and isoborreverine* (**70**) *from borrerine* (**66**) (*77*).

spectrum did not show the highly shielded methyl group of **68**; instead, two quaternary methyl groups were observed at 0.76 and 1.06 ppm. An olefinic methyl group was observed at 1.69 ppm and the corresponding vinylic proton at 5.40 ppm. Three protons were exchanged on the addition of D_2O, two secondary amine protons in the region 2.35–2.65 ppm and an indole NH at 4.05 ppm. Acetylation gave a diacetyl derivative (**67**) identical to that obtained from borreverine (**68**) (*77*).

IV. Corynanthe–Tryptamine Type

A. Roxburghines

The roxburghines A–E are an interesting group of alkaloids obtained from the leaves and stems of *Uncaria gambier* Roxb. (*80*). Some aspects of their chemical and physical properties have been described in a previous volume of this series (*81*). The alkaloids are isomers having the molecular formula $C_{31}H_{32}N_4O_2$, and displaying similar UV spectra, unchanged in acid or base. The IR spectra are also quite similar, showing both NH and α,β-unsaturated carbonyl absorptions, the former appearing as two signals at 9.72 and 10.08 ppm in the NMR spectrum. Also observed was a low-field vinylic proton at 7.53 ppm, a carbomethoxy singlet at 3.64 ppm, and a tertiary methyl singlet at 1.21 ppm.

Catalytic reduction (PtO_2–HOAc) of roxburghine D gave a dihydro derivative which did not show a vinylic proton and only had an indolic UV spectrum. By subtraction, the additional UV chromophore was attributed to a vinylogous urethane system similar to that found in vallesiachotamine.

Partial dehydrogenation of roxburghine D gave a pyridinium species showing two low-field aromatic protons at 8.43 and 8.78 ppm and a vinylic proton at 8.76 ppm. Two —CH_2CH_2— groups and eight aromatic protons were observed, together with singlets for the carbomethoxy group at 3.85 ppm, and the downfield shift indicates a close proximity to the plane of the new aromatic system.

Elegant double- and triple-resonance studies on the ^{1}H-NMR spectrum of roxburghine D established a piperidine ring substitution and stereochemical array as shown in **71**. Because work with the dehydro compound had indicated partial formulas **72** and **73**, together with a tertiary methyl group, two possible structures (**74** and **75**) were suggested for roxburghine D; structure **74** would be favored on biogenetic grounds (*80*).

In addition, the low-field shift of the C-14, C-17, and C-21 protons in the dehydro compound was attributed to delocalization of the pyridinium charge

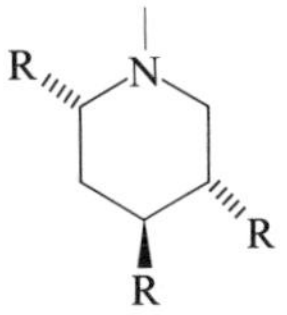

71 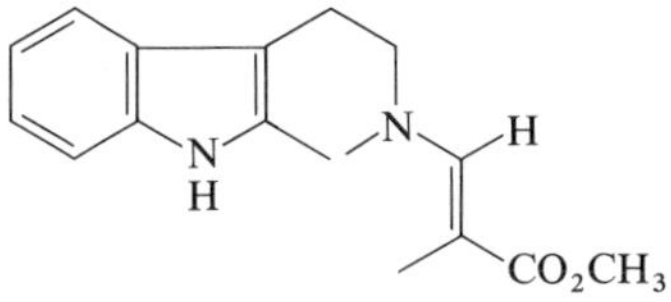 72 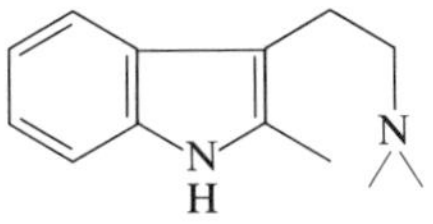 73

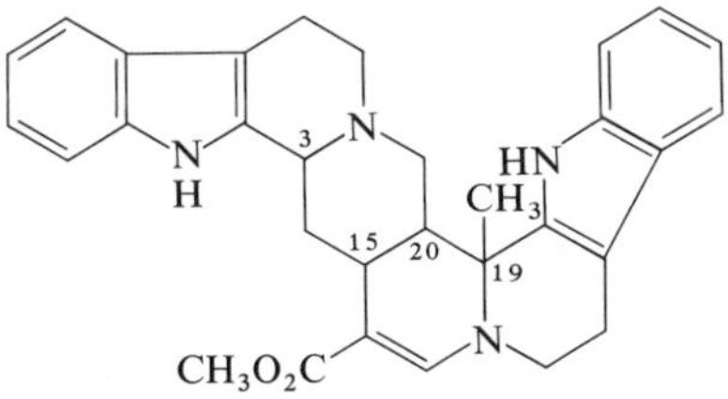

74 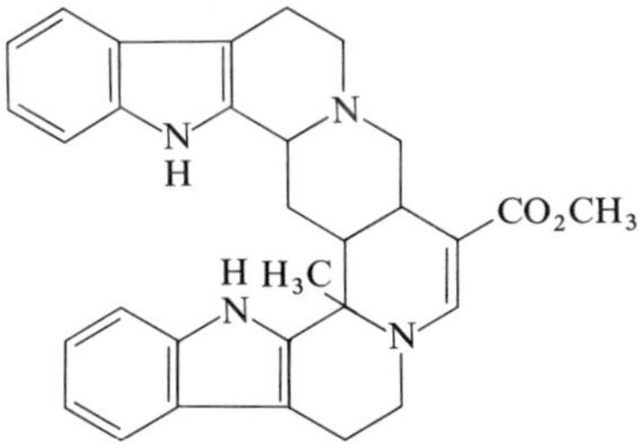 75

by N_b. This is only possible for dehydroroxburghine D derived from the skeleton **74**.

Formation of the dehydro derivatives of roxburghines B, C, and E gave indications of their interrelationships. Dehydroroxburghines B and E were identical, as were dehydroroxburghines C and D. As dehydroroxburghine has only one asymmetric center, these two groups of compounds must be diastereomeric at the C-19 methyl group (*80*).

When roxburghine E has heated with zinc and acetic acid, roxburghine B was produced (*80*). Thus, these two alkaloids differ in configuration only at C-3 (*82*), and H-3 is β in roxburghine E and α in roxburghine B (but this was later revised).

The stereochemistry of roxburghine D could now be deduced on the basis of the above reactions and the extensive decoupling experiments. Because H-3 is equatorial (δ 4.31) the C/D ring junction must be cis, and because of the large coupling ($J = 11.0$ Hz) established for H-15 and H-20, the D/E ring junction should be trans. The 15-H in all indole alkaloids having such a proton is α based on a biosynthesis from loganin (*83*); consequently, the two possibilities for the configurations of roxburghine D (and E) are 3β, 15α, 20β, 19α, and 3β, 15α, 20β, 19β (*80*).

The configuration at C-19 of roxburghine D was deduced as α by total synthesis (Scheme 3) (*84*). Using established procedures (*85*), addition of methyl *tert*-butyl malonate to the unsaturated ketone **76** gave, under conditions of thermodynamic control, a single product (**77**). Successive treatment

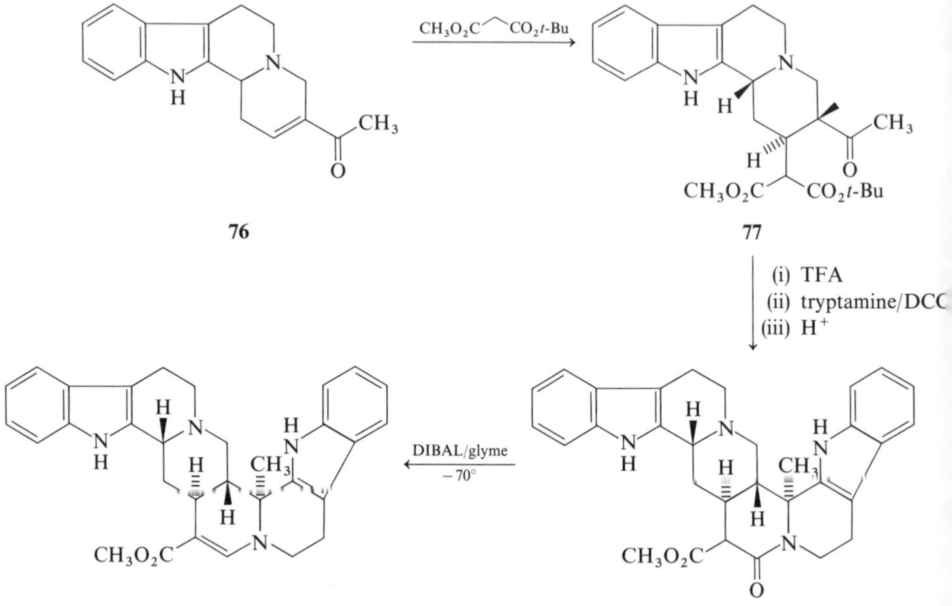

SCHEME 3. *Winterfeldt synthesis of roxburghine D* (**79**) (*84*).

with TFA, tryptamine, dicyclohexylcarbodiimide, and acid gave, stereospecifically, lactam **78**. Reduction of **78** with DIBAL in glyme at −70° afforded roxburghine D, which therefore has structure **79** (*84*).

Analysis of the ^{1}H-NMR spectrum of roxburghine E led to the conclusion that, like roxburghine D, the C/D ring junction was cis and the D/E junction trans (*86*, *87*), and consequently that the alkaloids differed only by their stereochemistry at C-19. Subsequently, the ^{1}H-NMR data for roxburghines B and C were reported (*87*). From these the complete stereochemical assignments could be deduced as follows.

In roxburghines C and E, as with roxburghine D, the coupling of H-15 and H-20 permits assignment of the trans configuration to the D/E ring junction; and recalling that roxburghines B and E differ only at C-3, the former should also be assigned a trans C/D junction. From the chemical shift (3.45 ppm) of H-3 in roxburghines B and C, the stereochemistry at C-3 in these compounds was deduced to be α, and the C/D ring junction trans. Interlocking evidence comes from the oxidation–reduction of roxburghine D and roxburghine C with $Pb(OAc)_4$ followed by $NaBH_4$. Taken with the interrelationships deduced from the dehydro compounds and the established stereochemistry for the C-19 methyl group in roxburghine D (*84*), roxburghines B–E were assigned (*87*) the following absolute configurations:

Compound	H-3	C-18
Roxburghine B	α	β
Roxburghine C	α	α
Roxburghine D	β	α
Roxburghine E	β	β

From the mass spectra of roxburghine D, decarbomethoxyroxburghine D, dihydroroxburghine D, and other derivatives, some interesting mass spectral fragmentation schemes were proposed. Two of the most important fragment ions in the mass spectrum of roxburghine D were at *m/e* 184 (**80**) and 279 (**81**), and these were suggested to be derived by reactions of the $M^+ - CH_3$ ion **82** as shown in Scheme 4. The second most important fragment ion from **79**, at *m/e* 321 (**83**), is dependent on the presence of the 16,17-double bond; the proposed pathway of formation is shown in Scheme 5.

More recently, additional data have become available which indicate that the stereochemistry of roxburghine B (**84**) should be revised to that in which H-3 has the β-stereochemistry (*88*, *89*).

The 300-MHz NMR spectrum of roxburghine B (*88*) was thoroughly analyzed by the INDOR technique to determine accurately the chemical shifts and coupling constants of as many aliphatic protons as possible. The results of these experiments are summarized in Table I.

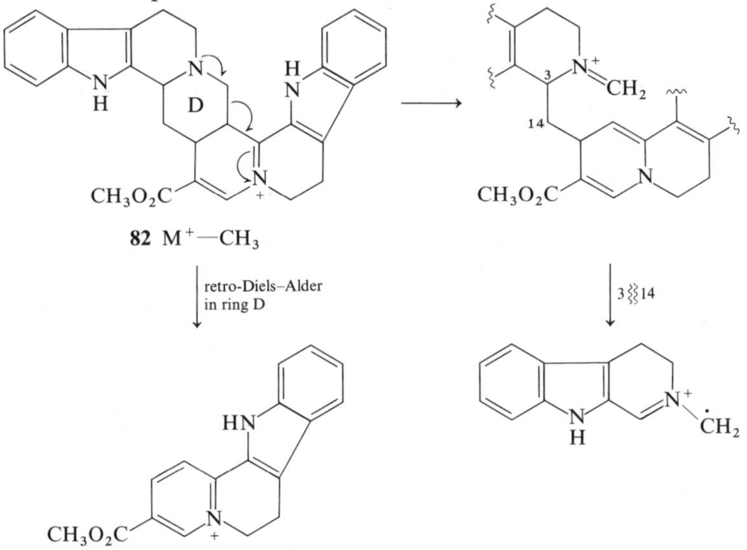

82 M^+—CH_3

retro-Diels–Alder in ring D

81 *m/e* 279

80 *m/e* 184

SCHEME 4.

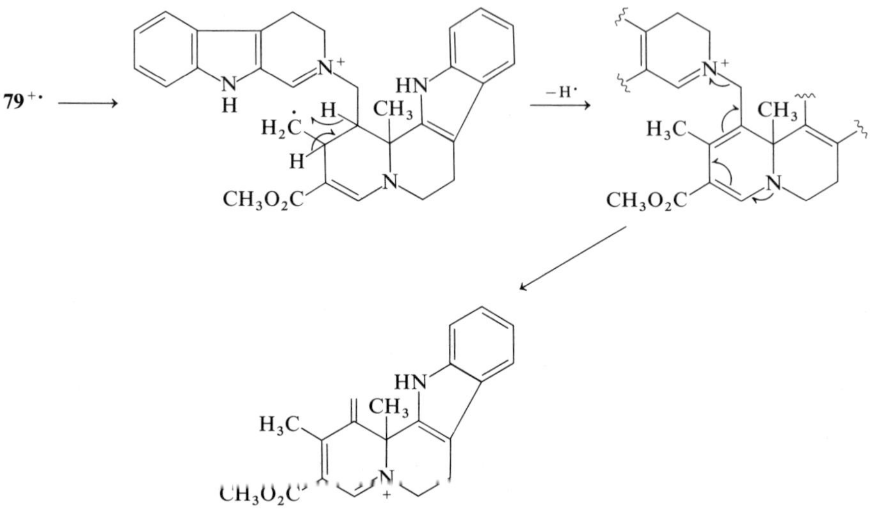

83 *m/e* 321

SCHEME 5.

TABLE I
^{1}H-NMR DATA FOR ROXBURGHINE B (**85**)[a]

Signal	Chemical shift		Coupling constants, J (Hz)			
	Acetone	BPA[b]				
H-3	3.19	3.73	3,14ax	11.5;	3,14eq	2.5
H-14ax	1.49	1.66	14ax,14eq	−13.0;	14ax,15	2.5
H-14eq	3.31	3.47	14eq,15	2.5		
H-15	2.74	~3	15,20	≤2;	15,17	2.0
H-20	2.90	~3	20,21eq	5.0;	20,21ax	11.0
H-21eq	3.10	3.17	21ax,21eq	−10.5		
H-21ax	2.60	2.67				
H-5′ax	3.75	3.58	5′ax,5′eq	−13.5;	5′ax,6′ax	11.0
H-5′eq	3.61	3.37	5′eq,6′ax	6.5;	5′eq,6′eq	≤1
H-6′ax	2.86	2.82	6′ax,6′eq	−15.0		
H-6′eq	2.71	2.62	5′ax,6′eq	5.0		
H-17	7.42	—	5′eq,17	≤1		

[a] Data from Cistaro *et al.* (88).
[b] Benzene:pyridine:acetone, 30:65:5 (v/v).

A trans C-15–C-20 ring junction for roxburghine B was excluded by the small coupling constants of H-15 with H-14 axial and H-20, and the large coupling of H-20 with H-21 axial, which must be coupling between diaxial protons. The H-15 is therefore equatorial and H-20 axial. Previous work (*90*) had demonstrated that roxburghine E could be converted to roxburghine B with zinc–acetic acid. This had been previously interpreted as an inversion of C-3, but must now be viewed in terms of an inversion at C-20. The large coupling of H-3 with H-14 axial establishes this proton to be axial and therefore β-oriented. Therefore, roxburghine B has structure **85**, and its preferred conformation is shown (*88*).

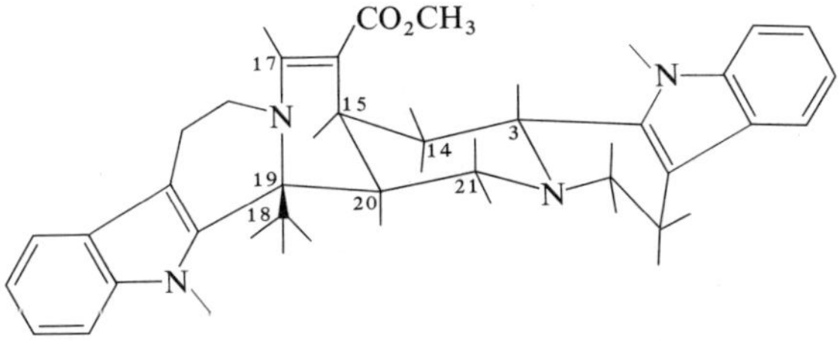

85 Preferred conformation of roxburghine B

Mondelli and co-workers, in collaboration with Wenkert's group (*89*), have also reported the ^{13}C-NMR spectra of the four established roxburghines. The data (Table II) confirmed the structures of roxburghines D (**79**), C (**86**), and E (**87**), and agreed with the revision of the structure of roxburghine B to **85**.

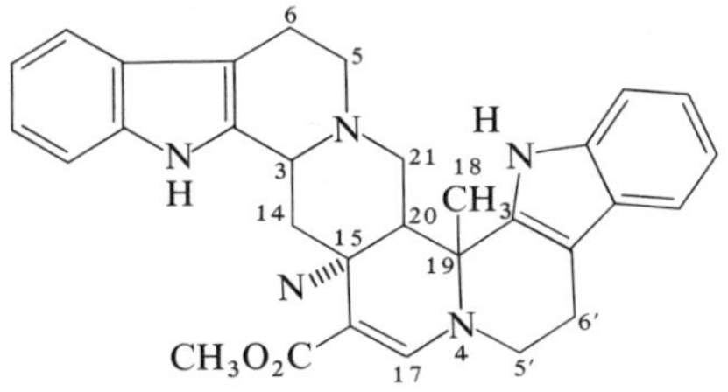

	H-3	H-20	C-18
79 Roxburghine D	β	β	α
84 Roxburghine B (original)	α	β	β
85 Roxburghine B (revised)	β	α	β
86 Roxburghine C	α	β	α
87 Roxburghine E	β	β	β

Crucial in establishing the carbon shifts was the availability (*91*) of data on the two ajmalicinoid systems, ajmalicine (**88**) and 3-iso-19-epiajmalicine

TABLE II
^{13}C-NMR DATA FOR THE ROXBURGHINES[a]

Carbon	Chemical shift, δ (ppm)			
	Roxburghine D (**79**)	Roxburghine E (**87**)	Roxburghine C (**86**)	Roxburghine B (**85**)
C-2	132.9	132.3	136.5[a]	136.5
C-3	54.0	55.1	60.6	56.0
C-5	51.4	52.1	54.0	54.2
C-6	17.6	17.5	22.7[b]	22.7[b]
C-7	108.1	109.7	109.9	108.1
C-14	32.9	32.8	35.3	32.0
C-15	30.6	29.5	36.1	30.3
C-16	95.1	105.7	96.2	102.3
C-17	144.8	149.1	146.9	149.0
C-18	18.1	26.9	18.7	26.6
C-19	57.7	58.1	58.5	57.7
C-20	49.4	49.3	50.0	42.7
C-21	48.1	48.6	57.9	52.9
$CO_2\underline{C}H_3$	49.4	50.8	50.1	50.8
$\underline{C}O_2CH_3$	165.4	167.8	167.7	168.2
C-2′	135.3	137.2[a]	137.2[a]	137.0[a]
C-5′	49.9	47.4	50.7	46.6
C-6′	23.0	23.0	23.3[b]	23.0[b]
C-7′	106.0	107.6	107.7	106.8
C-8, C-8′	125.5, 126.8	127.7, 128.7	127.3, 128.1	128.2
C-9, C-9′	116.4, 117.0	118.7, 118.8	118.1, 118.8	118.2, 118.5
C-10, C-10′	117.5, 118.0	119.8, 120.0	119.2, 119.7	119.3, 119.6
C-11, C-11′	119.5, 120.4	122.1, 122.4	121.1, 122.2	121.2, 121.9
C-12, C-12′	110.2, 110.4	112.2, 112.4	111.7, 111.9	111.8, 112.0
C-13-C-13′	135.3	137.4[a], 137.7[a]	137.3[a], 137.4[a]	137.5[a], 139.4[a]

[a] Superscripts *a* and *b* indicate signals may interchanged.

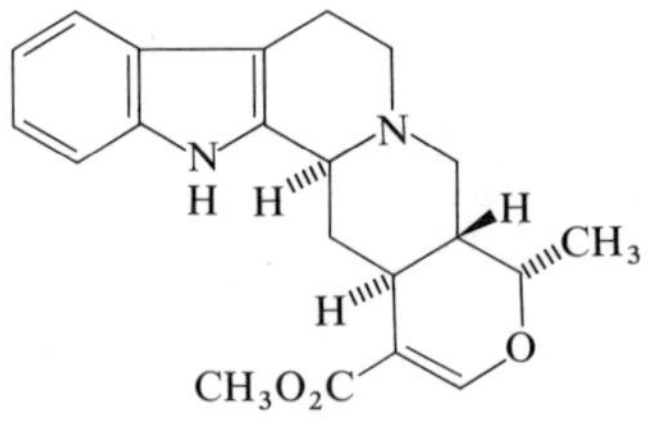

88 Ajmalicine

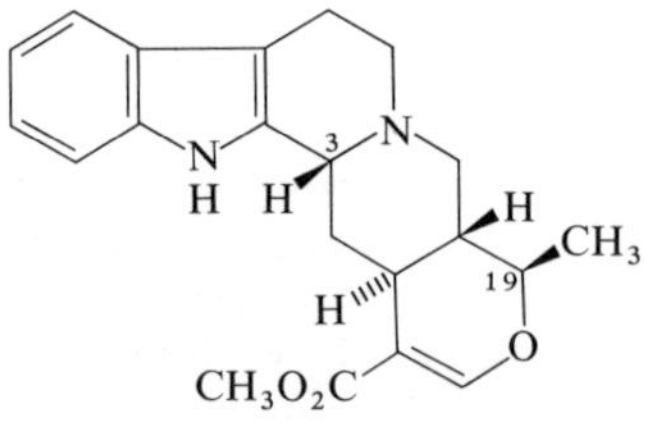

89 3-Iso-19-epiajmalicine

(**89**). These data are summarized and compared with roxburghines C (**86**) and E (**87**) in Table III.

TABLE III

COMPARISON OF THE ^{13}C-NMR DATA OF AJMALICINOID AND ROXBURGHINE SYSTEMS (PARTIAL DATA)

Carbon	Chemical shift, δ (ppm)			
	Roxburghine C[a] (**86**)	Ajmalicine (**88**)	Roxburghine E (**87**)	3-Iso-19-epiajmalicine (**89**)
C-2	136.5	134.0	132.2	132.4
C-3	60.6	59.8	54.2	53.8
C-5	54.0	52.7	51.3	50.9
C-6	22.7	21.3	17.0	16.8
C-7	109.9	106.1[b]	109.2	107.4[b]
C-14	35.3	32.1	31.9	31.2
C-15	36.1	30.1	28.8	30.8
C-16	96.2	106.5[b]	104.7	107.7[b]
C-17	146.9	154.5	148.3	155.9
C-18	18.7	14.5	26.6	18.0
C-19	58.5	73.3	57.3	75.3
C-20	50.0	40.2	48.4	43.8
C-21	57.9	56.2	46.7	46.8

[a] Data obtained in acetone; all other data in $CDCl_3$.
[b] Assignments may be reversed.

Roxburghines D (**79**) and E (**87**) belong to the pseudoajmalicinoid series, and each exhibits characteristic shifts of C-3 (54 ppm) and C-6 (17 ppm) for pseudo derivatives. Thus, these compounds have a close similarity to the shifts of C-3, C-5, C-6, C-14, and C-2 in 3-iso-19-epiajmalicine (**89**). Although roxburghine D (**79**) and E (**87**) differ in stereochemistry, C-21 must experience similar steric influences from C-18, C-2′, and N-1′, because the chemical shifts are extremely close. However, there is marked difference observed in the chemical shift of C-18 depending on the C-19 methyl group, the γ-effects of C-15 and C-17 are greatly reduced, and consequently C-18 is deshielded to 26.9 ppm.

Roxburghine C (**86**) has the normal ajmalicinoid system, and comparison with the data of ajmalicine (**88**) bear out this assignment. In addition, **86** shows the shifts of C-3 (60 ± 1 ppm) and C-6 (21.5 ± 0.5 ppm) characteristic of a normal system.

In roxburghines C (**86**) and D (**79**), both C-16 and C-17 are shielded relative to the shift in roxburghines B (**85**) and E (**87**). This shielding has been rationalized (*91*) in terms of increased overlap of the enamino nitrogen lone pair with the conjugated double bond.

For roxburghine B (**85**), the shifts of C-3 and C-6 indicate that it is not a member of the normal series, but rather a member of the C/D trans epi-allo series, having shifts in the ranges 54.5 ± 0.5 and 21.5 ± 0.5 ppm for C-3 and C-6, respectively. Both C-15 and C-20 are relatively shielded in roxburghine B compared with the normal system [e.g., roxburghine C (**86**)]; C-20 is particularly shielded because of the added γ-effect of N-1′. The shifts of C-18 in roxburghine B (**85**) and roxburghine E (**87**) indicate that they have similar spatial environments and that C-18 is β. The data, therefore, agree with the revised structure for roxburghine B.

The only point remaining to be discussed is the mechanism by which C-20 is epimerized in the conversion of roxburghine E (**87**) to roxburghine B (**85**) with zinc–acetic acid. Two mechanisms have been postulated (*91*) to explain this reaction and a modified version of one of these is shown in Scheme 6.

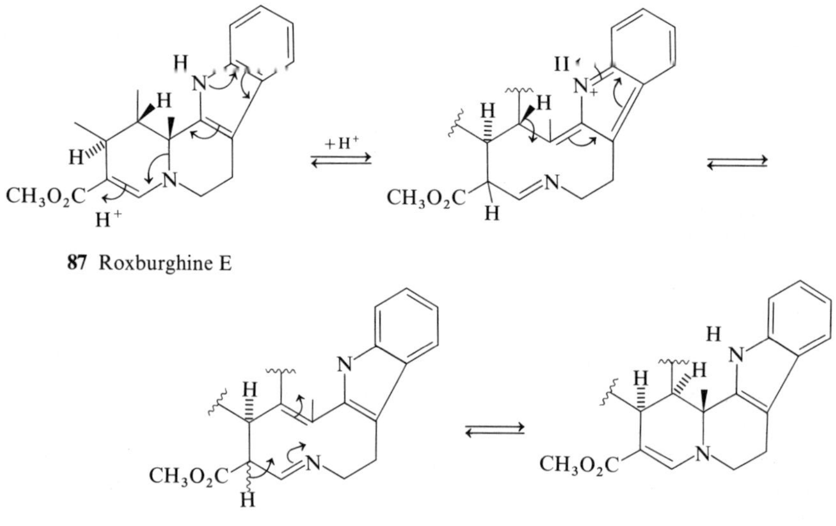

87 Roxburghine E

85 Roxburghine B

SCHEME 6.

B. THE OCHROLIFUANINES, THE USAMBARINES, AND RELATED ALKALOIDS

Like the roxburghines, the ochrolifuanines and the usambarines are formed from a corynanthe moiety and a tryptamine residue. Because of their very close structural relationship, they will be discussed together.

Usambarine I, mp 215°–218°, was isolated from the leaves of *Strychnos usambarensis* Gilg. (Loganiaceae), and from high-resolution analysis of the molecular ion (*m*/*e* 450) a molecular formula $C_{30}H_{34}N_4$ was deduced (*92*). The UV spectrum indicated the presence of two indolic chromophores, and two indolic NH protons were found in the NMR spectrum (6.35 and 7.71 ppm). Also revealed in the latter were an ethylidene group (doublet, $J =$ 7 Hz, at 1.22 ppm; quartet, $J = 7$ Hz at 5.22 ppm), an *N*-methyl group (singlet at 2.44 ppm), eight aromatic protons, and a one-proton multiplet at 3.65 ppm attributable to the H-3 proton of a corynane skelton in the β-configuration. The molecular ion was quite intense (77%) compared with the peak at *m/e* 185 (**90**), and the latter together with the fragment at *m/e* 265 (**91**) indicated a complete structure **92** for usambarine I. The C-17 stereochemistry was not determined (*92*).

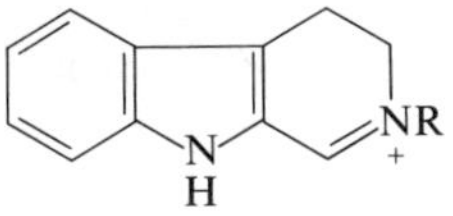

90 R = CH_3, *m/e* 185
93 R = H, *m/e* 171

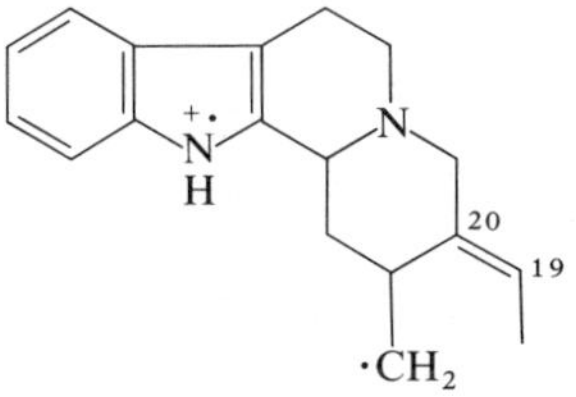

91 *m/e* 265
94 19,20-Dihydro, *m/e* 267

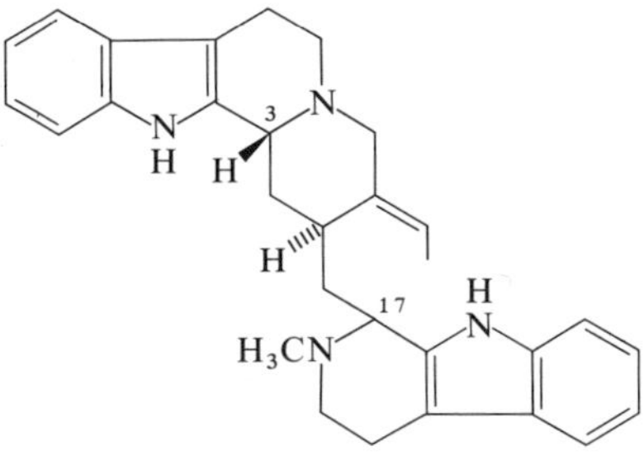

92 Usambarine I

From the leaves of *Ochrosia lifuana* Guill. (Apocynaceae) two isomeric, amorphous alkaloids, ochrolifuanines A and B, were isolated (*90, 93*). The UV spectra of these compounds indicated their dimeric indole nature, and in the IR spectrum Bohlmann bands were observed, indicating an H-3α stereochemistry. This was confirmed by the NMR spectrum, which exhibited a quartet at 4.15 ppm. Also observed in the spectrum of ochrolifuanine A were eight aromatic protons, two exchangeable indolic NH protons, a nonindolic NH proton (1.50 ppm), and the distorted methyl triplet (0.73 ppm) of an ethyl side chain. The mass spectral fragmentation of a molecular ion at 438 mu ($C_{29}H_{34}N_4$) was essentially the same as that of usambarine I, showing a base peak at *m/e* 171 (**93**) and a fragment ion at *m/e* 267 (**94**). As with **92**, cleavage of the 15–16 bond also occurs to afford ions at *m/e* 253 and 185.

Ochrolifuanine B displayed mass, IR, and UV spectra essentially identical to ochrolifuanine A, but the "triplet" of the ethyl side was now observed at 0.20 ppm. Therefore, structures **95** and **96** were suggested for ochrolifuanines A and B, respectively. They were considered to be isomers at C-20, and no C-17 stereochemistry was assigned (*90*, *93*).

Further isolation work yielded other closely related alkaloids (*93*). The first was an unstable, amorphous alkaloid, $C_{29}H_{32}N_4$, showing in acidic media the UV spectrum of an indole and a 1,2-dehydro-β-carboline. A strong band at 1550 cm^{-1} in the IR spectrum also suggested this group. The base peak in the mass spectrum at *m/e* 251 confined the oxidation to the corynane nucleus, and the alkaloid was assigned the structure 3-dehydroochrolifuanine (**97**) (*93*).

A more polar alkaloid showed a molecular ion at *m/e* 454 ($C_{29}H_{34}N_4O$) and also displayed UV and IR spectra very similar to those of ochrolifuanines A and B. The nature of the oxygen function was evident from the observation of important fragment ions at *m/e* 269 and 283 corresponding to **98** and **99**. This compound was, therefore, formulated as ochrolifuanine *N*-oxide (**100**) (*93*).

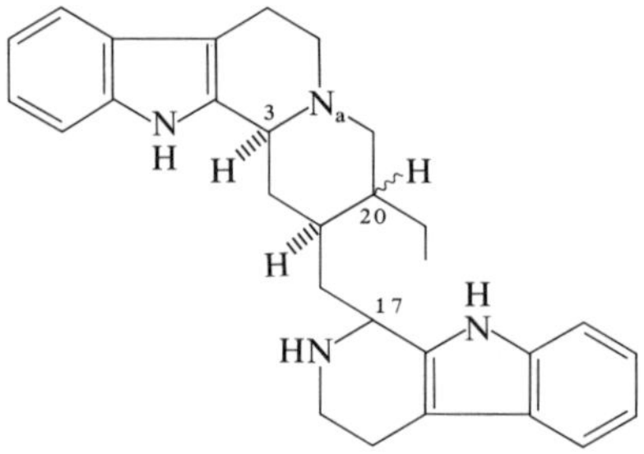

95/96 Ochrolifuanines A and B
100 N_a-Oxide

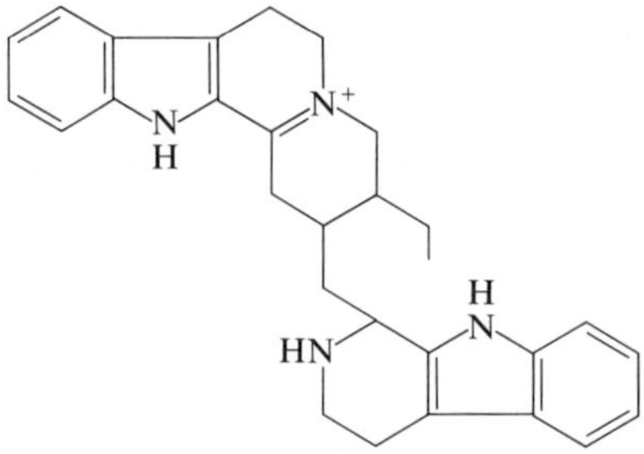

97 3-Dehydroochrolifuanine

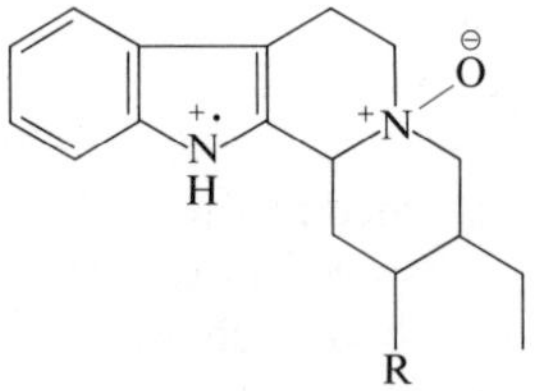

98 R = radical, *m/e* 269
99 R = $CH_2^{\cdot}$, *m/e* 283

The stereochemistry of ochrolifuanines A and B was determined by partial synthesis from precursors of known stereochemistry (*94*). Condensation of the dihydrocorynantheals **101** (H-20β) and **102** (H-20α) with tryptamine under Pictet–Spengler conditions gave a mixture of C-17 stereoisomers from each. Thus, four compounds were obtained in which H-3 and H-15 were both fixed and α, but which had the following configurations at positions 17 and 20: **103**, H-17β, H-20β; **104**, H-17α, H-20β; **105**, H-17α, H-20α; **106**, H-17β, H-20α. With H-3 fixed, the CD curve is dependent on the H-17 stereochemistry; in this way, **104** and **105** were assigned the 17*S*-configuration and **103** and **106** the 17*R*-configuration. By TLC and IR comparison, ochrolifuanine A was shown to be identical with **103** and ochrolifuanine B with **104** (*94*). Therefore, the isomers differ in configuration at C-17 and not at C-20, as was originally thought.

Subsequently, ochrolifuanine B (**104**) and dehydroochrolifuanine (of unknown stereochemistry) were isolated from the stem bark of *Ochrosia confusa* Pichon (*95*), and **103** and **104** were isolated from the bark, leaves, and twigs of *Ochrosia miana*. H. Bn ex Guill. (*96*). Also obtained from the leaves and twigs of the latter plant were two new isomeric ochrolifuanines. The molecular formula of these compounds was deduced to be $C_{29}H_{34}N_4O$, and from the shift in alkali the hydroxy group in each could be assigned to the 10-position. An alternative position (C-6′) could be ruled out because each compound displayed a pronounced ion at *m/e* 171 (**93**). The hydroxyochrolifuanines were therefore ascribed structures **107** and **108** in which the C-17 stereochemistry remains to be determined (*96*).

Additional evidence for the structures of ochrolifuanines A and B and their isomers was obtained by analysis of the ^{13}C-NMR spectra of these compounds, and by comparison with some corynanthe derivatives of known stereochemistry at C-20 (*97*). Some of the pertinent signals are summarized in Table IV. The C-3 shift confirms the trans nature of the C/D ring junction,

TABLE IV
^{13}C-NMR DATA OF PERTINENT CARBONS IN THE OCHROLIFUANINES AND RELATED COMPOUNDS

	Chemical Shift, δ (ppm)									
Compound	C-3	C-14	C-15	C-16	C-17	C-18	C-19	C-20	C-21	Ref.
Ochrolifuanine A (**103**)	59.3	34.3	35.8	38.1	48.4	11.0	23.2	42.2	59.9	(*97*)
Ochrolifuanine B (**104**)	59.5	36.4	37.8	38.4	51.9	11.2	23.8	42.5	60.1	(*97*)
109	60.2	33.8	38.7	—	—	11.3	24.4	39.3	61.3	(*97*)
Ochrolifuanine C (**105**)	60.3	32.4	36.1	37.8	50.0	12.4	17.5	38.3	57.5	(*97*)
Ochrolifuanine D (**106**)	59.4	31.1	35.1	38.4	49.8	12.5	18.6	41.3	57.3	(*97*)
110	61.2	29.8	40.8	—	—	12.8	19.1	40.0	57.9	(*97*)
Tchibangensine (**132**)	53.2	29.0	31.3	36.4	160.9	12.6	119.7	134.9	52.1	(*106*)
Ochrolifuanine C (**105**)	60.5	32.6	36.2	38.3	50.0	12.6	17.7	38.5	57.7	(*106*)
Ochrolifuanine D (**106**)	59.7	31.3	35.3	38.5	49.8	12.6	18.7	41.6	57.7	(*106*)

and the shifts of C-18, C-19, and C-21 distinguish the two alternative stereochemistries at C-20. Note that these shifts are quite analogous to those observed for the corresponding carbon atoms in the dihydrocorynantheines **109** and **110**.

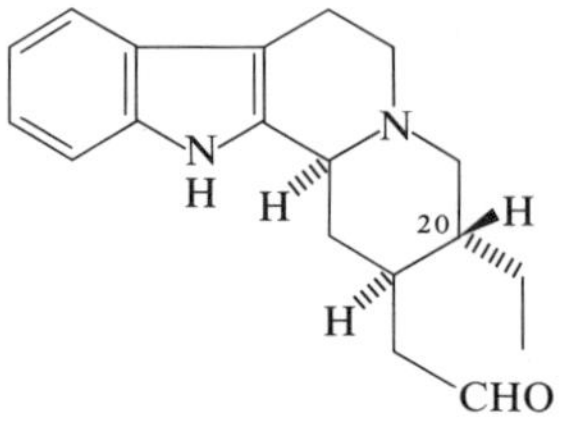

101

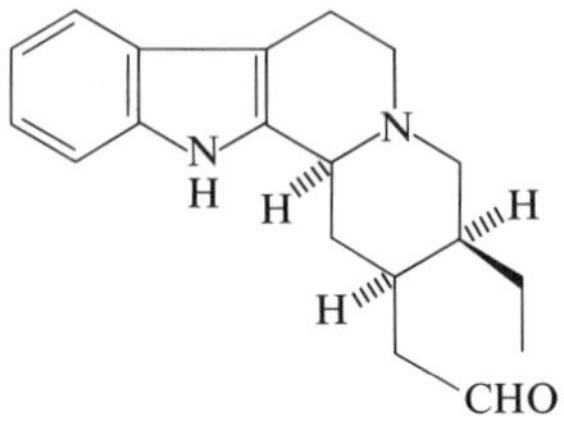

102

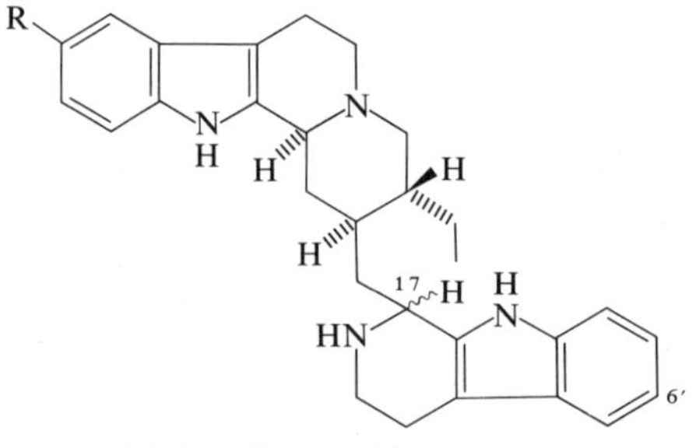

103 R = H Ochrolifuanine A, H-17β
104 R = H Ochrolifuanine B, H-17α
107/108 R = OH

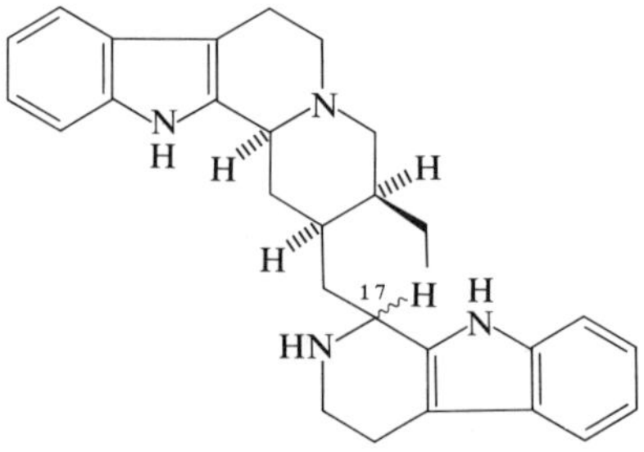

105 Ochrolifuanine C, H-17α
106 Ochrolifuanine D, H-17β

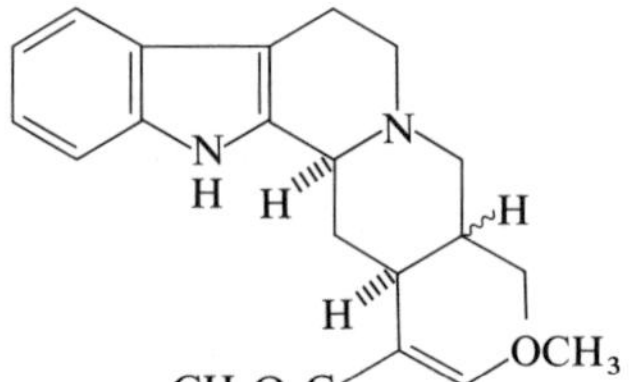

109 H-20β
110 H-20α

The Banyambo hunters in Rwanda have used preparations of the root bark of *Strychnos usambarensis* in arrow poisons. A group at the University of Liège (*98–102*) has investigated the alkaloids of this plant over several years from the point of view of both chemistry and pharmacology, and as a result two groups of alkaloids have been isolated and characterized: one in the ochrolifuanine series and another group in the curarine series, which will be discussed below. Four ochrolifuanine-type bases were isolated from the root, two tertiary bases and their corresponding N_b-methyl salts.

Usambarensine gave a molecular ion at *m/e* 432 which analyzed for $C_{29}H_{28}N_4$; its UV spectrum indicated the presence of a harmane (**111**) nucleus and an indole nucleus showing λ_{max} 233, 291, 337, and 349 nm. The IR spectrum displayed bands at 3415 cm^{-1} for an NH group and at 1634 cm^{-1} for the pyridine nucleus.

The mass spectrum was particularly informative, showing ions resulting from the harmane and tetrahydro-β-carboline nuclei. The former unit showed a prominent ion at *m/e* 182 suggested to have the structure **112**, whereas the latter was responsible for the base peak in the spectrum at *m/e* 251 (**113**) and for important fragment ions at *m/e* 250, 249 (**114**), 185, 169, and 144.

The NMR spectrum of usambarensine also indicated the presence of a harmane moiety showing the 3′ and 4′ protons at 8.52 and 7.86 ppm as doublets ($J = 5$ Hz), and showing the 8′ proton as a doublet ($J = 7.5$ Hz) at 8.08 ppm. Besides eight aromatic protons, indicating unsubstituted indole nuclei, only an ethylidene group (quartet at 5.43 ppm, doublet at 1.67 ppm) and two NH protons at 8.09 and 10.8 ppm, were readily assigned. Two other signals were noted at 4.35 and 3.80 ppm but were not assigned. On this basis, structure **115** was suggested for usambarensine without stereochemistry (*98*).

The configuration at C-15 could be assigned by analogy with all other corynane indole alkaloids and the C-3 stereochemistry deduced chemically. Oxidation with mercuric acetate followed by reduction with zinc and HCl gave a product different from the starting material. The product was regarded as a compound having a C-3α stereochemistry with a trans C/D ring junction; consequently, usambarensine was regarded as having a C-3β stereochemistry (*98*).

The second alkaloid isolated was shown by high-resolution mass spectrometry to have a molecular formula $C_{29}H_{30}N_4$, and from the UV spectrum (λ_{max} 229, 284, 291, and 318 nm) it was shown to be comprised of indole and harmalane nuclei. The IR spectrum gave indications of NH (ν_{max} 1390 cm^{-1}) and imine (1620 cm^{-1}) functionalities. Overall, the mass spectrum was quite similar to that of usambarensine showing a molecular ion at *m/e* 434 and major fragment ions at *m/e* 250 (base peak), 249, 183, 169, 155, and 144.

The ion at *m/e* 183 can be assigned structure **116**, and the ion at *m/e* 250 structure **117**. The NMR spectrum was quite similar to that of usambarensine, but clearly demonstrated an absence of C-3′ and C-4′ protons in the aromatic region. Instead, the aliphatic region of the spectrum was more complex. With these data on hand, the structure 3′,4′-dihydrousambarensine (**118**) was given to this compound. No stereochemical assignments were made at this time (*98*).

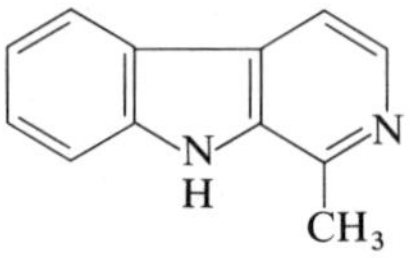

111
112 *m/e* 182

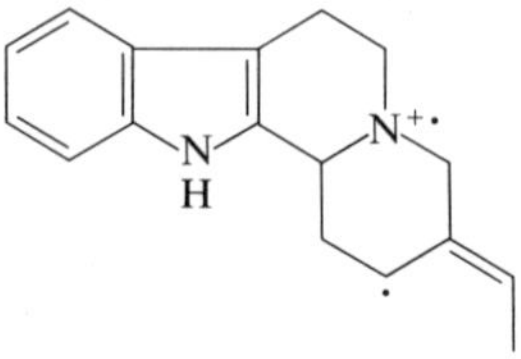

113 *m/e* 251

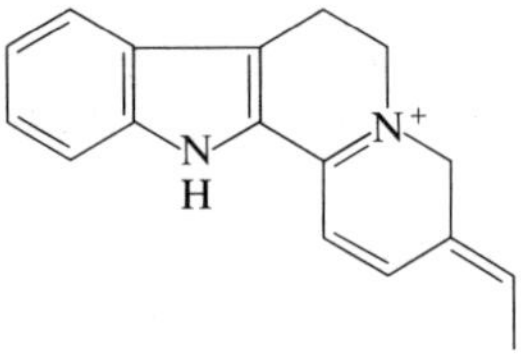

114 *m/e* 249

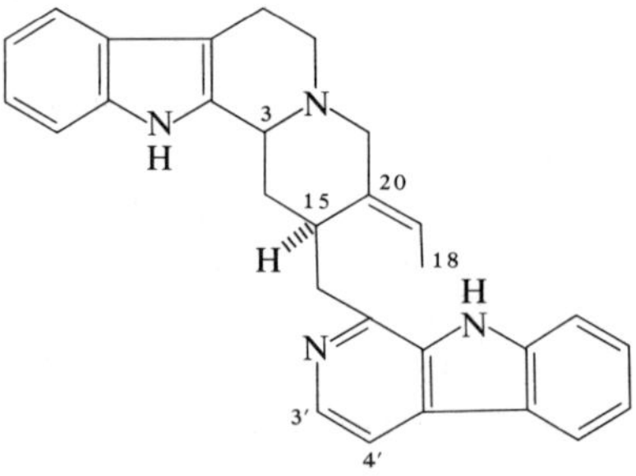

115 Usambarensine
123 N_b-Methyl

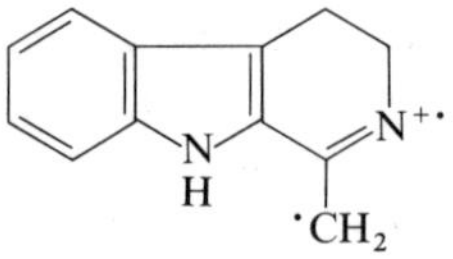

116 *m/e* 183

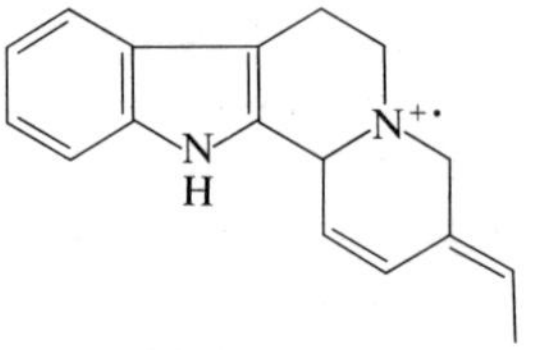

117 *m/e* 250

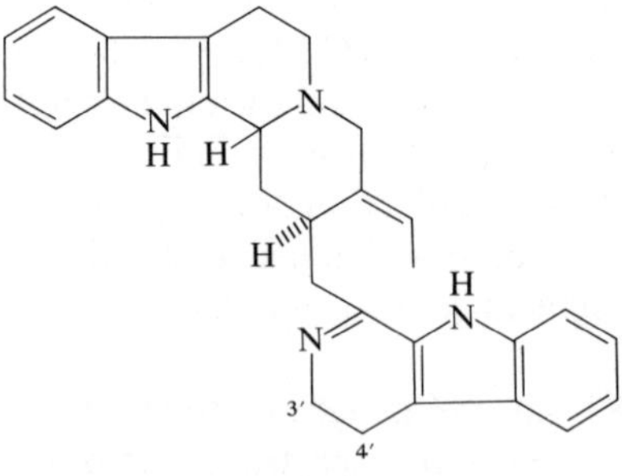

118 3′,4′-Dihydrousambarensine
124 N_b-Methyl

The stereochemistry of 3′,4′-dihydrousambarensine was deduced by partial synthesis (*103*). Condensation of geissoschizoic acid (**119**) with tryptamine in the presence of dicyclohexylcarbodiimide followed by cyclization of the intermediate amide with phosphorus oxychloride gave 3′,4′-dihydrousambarensine (**120**) identical with the natural product. To confirm the C-3 stereochemistry, 3-epi-3′,4′-dihydrousambarensine (**121**) was prepared in analogous fashion from 3-epigeissoschizoic acid (**122**).

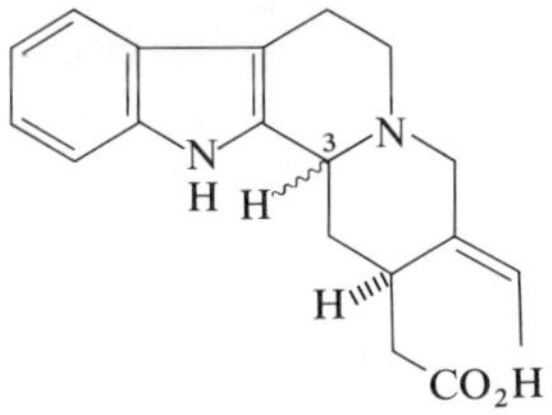

119 H-3α
122 H-3β

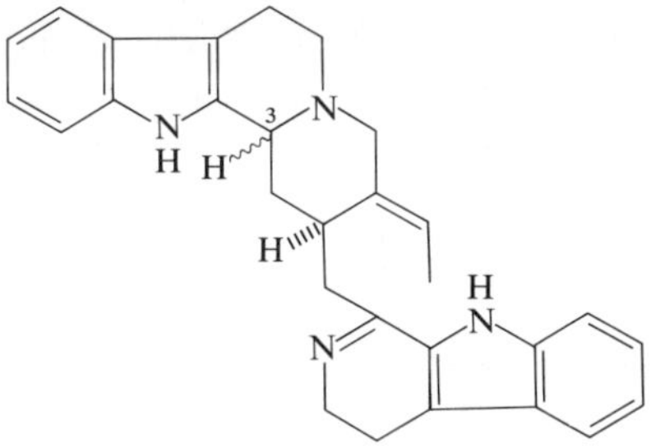

120 3′,4′-Dihydrousambarensine, H-3α
121 H-3β

The two remaining alkaloids isolated were quaternary alkaloids, and on the basis of spectral properties they were assigned the structures N_b-methylusambarensine (**123**) and N_b-methyl-3′,4′-dihydrousambarensine (**124**) (*98*).

N_b-Methylusambarensine (**123**) showed a molecular ion at *m/e* 447 and important fragment ions at *m/e* 265, 264, 248, 185, 182, and 143. Methylation of usambarensine (**115**) gave a product identical with the natural product (*98*).

Only very small quantities of the final compound were obtained, showing a molecular ion at *m/e* 449 and major fragment ions at *m/e* 264, 263, 250, 249, 185, (base peak) 154, and 143. Methylation of 3′, 4′-dihydrousambarensine (**118**) afforded a product identical with the natural product (*98*).

After establishment of the structure of dihydrousambarensine as **120** by partial synthesis (*103*), attention was turned to usambarine (*104*), which had originally (*92*) been suggested to have structure **92**.

When geissoschizal (**125**) or 3-epigeissoschizal (**126**) was condensed with *N*-methyltryptamine (**30**) in 0.3 *M* sulfuric acid at 103°, none of the diastereomeric products was identical with natural usambarine (*104*). This observation led to the reisolation of usambarine from the leaves of *Strychnos usambarensis* by Angenot and co-workers and to a redetermination of its physical properties. The molecular formula was verified as being $C_{30}H_{34}N_4$ and the UV spectrum was purely indolic. However, this time the IR spectrum displayed Bohlmann bands at 2840, 2795, and 2780 cm^{-1}, together with an absorption at 918 cm^{-1} interpreted as being due to a vinyl group.

The presence of a vinyl group was verified by the ^{1}H-NMR spectrum which showed a complex three-proton multiplet for the C-18 and C-19 protons in the region 5.46–4.95 ppm. An *N*-methyl group was observed at 2.42 ppm and two indolic NH protons at 6.40 and 7.74 ppm. The stereochemistry at C-15 was assumed on biogenetic grounds, leaving C-3 and C-17 to be determined. Examination of the CD spectrum of usambarine indicated a strong positive Cotton effect at 276 nm analogous to that observed for ochrolifuanine B (**104**). On this basis, usambarine was assigned the complete structure shown in **127**, in which C-3 and C-17 both have the *S*-configuration (*105*). The stereochemistry at C-20 was inferred by considering a biogenetic derivation from strictosidine, the probable biogenetic precursor of all indole alkaloids (*105*).

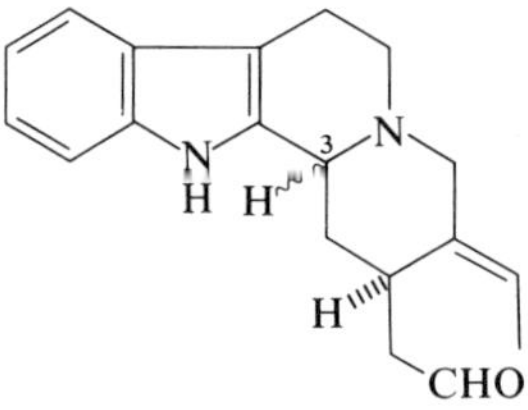

125 Geissoschizal, C-3Hα
126 3-Epigeissoschizal, C-3Hβ

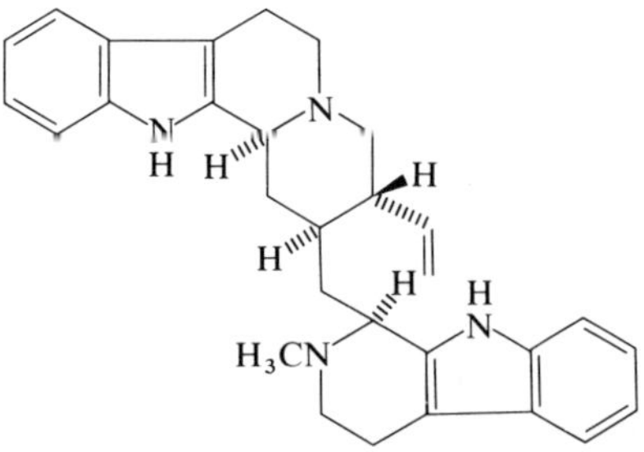

127 Usambarine

C. Tchibangensine

Tchibangensine (**132**) was obtained by Le Men and co-workers from the root bark and stem bark of *Strychnos tchibangensis* Pellgr., a plant native to Zaire (*106*).

The alkaloid was the major alkaloid of the plant (90% of the total alkaloids), and although other alkaloids were detected, they have not yet been isolated and characterized.

Tchibangensine, mp 143°–145°, $[\alpha]_D + 109.7°$ in $CHCl_3$, showed a complex UV spectrum (λ_{max} 227, 286, 293, 319 nm) corresponding to an indole chromophore superimposed on that of an indole further conjugated with an enamine or imine. The spectrum was changed markedly on the addition of acid to λ_{max} 220, 250, 291, and 358 nm with shoulders at 274 and 283 nm. In the IR spectrum important absorptions were noted at 3480 cm^{-1} for an NH group and at 1620, 1600, and 1545 cm^{-1} for aromatic and imine groups.

The molecular ion at *m*/*e* 434 analyzed for $C_{29}H_{30}N_4$, and the fragmentation pattern indicated that tchibangensine belonged to the usambarine group of alkaloids with no additional oxygen or methyl substituents. Thus, principal fragment ions were observed at *m/e* 250, 249, 247, 235, 184, and 183; these were, respectively, assigned structures **117**, **114**, and **128**–**131**.

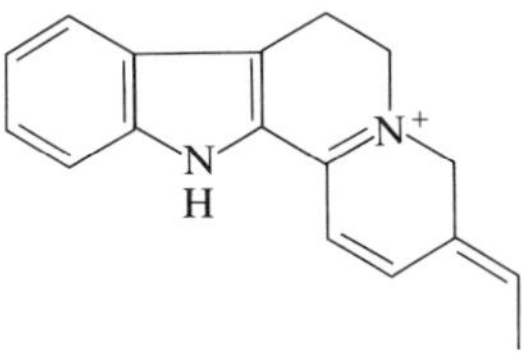

114 *m/e* 249 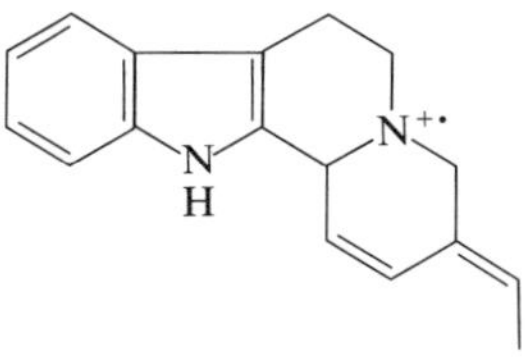 **117** *m/e* 250 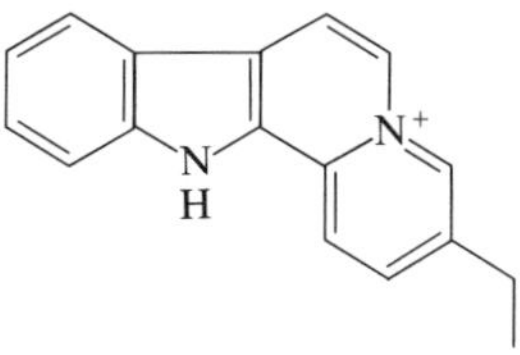 **128** *m/e* 247

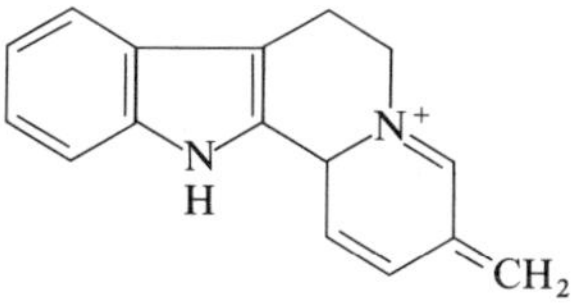

129 *m/e* 235 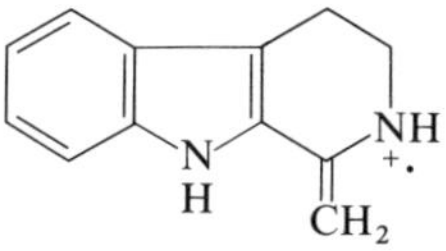 **130** *m/e* 184 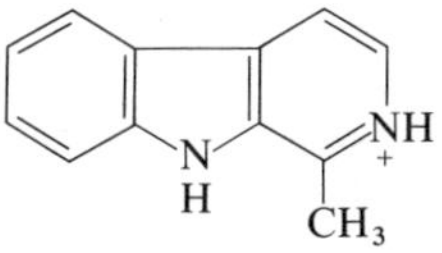 **131** *m/e* 183

The ^{1}H-NMR spectrum revealed a number of important features, including the presence of an ethylidene group with a doublet ($J = 6$ Hz) at 1.62 ppm and a quartet at 5.45 ppm, two indolic NH peaks at 10.9 and 8.2 ppm, and eight aromatic protons in the region 6.9–7.8 ppm. No other olefinic protons were observed.

On this basis, tchibangensine was assigned structure **132**, in which it remained to determine the C-3 stereochemistry. Catalytic reduction of tchibangensine using a palladium catalyst gave a mixture of two products, ochrolifuanine C (**105**) and ochrolifuanine D (**106**), identical in their spectroscopic and optical properties to the synthetic products (*94*). Because **105** and **106** differ only in their stereochemistry at C-17, the stereochemistry at C-3 in **132** must be α and the C/D ring junction cis (*106*).

Several other factors also support structure **132** for tchibangensine. For example, tchibangensine does not show Bohlmann bands in the IR spectrum, and the chemical shift of C-3 (53.2 ppm) is close to that of geissoschizine (53.6 ppm) which also has a cis C/D ring junction. In a compound with a trans C/D ring junction, such a signal would be expected at about 58–59 ppm. In addition, a highly deshielded (160.9 ppm) quaternary carbon was observed, suggesting the presence of an imine grouping. The pertinent ^{13}C-NMR data are shown in Table IV. Finally, mild selective oxidation of tchibangensine with cupric chloride gave usambarensine (**133**) (1.7% yield) identical with the natural product.

Although these data establish the relationship of tchibangensine (**132**) with usambarensine (**133**), the possible identity of **132** with 3′,4′-dihydrousambarensine (**120**) remains to be established. No direct comparison of the

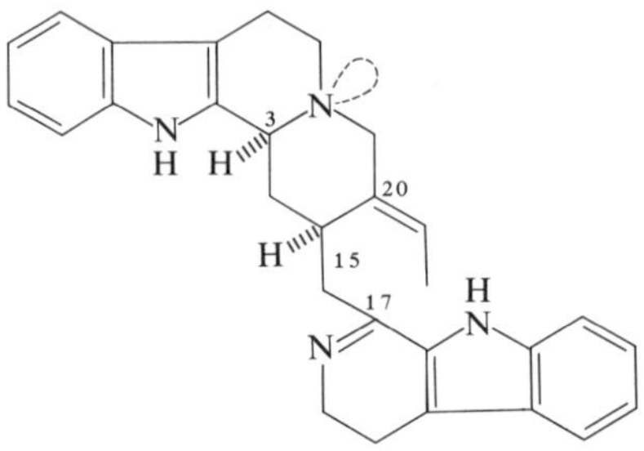

132 Tchibangensine

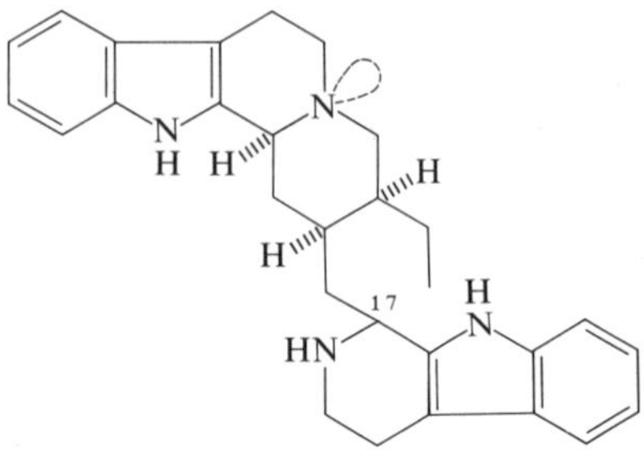

105 Ochrolifuanine C, H-17α
106 Ochrolifuanine D, H-17β

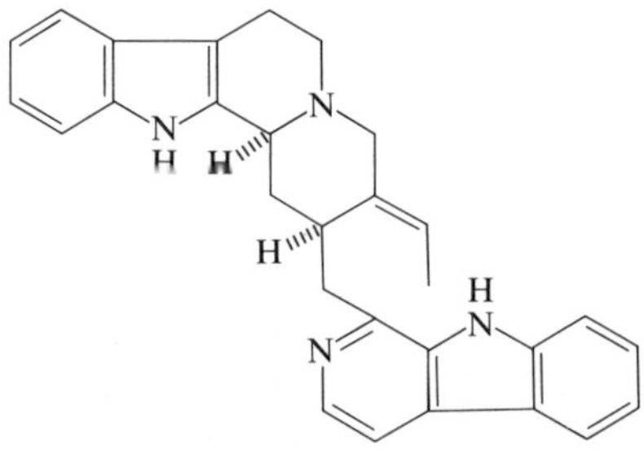

133 Usambarensine

samples appears to have been made. In addition, there is some confusion concerning the C-3 stereochemistry of usambarensine. In the original publication (*98*) H-3 was assigned a β-stereochemistry on the basis of an oxidation–reduction sequence to give a C-3 epimer thought to be the α-trans isomer. Subsequent publications (*99*, *101*) give the C-3 stereochemistry as α without any rationale for the need of a stereochemical reassignment.

D. 18,19-Dihydrousambarine (**134**)

Angenot and co-workers have recently reported (*102*) on their extensive studies of the leaves of *Strychnos usambarensis*, and besides usambarine (**127**) (*104*) and strychnopentamine (Section IV,H) (*107*), several new alkaloids in this series were isolated.

The first alkaloid was closely related to usambarine, but had a molecular weight of 452 ($C_{30}H_{36}N_4$), two mass units higher than **127**. Spectral data suggested the structure 18,19-dihydrousambarine (**134**) for this compound. Thus, the IR spectrum did not show ν_{max} 920 cm^{-1} typical of a vinyl group, and the NMR spectrum indicated the presence of an ethyl group (methyl triplet at 0.82 ppm). The mass spectrum displayed a characteristic base peak

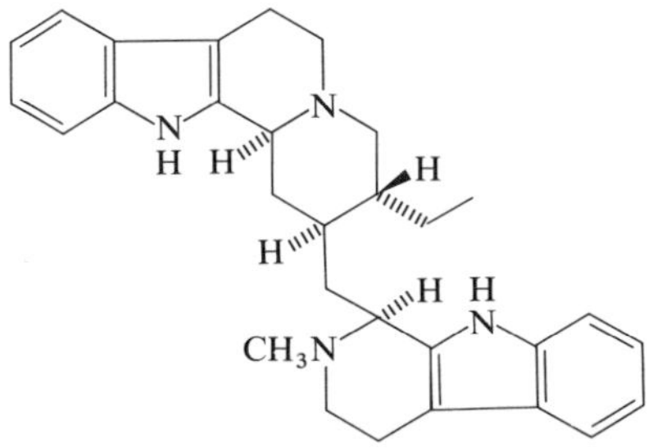

134 18,19-Dihydrousambarine

at *m/e* 185 (**90**) analyzing for $C_{12}H_{13}N_4$, demonstrating that the additional two mass units were in the corynane unit. The remaining NMR spectral features were very similar to those of usambarine, with an *N*-methyl singlet at 2.39 ppm, eight aromatic protons in the region 6.84–7.53 ppm, and two exchangeable protons at 6.52 and 7.90 ppm.

The stereochemistry of the product was suggested to be the same as that for usambarine (3*S*, 4*R*, 15*S*, 17*S*, and 20*R*). Observation of Bohlmann bands in the IR spectrum and a strong positive Cotton effect indicated 3*S*-, 4*R*-, and 17*S*-configurations, and the 15*S*-configuration was based on biogenetic grounds. The 20*R*-configuration (equatorial) was assigned on the basis of the asymmetric nature of the methyl triplet at 0.82 ppm. The "new" alkaloid was, therefore, assigned the structure 18,19-dihydrousambarine (**134**) (*102*). The alkaloid was not correlated with usambarine (**127**).

However, the shift of the methyl triplet would appear to be more in agreement with a 17β-H stereochemistry, since ochrolifuanine A (**103**) displayed this signal at 0.73 ppm (*93*).

E. Hydroxyusambarine Derivatives

Usambaridine was obtained by Koch and co-workers (*108*) from the leaves of *Strychnos usambarensis* together with usambarine. The UV spectrum was mainly indolic, with a shoulder at 312 nm shifting to λ_{max} 325 nm on the addition of base. These data suggested that the alkaloid was dimeric, having indole and hydroxyindole chromophores.

In the mass spectrum a molecular ion was observed at *m/e* 466, 16 mass units higher than usambarine. The base peak was at mass 185 (**90**) as it was in **127**, indicating that the phenolic group should be in the A ring. The main features of the ^{1}H-NMR spectrum were very similar to those of usambarine except that three signals were removed on the addition of D_2O. On this basis, usambaridine was assigned structure **135**, in which the hydroxy group was not precisely located (*108*).

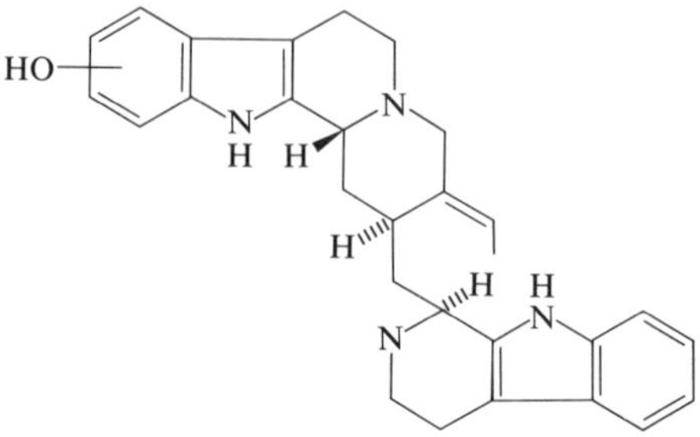

135 Usambaridine (incorrect)

Several phenolic alkaloids were also encountered by the Belgian group (*102*). Two of these were given the names usambaridine Br and usambaridine Vi,* the latter being the more abundant. Usambaridine Br gave a molecular ion at *m/e* 466 which analyzed for $C_{30}H_{34}N_4O$, and the alkaloid gave a bathochromic shift in alkali (305 to 320 nm), indicating the presence of a phenolic hydroxy group. The fragmentation pattern displayed a base peak at *m/e* 185, limiting the substitution to ring A. In the aromatic region of the ^{1}H-NMR spectrum, a doublet ($J = 2$ Hz) at 6.22 ppm and a doublet of doublets ($J = 8, 2$ Hz) at 6.45 ppm indicated the phenolic group to be located at C-11. A three-proton complex for a vinyl group was also observed in the region 5.46–4.96 ppm, together with an *N*-methyl group at 2.37 ppm.

Bohlmann bands were observed in the IR spectrum, and because of the strong positive Cotton effect in the CD spectrum, usambaridine Br was assigned structure **136** (*102*). A preferred name for this alkaloid is 11-hydroxyusambarine.

Usambaridine Vi was very similar to usambaridine Br in all spectral properties. The mass spectrum again showed an M^+ at 466 mu, analyzing for $C_{30}H_{34}N_4O$; similarly, the oxygen function was traced to a phenolic group (bathochromic shift in the UV from 312 to 328 nm). A base peak was observed at *m/e* 185 (**90**), confining the phenolic group to ring A. But in this instance, two doublets ($J = 9$ Hz) were found at 6.58 and 6.87 ppm for H-9 and H-11, suggesting that the phenolic group was located at C-12. The Cotton effect of usambaridine Vi was similar to that of both usambarine (**127**) and usambaridine Br (**136**); consequently, the alkaloid was assigned structure **137** (*102*). The name 12-hydroxyusambarine is to be preferred.

Two other closely related alkaloids were also obtained and shown to be dihydro derivatives of **136** and **137**. In the NMR spectrum of one of the isolates the signals for the vinyl group were absent, but the remaining data indicated that all other aspects of the structure were the same as **137**, including the CD curve. This alkaloid was, therefore, assigned the structure

* Br = brown, Vi = violet; based on a color reaction with *sel de bleu solide* B.

18,19-dihydro-12-hydroxyusambarine (**138**) (*102*). The second alkaloid gave similar UV spectral properties to **136** and was consequently proposed to be 18,19-dihydro-11-hydroxyusambarine (**139**) (*102*).

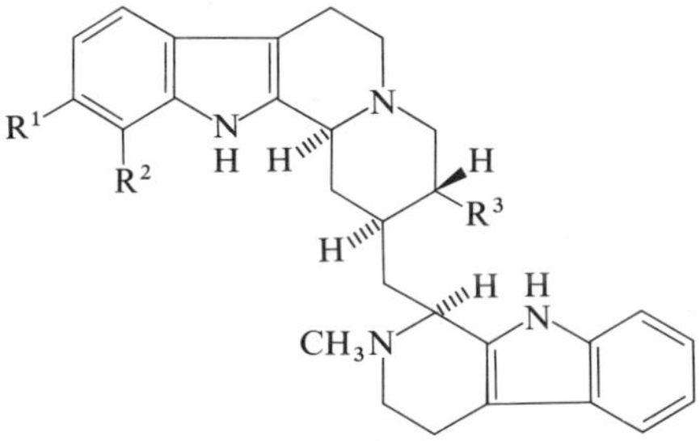

	R^1	R^2	R^3	
136	OH	H	$-CH{=}CH_2$	11-Hydroxyusambarine
137	H	OH	$-CH{=}CH_2$	12-Hydroxyusambarine
138	H	OH	$-CH_2CH_3$	18,19-Dihydro-12-hydroxyusambarine
139	OH	H	$-CH_2CH_3$	18,19-Dihydro-11-hydroxyusambarine

The last of the "simple" bisindole alkaloids to be obtained from the leaves of *S. usambarensis* was strychnobaridine. A molecular ion was observed at *m/e* 482 which analyzed for $C_{30}H_{34}N_4O_2$. A shift similar to that observed for **137** was noted in the UV spectrum on the addition of base and, from the appearance of ions at *m/e* 201 and 187 (?), the alkaloid was proposed to have a hydroxy group in each indole nucleus. Because both the ^{1}H-NMR and CD spectra were similar to those of **137**, the isolate was suggested to have structure **140** in which the phenolic groups were not precisely placed (*102*).

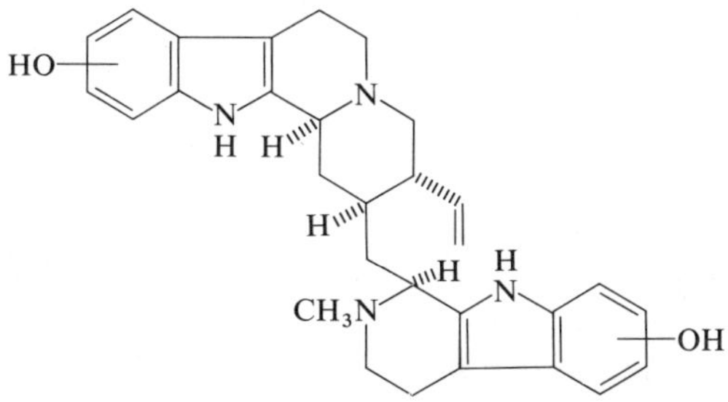

140 Strychnobaridine

These data, however, are not in agreement with an observed base peak at *m/e* 185 (**90**). Rather, they suggest that strychnobaridine could be a mixture of two compounds. Bearing in mind the similar color reaction and UV spectral properties with **137**, strychnobaridine could well be a mixture of compounds **141** and **142**.

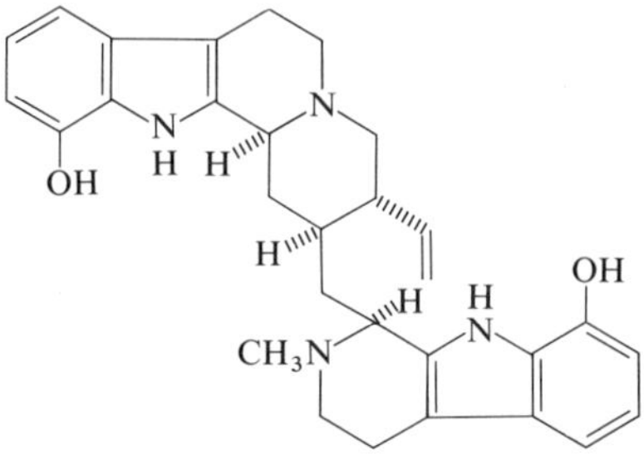

141 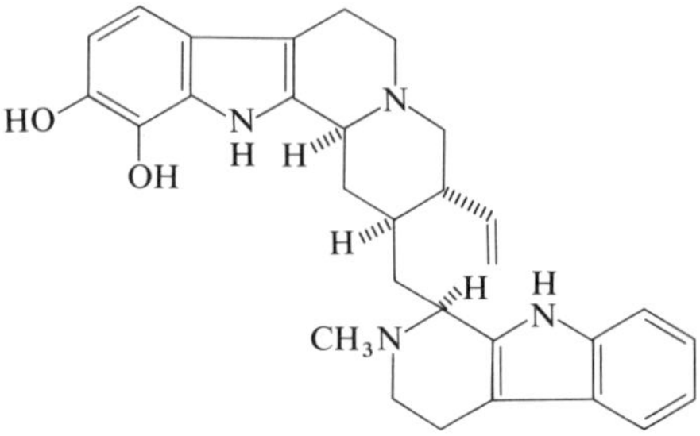 **142**

F. Alkaloids of *Rauwolfia obscura*

The roots of *Rauwolfia obscura* K. Schum. native to Zaire have been investigated over many years and several monomeric indole alkaloids obtained. Timmins and Court have reported on the alkaloids of the leaves (*109*) with some interesting results. In particular, the leaf alkaloids were found to be quite different, and no oxindoles or reserpiline type alkaloids could be detected. 10-Methoxygeissoschizol (**143**), α-yohimbine (**144**), and alstonine (**145**) were the only monomeric alkaloids detected, but four dimeric alkaloids, B_1, C_2, D_1, and D_2 in the ochrolifuanine series, were obtained.

Alkaloid B_1, mp 160°–162°, showed a typical indole chromophore (λ_{max} 226, 280, and 289 nm) in the UV spectrum and NH absorption (λ_{max} 3460 cm^{-1}) in the IR spectrum. An olefinic proton (quartet) was observed at 5.45 ppm indicating the presence of an ethylidene group, but the expected methyl doublet was not noted. In the aromatic region of the NMR spectrum nine protons were found, two of which exchanged on the addition of D_2O. An aromatic methoxy group was detected at 3.80 ppm, and although its location was not proven it was suggested to be at position 10 of the corynane nucleus. No alkyl *N*-methyl group was observed.

The mass spectrum of alkaloid B_1 displayed a molecular ion at *m/e* 466 which analyzed for $C_{30}H_{34}N_4O$, and major fragment ions at *m/e* 326, 325, 282 ($C_{18}H_{22}N_2O$), 281, 238, 185 ($C_{12}H_{13}N_2$), 169, 168, and 143. Such a fragmentation is typical of alkaloids in this series; indeed, the spectrum of B_1 resembled that of usambarine (**127**). The ion at *m/e* 281 was suggested to have structure **146** and the base peak at *m/e* 185 was formulated as **147**. On this basis and owing to the similarity of the IR spectra of **143** and alkaloid B_1, the latter was assigned structure **148** (*109*). No stereochemistry was assigned to C-17 and no proof of the C-3 stereochemistry was carried out.

Alkaloid C_2, mp 142°–144°, also showed a purely indolic UV spectrum (λ_{max} 3450 and 3300 cm^{-1}). In the NMR spectrum a single olefinic proton was observed at 5.3 ppm, together with both aromatic methoxy (3.8 ppm)

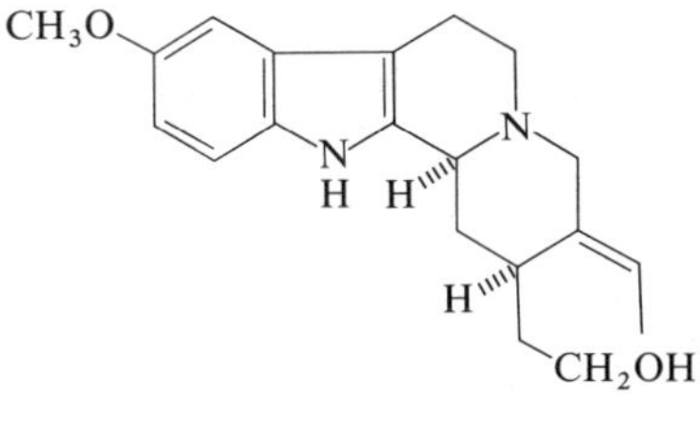

143 10-Methoxygeissoschizol

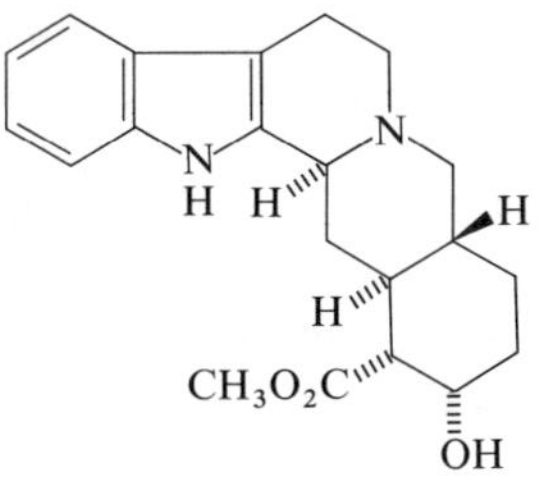

144 α-Yohimbine

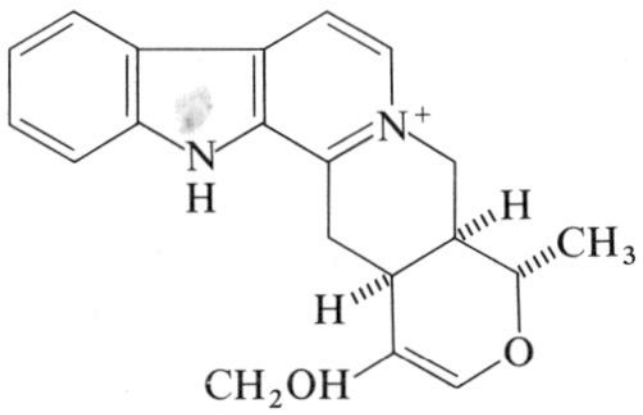

145 Alstonine

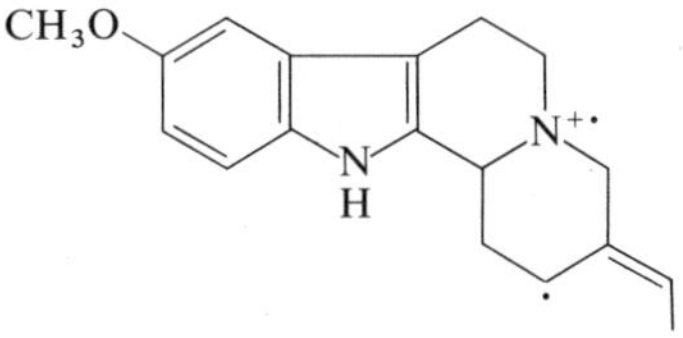

146 *m/e* 281

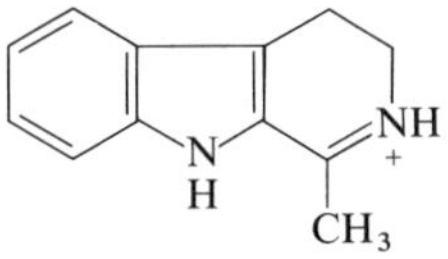

147 *m/e* 185

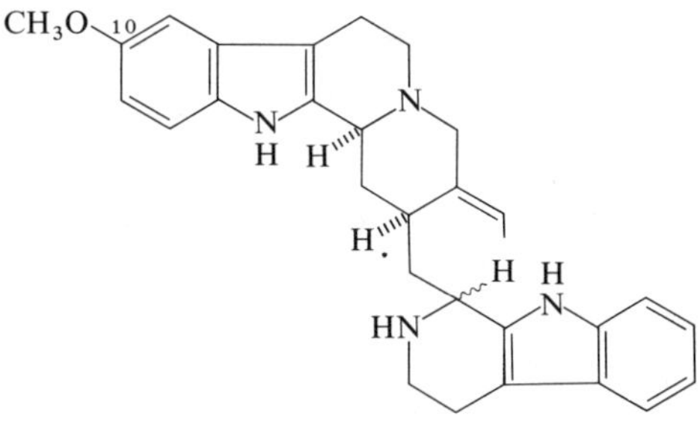

148 Alkaloid B_1

and aliphatic *N*-methyl (2.5 ppm) groups. However, these data are not in agreement with the mass spectrum, which showed a molecular ion at *m/e* 466 analyzing for $C_{30}H_{34}N_4O$ isomeric with alkaloid B_1. Two characteristic fragment ions were observed at *m/e* 281 ($C_{18}H_{21}N_2O$) and *m/e* 185 ($C_{12}H_{13}N_2$) (*109*). A compound containing both N—CH_3 and O—CH_3 groups in this series should have a molecular weight of 480 and contain 31 carbon atoms. No structure was suggested for alkaloid C_2.

Alkaloid D_1, mp 190°–192°, similarly showed a purely indolic UV spectrum (λ_{max} 225, 281, and 289 nm), and the presence of two NH groups (λ_{max} 3450, 3350 cm^{-1}) was also noted. No NMR data were presented, but a molecular ion was observed at *m/e* 466, isomeric with alkaloids B_1 and C_2. The mass spectrum was now somewhat different, showing major fragment ions at *m/e* 280, 265, 233 (base peak), 200, 199, 185, 184, 183,

169, and 143. The compound was suggested to be an isomer of alkaloids B_1 and C_2, but no structural details were suggested except the possibility of an aromatic methoxy group in the corynane unit (*109*).

Alkaloid D_2 was obtained as an amorphous yellow powder showing λ_{max} 226, 280, and 290 nm. A molecular ion was observed at *m/e* 480 and major fragment ions at *m/e* 280, 279, 278, 277, 251, 250, 249, 185, 184 (base peak), 183, 171, and 169. No further information is available (*109*).

G. Cinchophyllamine and Isocinchophyllamine

Cinchophyllamine and isocinchophyllamine were obtained by a group at the C.N.R.S. in Gif (France) from the leaves of *Cinchona ledgeriana* Moens in 1965 (*110*), and in collaboration with Budzikiewicz and Djerassi the proposed structures for these alkaloids were published the following year (*111*). The alkaloids co-occurred with quinamine (**149**) but not with any of the quinoline alkaloids present in the bark. Details of the structure elucidation of these alkaloids were discussed previously in this series (*8*), where it was indicated that of the two suggested structures for cinchophyllamine (**150** and **151**), the latter was regarded as more likely based on the dissimilarity of the mass spectra in comparison with a model compound (**152**) prepared from dihydrocorynantheal. It was suggested (*111*) that isocinchophyllamine was a stereoisomer of cinchophyllamine, although no indication was made as to the location of the isomerization.

In an attempt to confirm the structure analysis, isocinchophyllamine was submitted to single-crystal X-ray crystallography (*112*). Somewhat surprisingly, the analysis indicated that the alternative, previously dismissed structure for isocinchophyllamine was in fact correct, and that the alkaloid should be represented by structure **153**. At present, no additional information is available on the structure of cinchophyllamine and its relationship to isocinchophyllamine (**153**).

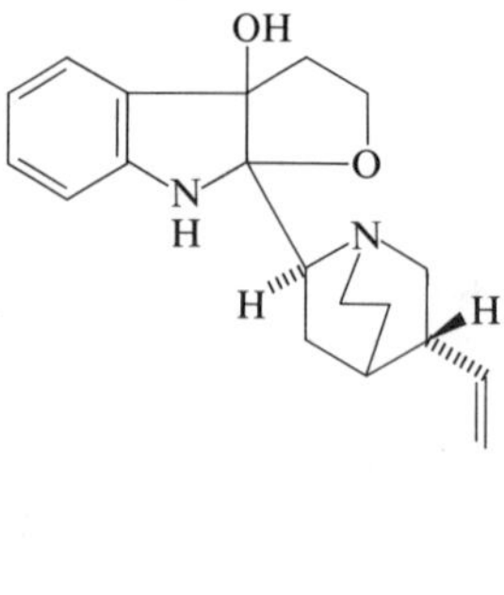

OH O N H N H H

149 Quinamine

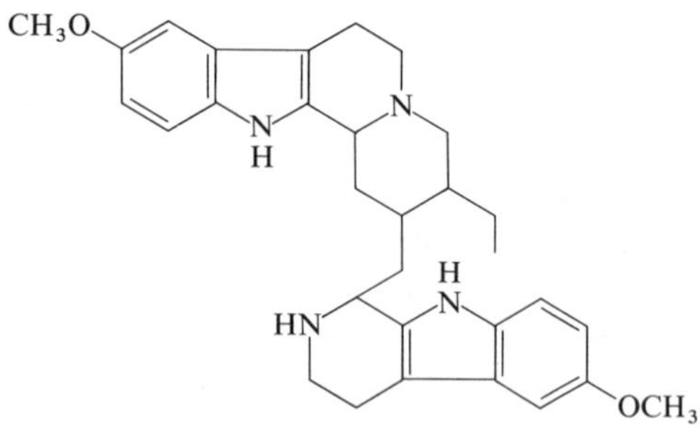

CH_3O N H N H N HN OCH_3

150

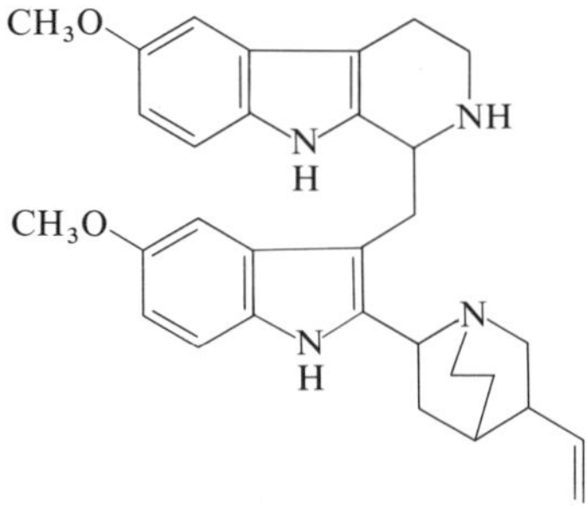

151

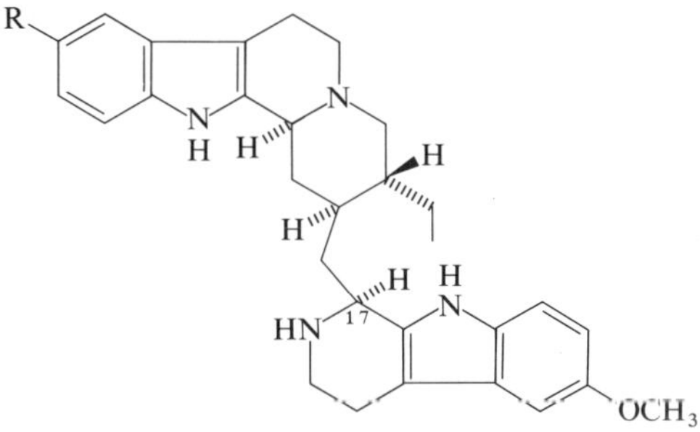

152 R = H, H-17 not defined
153 R = OCH_3

Potier has indicated (*113*) that the probable structure of cinchophyllamine is that of the C-17 isomer. Material is apparently lacking in order to settle this point.

H. Strychnopentamine and Related Isomers

The most novel alkaloid to be isolated in this series is strychnopentamine, which was obtained by Angenot and co-workers from the leaves of *Strychnos usambarensis* Gilg. (*102*, *107*) together with two other isomers (*102*).

The spectral data of strychnopentamine indicated the presence of five nitrogen atoms, a phenolic hydroxy group (which was not affected by the addition of alkali!), two *N*-methyl groups (2.21 and 2.31 ppm), a vinyl group, and an aromatic proton doublet at 6.49 ppm. The molecular ion at *m/e* 549 ($C_{35}H_{43}N_5O$) gave principal fragment ions at *m/e* 348 ($C_{22}H_{26}N_3O$) and 185 (base peak). Again, an unsubstituted lower indole moiety was indicated. The structure could not be rationalized from the data (*102*); consequently, an X-ray crystallographic analysis was undertaken (*107*).

Strychnopentamine crystallized with 3.5 molecules of water in the space group $P4_12_12$ with dimensions $a = b = 13.895$ Å and $c = 36.105$ Å. The crystal structure was solved by direct methods to an *R* factor of 8.4% after 2125 reflections. The absolute configuration was not determined but inferred on biogenetic grounds, considering the fixed stereochemistry at C-15 in all corynane alkaloids. Strychnopentamine was therefore assigned structure **154** (*107*) and joins the small group of alkaloids, such as nicotine, brevicolline, and ficine, which contain an *N*-methylpyrrolidine group attached to an aromatic ring.

The two remaining compounds, designated isostrychnopentamine A and isostrychnopentamine B, were isomeric with strychnopentamine (*102*). The UV, IR, mass, and NMR spectral data were very close for the isomer A and strychnopentamine, suggesting that they might be stereoisomers at the linkage to the *N*-methylpyrrolidine unit and that isomer A had structure **155**.

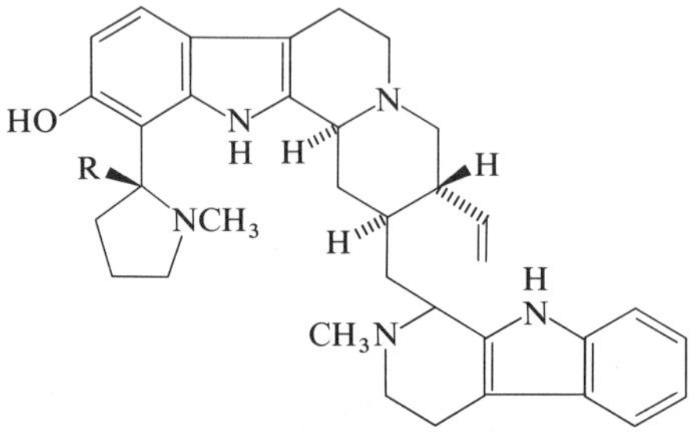

154 R = β-H Strychnopentamine
155 R = α-H Isostrychnopentamine A (proposed)

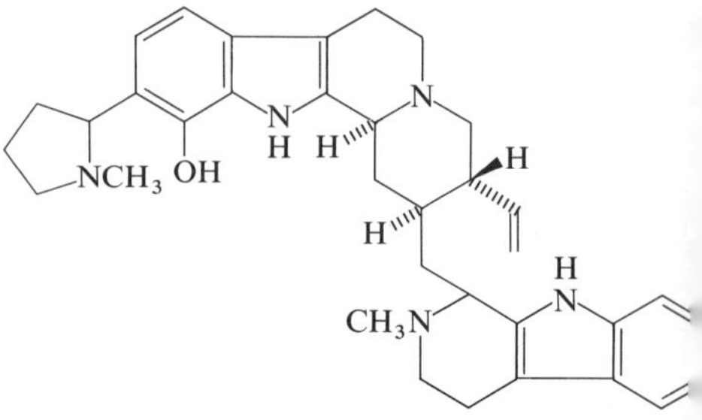

156 Isostrychnopentamine B (proposed)

Isostrychnopentamine B showed a UV spectrum close to that of 12-hydroxyusambarine (**137**). Therefore, although not inferred by the authors, it is suggested that this compound has structure **156**, in which the stereochemistry at C-2 of the pyrrolidine ring is unknown.

I. Pharmacology of the Alkaloids of *S. usambarensis*

Ten alkaloids, eight of them dimeric, were obtained from the root bark of *Strychnos usambarensis*, and their pharmacologic actions were studied. In particular, their actions were studied on noradrenergic, muscarinic, and nicotinic receptors (*99*, *101*).

At 3×10^{-6} *M* usambarensine was antagonistic to the effects of carbachol on the isolated rat intestinal muscle (muscarinic receptor). The other tertiary alkaloids did not show activity. None of the alkaloids had any potentiating or inhibiting effects on the isolated rat *vas deferens* (noradrenergic receptor).

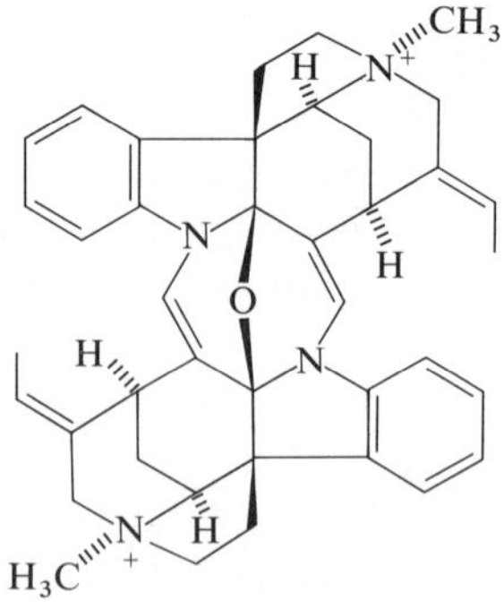

157 C-curarine

As expected, the bisquaternary alkaloids such as calebassine, dihydrotoxiferine, afrocurarine, and C-calebassine were all found to possess potent neuromuscular blocking action at doses of about 5×10^{-7} *M*/kg. Inhibition of acetylcholinesterase (with physostigmine) antagonized the curarizing properties of the alkaloids. Using (+)-tubocurarine as a standard, the decreasing order of potency (factor) was (+)-tubocurarine (1.00), C-curarine (0.94), dihydrotoxiferine (0.38), C-calebassine (0.30), and afrocurarine (0.05).

C-Curarine (**157**), one of the major bisquaternary alkaloids of the root bark, is therefore probably principally responsible for the curarizing effects of the arrow poisons (*101*).

V. Corynanthe–Corynanthe Type

Serpentinine (164)

The alkaloid serpentinine was first obtained from *Rauwolfia serpentina* Benth. ex Kurz (Apocynaceae) by Djerassi and co-workers (*114*). It is a deep yellow base possessing the molecular formula $C_{42}H_{44}N_4O_6$. The UV spectrum corresponds to an addition of the absorptions of serpentine (**158**) and yohimbine (**144**). In agreement with this, serpentinine exhibits two pK'_a values at 6.0 (for the tetrahydro-β-carboline part) and 10.6 (quaternary ammonium part) in aqueous DMF. A saturated ester group (1730 cm^{-1}) and a β-alkoxy-α,β-unsaturated ester (1705 and 1616 cm^{-1}) were observed in the IR spectrum.

The NMR spectrum of serpentinine confirmed the presence of two carbomethoxy groups with singlets at 3.45 and 3.75 ppm (*115*), and also indicated the presence of two doublets at 1.23 and 1.90 ppm. Each of these signals was originally assigned to secondary methyl groups adjacent to oxygen.

Selenium dehydrogenation gave alstyrine (**159**) and Pd–C gave deethylalstyrine (**160**), flavopereirine (**161**) and 5,6-dihydroflavopereirine (**162**), and structure **163** was suggested for serpentinine (*115*).

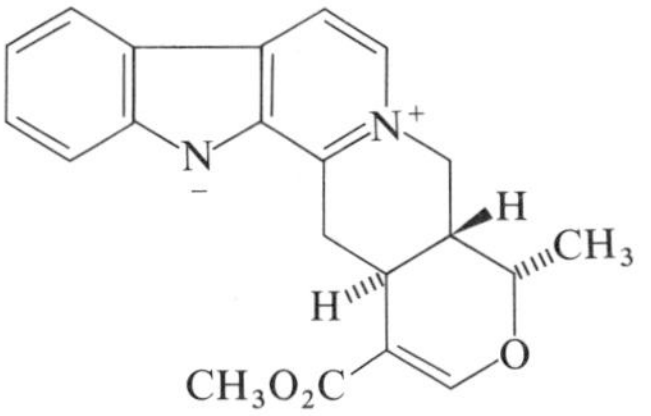

158 Serpentine

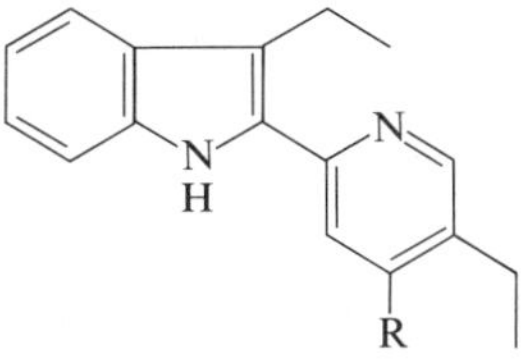

159 R = C_2H_5
160 R = H

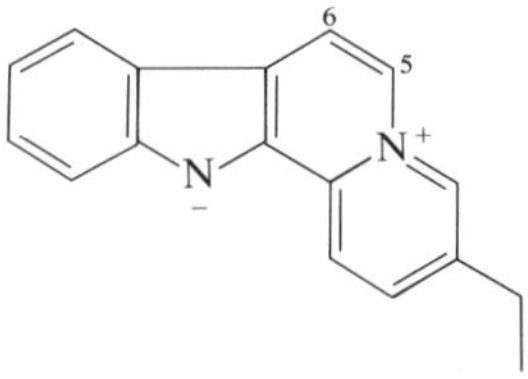

161
162 5,6-Dihydro

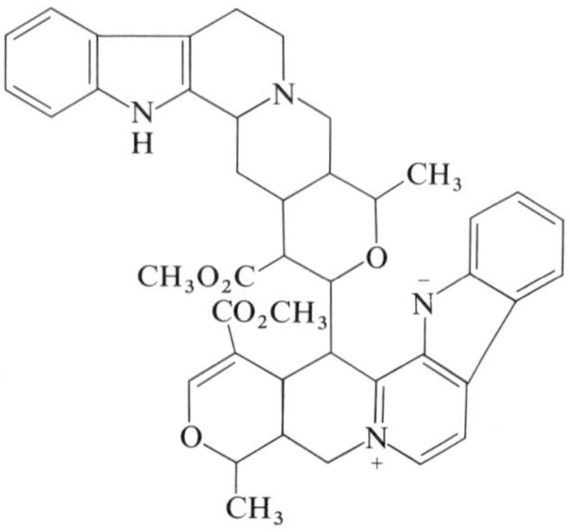

163

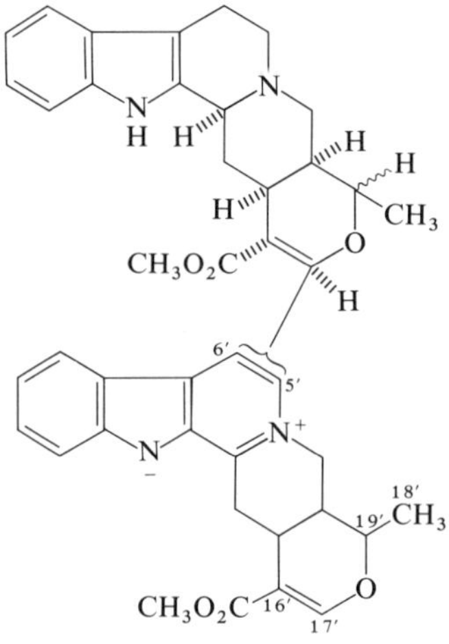

164

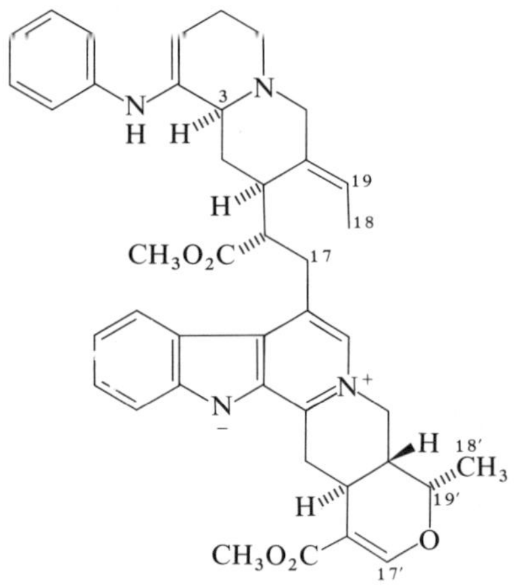

165 Serpentinine

The Swiss group (*116*) did not believe this formulation was correct and investigated further. Sodium borohydride reduction of serpentinine gave two tetrahydroserpentinines whose UV spectra indicated summation of two tetrahydro-β-carboline and a β-alkoxy-α,β-unsaturated ester units. Serpentinine gave rise only to thermal decomposition peaks in the mass spectrum, but the tetrahydroderivatives gave molecular ions at *m*/*e* 704.

To try to deduce an improved structure, a careful study of the mass spectra of these tetrahydroserpentinines was performed. One ion of particular interest was observed at *m*/*e* 438 ($C_{25}H_{30}N_2O_5$), formulated as the lower "half" of the molecule plus a unit $C_4H_7O_2$ from the upper half. On performing the reduction of serpentinine with $NaBD_4$ in CH_3OD, the ion at *m*/*e* 438 shifted to *m*/*e* 441. The secondary nature of the saturated ester was established by treating the tetrahydroserpentinine with $NaOCH_3$–CH_3OD to give a monodeuterated product in which the *m/e* 438 ion now appeared at *m/e* 439 (*116*).

Important ions were also observed at *m/e* 351 and 352 in the mass spectrum of tetrahydroserpentinine, and in the borodeuteride-reduced product these are shifted to *m/e* 354 and 355. In the base-catalyzed deuterium-exchange product, ions were observed at *m/e* 351, 352, and 353. The ion at *m/e* 352 in the original compound therefore involves transfer of the proton α to the secondary carbomethoxy group to the lower half of the molecule.

On this basis, a number of conclusions about the nature of the tetrahydroserpentinines could be made: (a) the lower half is bound to C-17 of the upper half, (b) the possible points of linkage to the lower half can be limited to C-5′, C-6′, C-15′, or C-21′.

The 220-MHz NMR spectrum of serpentinine in d_6-DMSO indicated the presence of ten low-field protons, eight of which could be assigned to the two benzene nuclei. Because the two remaining protons, at 7.6 and 7.3 ppm, in this region appeared as singlets, C-17 of the upper half must be joined to either C-5′ or C-6′ of the lower half. Therefore, one of these singlets should be C-5′ or C-6′ depending on the substitution, and the other on C-17′. Several other of the features observed previously were confirmed; in addition, new facets of the spectrum came to light. A "quartet-like" signal at 5.85 ppm was found, by double-resonance studies, to be the C-19 or C-19′ proton. A one-proton doublet of doublets ($J = 4.5$ and 18.5 Hz) at 4.75 ppm was assigned to C-21′, and a two-proton multiplet at 4.55 ppm contained the second C-19 or C-19′ proton by double resonance. It was suggested that the second proton at 4.55 ppm corresponded to C-17 or C-3 (*13*). With this supporting evidence, structure **164** was proposed for serpentinine in which the position of the linkage of C-17 to the lower unit (C-5′ or C-6′) was left unassigned (*13*). The cis D–E ring stereochemistry was deduced on the basis of a strongly hydrogen-bonded NH observed in both the IR and NMR spectra.

In spite of this compelling evidence for both the nature of the two units and the attachment points of the two halves, several problems remained. Not the least of these was the chemical shift of the C-19 or C-19′ proton to 5.85 ppm and the origin of the other proton at 4.55 ppm. This latter proton was shifted only to 4.3 ppm in tetrahydroserpentinine, which is difficult to understand if it is the C-17 proton.

Some of these problems were resolved (although others were created) when an X-ray crystallographic analysis of serpentinine dihydrobromide dihydrate (mp 280°–290°) was carried out (*117*). These data established that serpentinine actually has structure **165**. Note that the molecular formula of the base should now be revised to $C_{42}H_{44}N_4O_5$.

VI. Corynanthe-*Strychnos*-Type

GEISSOSPERMINE (**166**)

Geissospermine was first isolated from the bark of the Brazilian apocynaceous tree *Geissospermum laeve* (Vellozo) Baillon by Hesse in 1877 (*118*). In 1959 it became one of the first dimeric indole alkaloids to have its structure deduced (*119*). However, two stereochemical points—the nature of the 16′–17′ linkage and the stereochemistry of the C′/D′ ring junction—remained to be established. These problems have now been resolved by single-crystal X-ray analysis (*120*).

The alkaloid crystallized with two molecules of water in a cell which belonged to space group $P2_12_12_1$ and had dimensions $a = 10.274$, $b = 10.673$, $c = 33.087$ Å. The structure was resolved by direct methods after analyzing 2752 reflections.

Carbon atoms 16′ and 17′ were found to have the configurations *R* and *S*, respectively. But of greater interest was the observation that the N-4′ lone pair and the C-3′ proton are cis and α. Furthermore, the conformation of ring D′ is such that the 15′–16′ bond is axially oriented. Greissospermine is therefore represented by the complete structure **166**.

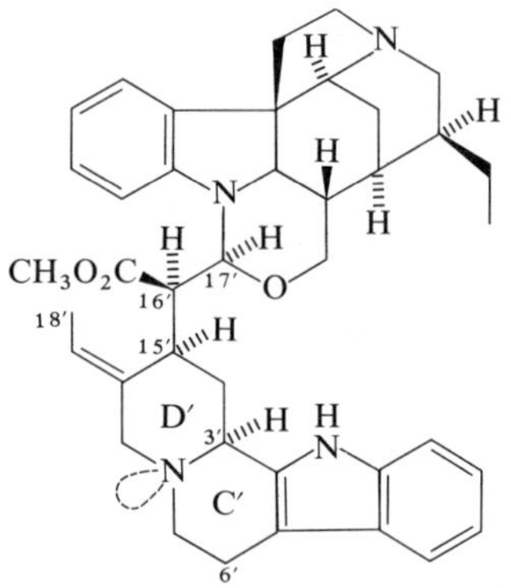

166 Geissospermine

More recently (*121*), Wenkert and co-workers have analyzed the ^{13}C NMR of geissospermine, as shown in **167**. In a typical corynanthoid indole alkaloid, the cis relationship H-3 and H-15 normally leads to a *trans*-quinolizidine configuration in which the C-15 substitutuent is equatorial. Geissospermine, however, has C-3′ and C-6′ shifts indicative of a *cis*-quinolizidine unit, which is in agreement with both the IR and ^{1}H-NMR spectral data. The C-18′ (13.4 ppm) and C-15′ (52.2 ppm) chemical shifts also indicate a δ interaction between C-18′ and H-15′, indicating that C-16′ is axial with

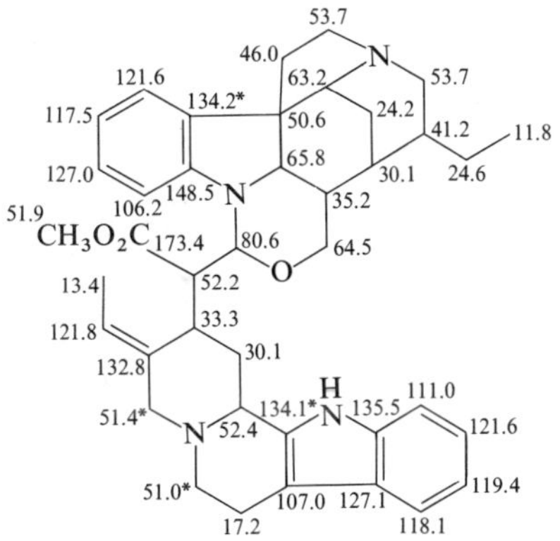

167

respect to ring D′. These data support the contention from the X-ray data that ring D′ adopts a chair conformation in which C-2′ and C-16′ are axially oriented toward ring D′ (*121*). This contrasts with the parent alkaloid geissoschizine, which exists in the *cis*-quinolizidine form in solution but as the *trans*-quinolizidine in the crystal state.

VII. Vobasine–Vobasine Type

A. Accedinisine (**168**)

Achenbach and Schaller have reported (*122*) the isolation of several dimeric indole alkaloids from the root bark of *Tabernaemontana accedens* Muell.-Arg. (Apocynaceae), including a new voacamine derivative, which is described in Section XVI, and two new dimeric alkaloids, accedinisine (**168**) and accedinine (**169**).

The UV spectrum of accedinisine indicated the presence of a purely indolic chromophore. Treatment with mineral acid gave cycloaffinisine (**170**), a compound also available in a similar manner from affinisine (**171**) (*122*).

The nature of the cyclizing groups in **168** was demonstrated from the ^{1}H-NMR spectrum, which indicated the presence of two ethylidene groups, one in each monomeric unit (quartet at 5.18 ppm, doublet at 1.67 ppm). In addition, accedinisine has four other important functional groups, an *N*-methyl group at 2.58 ppm, an indolic *N*-methyl group at 3.54 ppm, a shielded carbomethoxy group at 2.36 ppm, and a hydroxymethyl group as shown by the formation of a monoacetyl derivative. The aromatic region of the NMR spectrum indicated the probable presence of a vobasinol (**172**) unit, and this was supported by the mass spectrum.

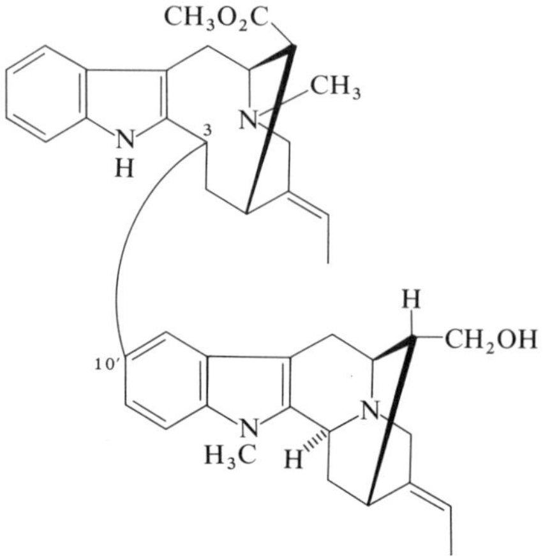

168 Accedinisine

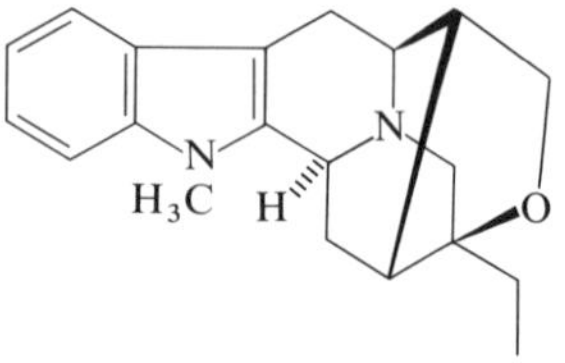

170 Cycloaffinisine

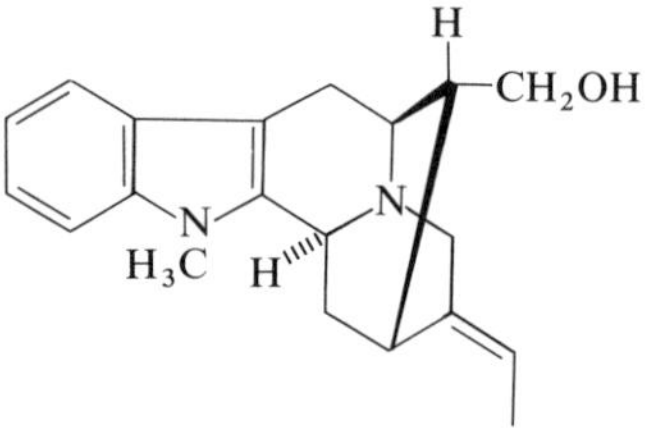

171 Affinisine

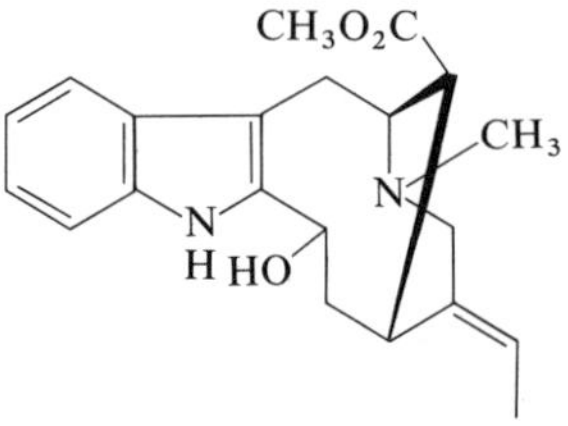

172 Vobasinol

The ^{1}H-NMR spectrum also indicated the point of attachment of the two units, with H-3 in the vobasinyl unit as a doublet ($J = 10$ Hz) at 4.64 ppm and an aromatic region in which the C-9′ and C-12′ protons were observed at 7.20 or 7.14 ppm. Attachment of the vobasinyl unit should, therefore, be at either C-10′ or C-11′. The structure was established by synthesis.

Condensation of vobasinol (**172**) with affinisine (**171**) in 3 *N* aqueous methanol under reflux for 2 hr gave a low yield of accedinisine (**168**), identical with the natural product (*122*).

B. Accedinine (**169**)

The second novel compound isolated from *Tabernaemontana accedens* by Achenbach and Schaller (*122*) was given the name accedinine. Its molecular formula was established as $C_{41}H_{48}N_4O_4$. The UV spectrum was very similar to that of accedinisine (**168**), and the alkaloid on treatment with acid gave cycloaccedine (**173**), thereby indicating the C-16 stereochemistry.

However, whereas accedinisine (**168**) gave a monoacetate derivative, accedinine gave a diacetate derivative having λ_{max} 320 nm.

The molecular ion of accedinine was found at *m/e* 660, with major fragmentations at *m/e* 629, 480, 467, 324, 181, 180, and 122.

In the NMR spectrum, signals characteristic of accedine (**174**) were observed, including an aromatic proton at 7.54 ppm, six additional aromatic protons in the region 6.84–7.11 ppm, and two ethylidene groups (quartet at 5.30 ppm, doublet at 1.67 ppm). An indole *N*-methyl group was observed at 3.45 ppm together with the *N*-methyl of a vobasinyl unit at 2.60 ppm. Substitution at the 3-position of a vobasinyl unit was established by a doublet ($J = 10$ Hz) at 4.60 ppm; and the stereochemistry at C-16′ was determined by a singlet for the carbomethoxy group at 2.43 ppm. Of particular significance from a stereochemical point of view was a doublet ($J = 15$ Hz) at 4.28 ppm for the C-21 α-H of the accedine (**174**) unit. Therefore, accedinine was assigned structure **169**, although the position of attachment to the accedine unit could not be determined with certainty (*122*).

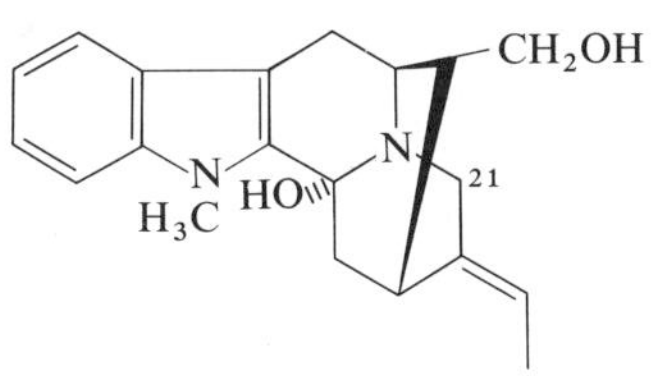

174 Accedine

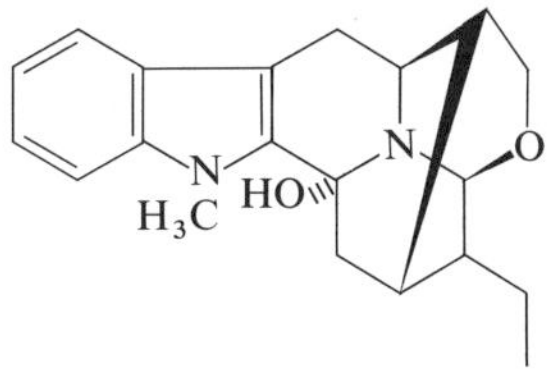

173 Cycloaccedine

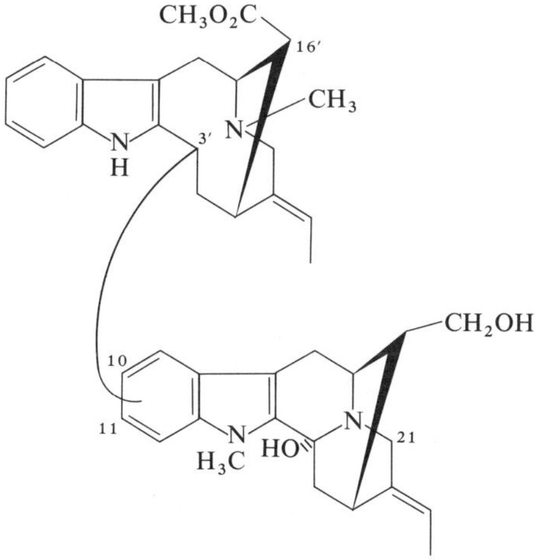

169 Accedinine

C. GARDMULTINE (**180**)

In addition to several monomeric indole alkaloids (*123*, *124*), the Japanese plant *Gardneria multiflora* Makino (Loganiaceae) has afforded a dimeric indole alkaloid (*125*) whose structure was subsequently deduced (*126*).

Gardmultine, mp 283°–285°, crystallized with half a molecule of water and was found to have the molecular formula $C_{45}H_{54}N_4O_{10}$. The NMR spectrum indicated the presence of seven methoxyl groups, six being aromatic and one aliphatic. An oxindole unit was observed in the IR spectrum, displaying absorptions at 3420 and 1713 cm^{-1}. The quite novel UV absorption curve, λ_{max} 213, 247, and 305 nm, was found to be due to a summation of 4,5,7-trimethoxyindoline and 4,5,7-trimethoxyoxindole chromophores; hence, the nuclei of the two halves of the molecule could be defined.

The nature of the side chains in the two units was revealed by the NMR spectrum. A doublet methyl group at 0.88 ppm ($J = 6.8$ Hz) coupled to a quartet olefinic proton at 4.95 ppm indicated the presence of an ethylidene group in half the molecule. The pronounced shielding of the olefinic methyl group will be discussed later. A second olefinic proton (at 5.26 ppm) appeared as a triplet coupled to a methylene at 3.84 ppm. This methylene, from its chemical shift, should be further substituted by the aliphatic methoxy group. Taken together, these data suggested that the dimer was composed of two monomer units, such as gardneramine (**175**) (*123*) and chitosenine (**176**) (*124*), compounds which co-occur with gardmultine.

From the mass spectrum, which showed a molecular ion at *m/e* 810, additional evidence for the proposed formulation of the dimer was obtained. Two prominent ions at *m/e* 412 (22%) and *m/e* 398 (100%) correspond to gardneramine (**175**) and dehydrochitosenine units, and together comprise the molecular ion. Intense fragments at *m/e* 356 and 355 were ascribed to ions **177** and **178**, respectively, of the chitosenine unit (*124*). With these data, it became necessary to deduce the mode of linkage of the two units.

Gardmultine was untouched by bases or acylating agents but reacted with dilute HCl to give a product showing a molecular ion at *m/e* 845 ($C_{44}H_{55}N_4O_{10}Cl$) and a pure oxindole UV chromophore. Such an addition of HCl to give an oxindole was also found when gardneramine (**175**) afforded compound **179** on reflux with dilute HCl.

The NMR spectrum of the dimeric acid-cleaved product now revealed an olefinic methyl group as a doublet at 1.43 ppm. Thus, the ether linkage between C-2 and C-17 is vital in forcing the olefinic methyl group into a highly shielding environment.

When the hydrochloric acid was replaced by formic acid and the intermediate formate ester hydrolyzed with ethanolic potassium hydroxide, a diol was obtained which on mesylation and treatment with potassium *tert*-butoxide in *tert*-butanol regenerated gardmultine. The remaining

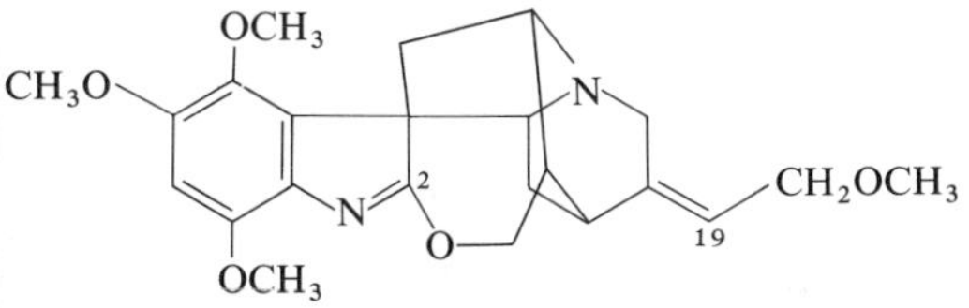

175 Gardneramine

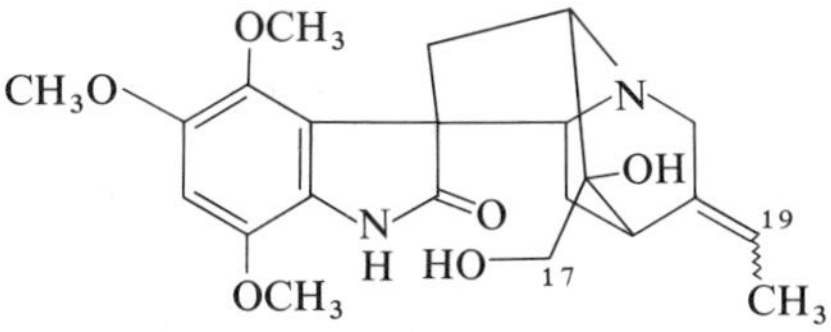

176 Chitosenine

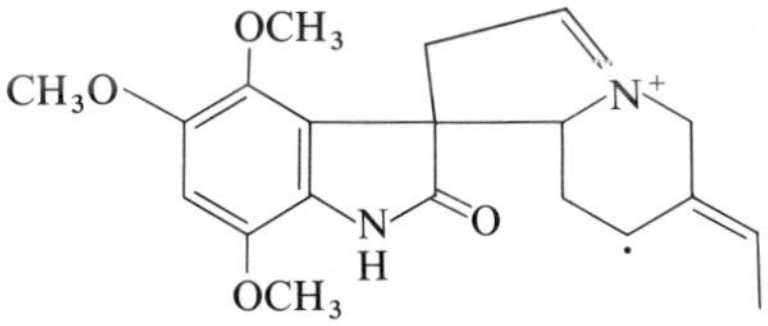

177 *m/e* 356

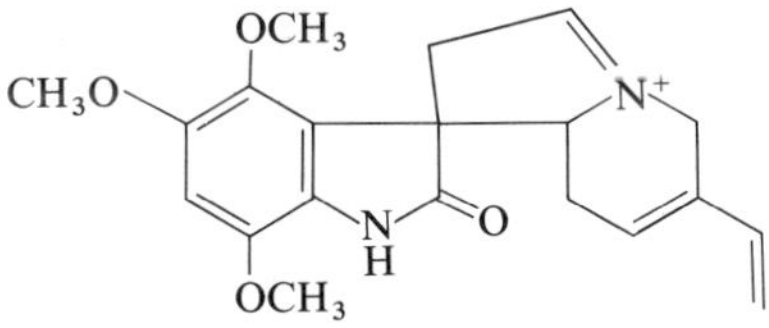

178 *m/e* 355

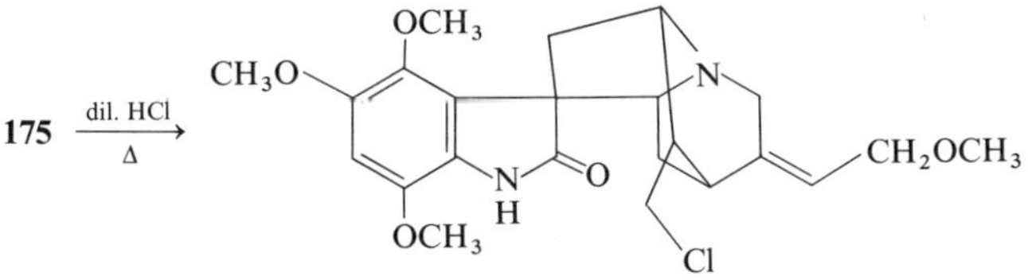

175 $\xrightarrow[\Delta]{\text{dil. HCl}}$ **179**

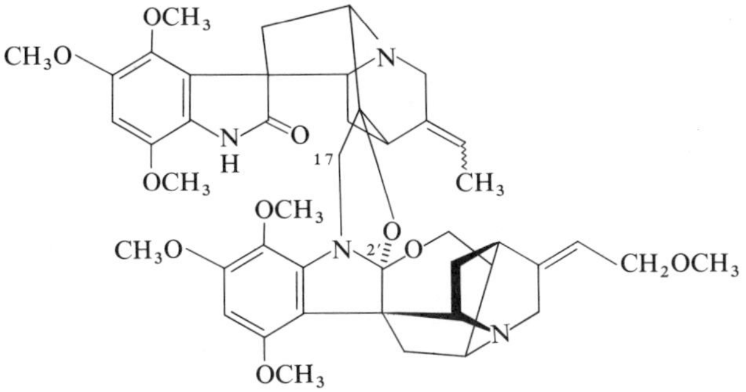

180 Gardmultine

linkage between the two units is, therefore, resistant to both acid and base; because of this, the indoline chromophore was envisaged as linkage between N_a of gardneramine and C-17 of chitosenine. Therefore, the acid-cleaved linkage must be between the C-16 hydroxy group of chitosenine and C-2 of gardneramine, and gardmultine has structure **180**. A proposed biosynthesis of gardmultine (**180**) from **175** and **176** follows these lines (Scheme 7).

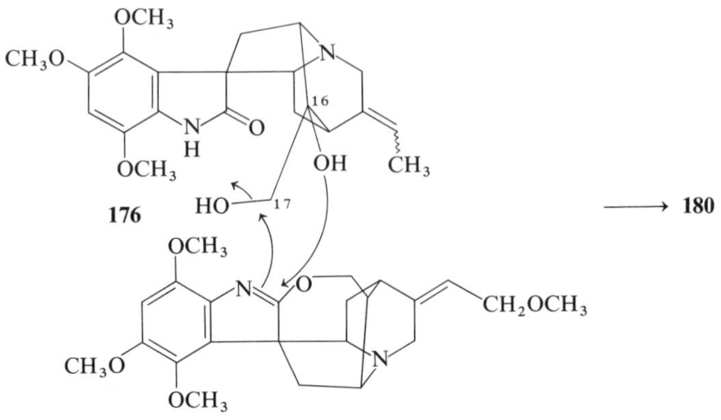

SCHEME 7.

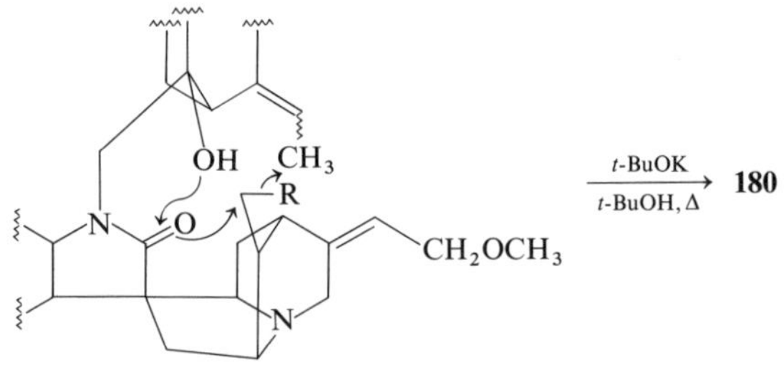

181 R = Cl
182 R = OMs
183 R = OAc

SCHEME 8.

The acid cleavage with HCl can now be rationalized as affording the chloro compound **181**, and the recyclization of the mesylate **182** can be considered as shown in Scheme 8. Reaction of **180** with hot acetic acid gave the acetate **183** which, as expected, had an oxindole UV spectrum.

VIII. Bisindole Alkaloids of *Alstonia* Species

Since the last summary of the alkaloids of *Alstonia* species in this series (*127*), significant progress has been made in the structure elucidation and synthesis of this interesting alkaloid group. For convenience, all the dimeric alkaloids isolated from *Alstonia* species are summarized later in Table XXI.

A. DES-N_a'-METHYLANHYDROMACRALSTONINE (190)

The bark of *Alstonia muelleriana* Domin. was first studied by Elderfield and Gilman (*128*) and was shown to contain a complex indole alkaloid mixture from which the dimers villalstonine (**184**) (*128*), alstonisidine (**185**), (*128*) and later macralstonine (**186**) (*129*) were isolated.

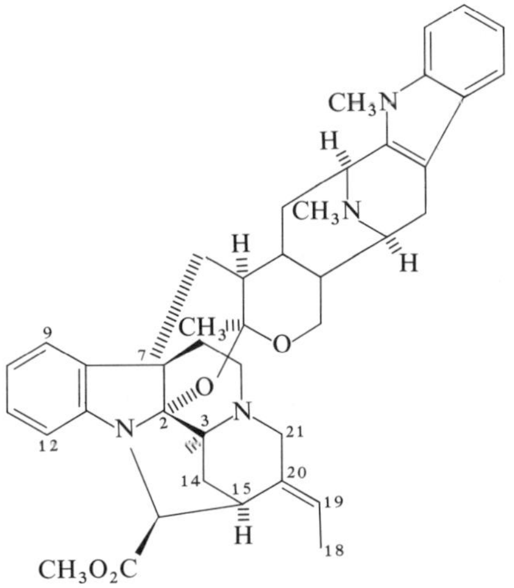

184 Villalstonine

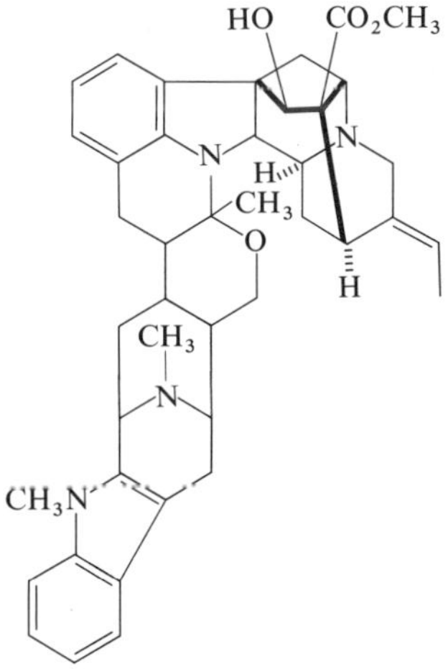

185 Alstonisidine (original)

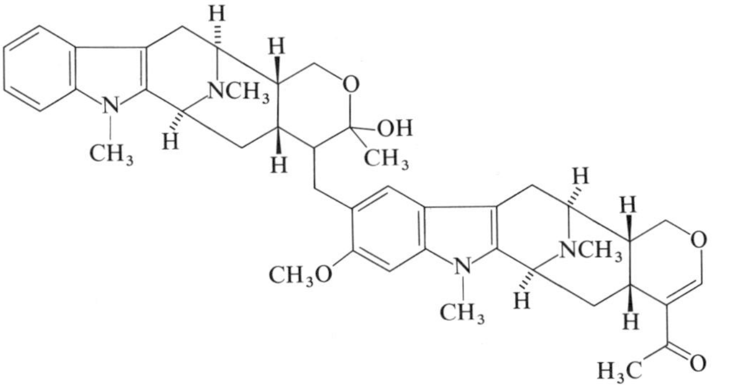

186 Macralstonine

The same group has also investigated the alkaloid fraction of *A. muelleriana* bark and obtained several monomeric and dimeric alkaloids (*130*). The crude alkaloid fraction contained 85% villalstonine (**184**), and alstonisidine (**185**) was also obtained together with a new dimeric alkaloid.

Des-N_a'-methylanhydromacralstonine, $C_{42}H_{48}N_4O_4$, mp 240°–248° (dec.), $[\alpha]_D$ +10.2°, displayed a UV spectrum in agreement with the summation

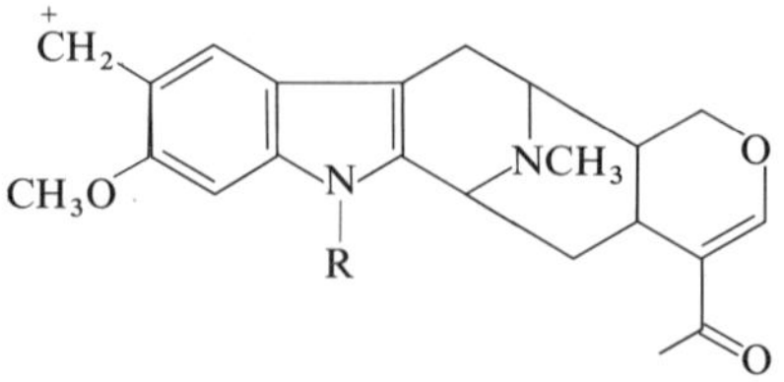

187 *m/e* 365, R = H
188 *m/e* 379, R = CH_3

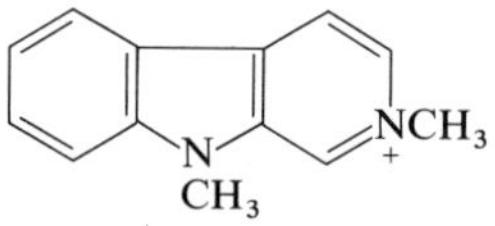

189 *m/e* 197

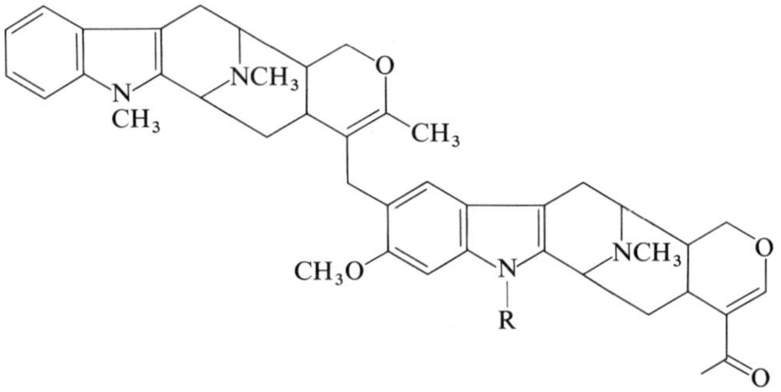

190 des-N_a'-Methylanhydromacralstonine, R = H
191 Anhydromacralstonine, R = CH_3

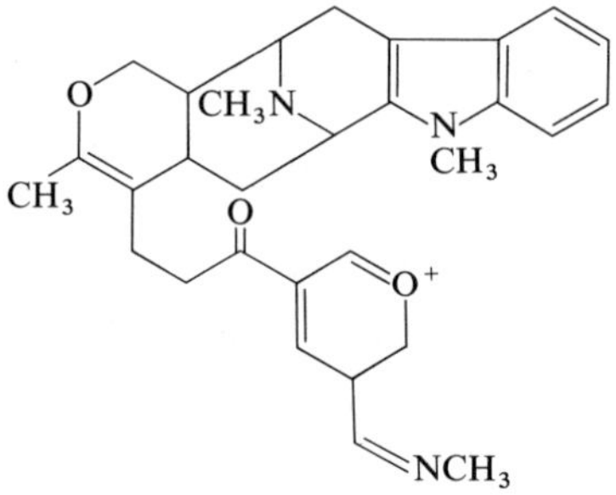

192 *m/e* 486

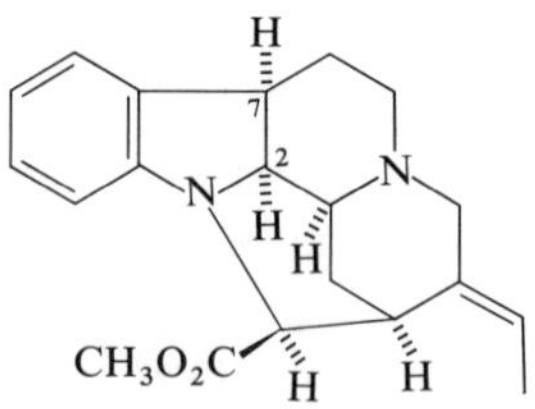

193 2,7-Dihydropleiocarpamine

of indole and 10-methoxyindole chromophores (λ_{max} 230, 285, 305, and 318 nm). The IR spectrum suggested the presence of hydroxyl or amino groups and a β-acetyl enol ether (1650 cm^{-1}). The latter group's presence was substantiated by the ^{1}H-NMR spectrum, that showed a low-field one-proton singlet at 7.63 ppm and a three-proton singlet at 2.15 ppm.

Other features of the NMR spectrum included two aromatic singlets and a four-proton complex around 7.0 ppm. Two aliphatic (2.25 and 2.21 ppm) and one indolic (3.58 ppm) *N*-methyl groups were observed together with a shielded allylic methyl group at 1.41 ppm.

The mass spectrum was crucial in deciding the location of the indolic NH group. The molecular ion at *m/e* 672 gave rise to two ions for the halves of the molecule at *m/e* 307 ($C_{20}H_{23}N_2O$) and *m/e* 365 ($C_{22}H_{25}N_2O_3$) (**187**). In macralstonine (**186**), this latter ion appears at *m/e* 379 (**188**), and consequently it is this unit which is lacking the indolic *N*-methyl group in the new dimer. The base peak in the spectrum was observed at *m/e* 197 ($C_{13}H_{13}N_2$) derived from the *m/e* 307 ion and having structure **189**. The new dimer was therefore assigned structure **190**, the des-N_a'-methyl derivative of **191**.

An important ion was also observed at *m/e* 472, analyzing for $C_{29}H_{34}\cdot N_3O_3$. In macralstonine (**186**) this ion appears at *m/e* 486, and Schmid and co-workers assigned structure **192** to this ion. Such a formulation does not agree with the revised structure of macralstonine (**186**) or with the shift to *m/e* 472 (74% of base peak) in des-N_a'-methyl macralstonine. The Michigan group did not suggest a probable structure for this ion and, indeed, its formulation is not immediately apparent.

Also isolated at this time was a biogenetically interesting monomeric alkaloid, 2,7-dihydropleiocarpamine (**193**) (*130*), an important structural unit of several dimeric alkaloids.

B. MACROCARPAMINE (202)

Alstonia macrophylla Wall. is the source of several dimeric alkaloids, including villalstonine (**184**) (*131*, *132*), macralstonine (**186**) (*133*), and macralstonidine (**194**) (*134*). Mayerl and Hesse (*135*) have recently described the isolation and structure elucidation of a new dimeric alkaloid, macrocarpamine, from the bark of *A. macrophylla*.

Macrocarpamine, $C_{41}H_{46}N_4O_3$, amorphous, $[\alpha]_D - 16°$ ($CHCl_3$), had a UV spectrum (λ_{max} 230, 254, 284, and 291 nm) characteristic of the addition of an indole and an indoline chromophore. The IR spectrum showed no OH or NH absorption and, in agreement with this, macrocarpamine could not be acetylated. Two ester carbonyl bands were observed at 1765 and 1735 cm^{-1}, characteristic of the carbomethoxy group of a pleiocarpamine (**195**) unit

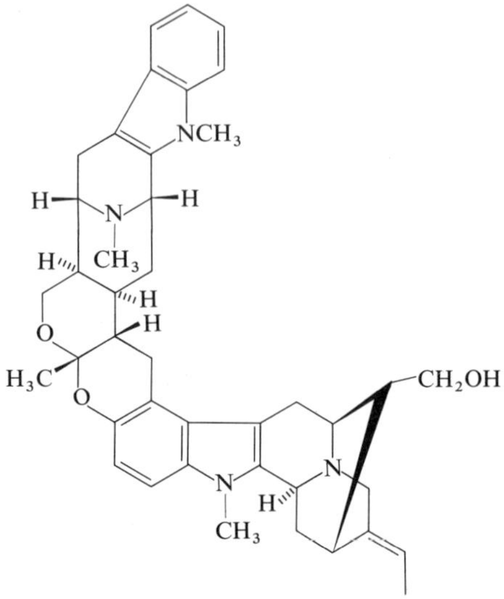

194 Macralstonidine

(*132*). Also seen in the IR spectrum were an intense enol ether band at 1645 cm^{-1} and an indoline absorption at 1615 cm^{-1}.

Catalytic hydrogenation (10% Pd/C in 99.5% ethanol) of macrocarpamine gave a mixture of dihydro and tetrahydro derivatives. The NMR spectrum of this mixture did not show two doublets ($J = 16$ Hz) at 5.44 and 4.58 ppm present in the original compound, and in the IR spectrum the enol ether band was now shifted to 1665 cm^{-1}.

The pleiocarpamine portion of the alkaloid was determined by direct acid-catalyzed ($CF_3CO_2H/(CF_3CO)_2O/CH_2Cl_2$) cleavage of macrocarpamine at room temperature for 10 min to give pleiocarpamine (**195**) identical to the authentic material.

The second half of the dimer was determined by pyrolysis (280°, 0.01 mm) to give (+)-pleiocarpamine (**195**) and (−)-anhydromacrosalhine–methine (**196**). The latter compound was identified by comparison of the mass, ^1H-NMR, UV, and IR spectra with those of an authentic sample.

Summation of the molecular weights (322 and 320) of the two component halves of macrocarpamine indicated that no other fragments had been lost on pyrolysis and suggested that either a Cope or Diels–Alder type reaction had taken place. It remained only to deduce the nature of the linkage between the two units.

The previously indicated ^1H-NMR data suggested that C-18′ and C-19′ of the anhydromacrosalhine methine unit were trans-substituted, and because an indoline must be present, the linkage of this unit should be to either C-2 or C-7 of the pleiocarpamine part.

The mass spectrum of macrocarpamine, besides giving rise to simple losses from the molecular ion (M^+, 642 mu), afforded ions at 322 ($C_{20}H_{22}N_2O_2$) and 320 ($C_{21}H_{24}N_2O$), and two series of ions probably derived from each of these ions. Characteristic ions for the pleiocarpamine unit were observed at *m/e* 135 (**197**) and *m/e* 107 (**198**) together with the series *m/e* 197 (**189**), *m/e* 182 (**199**), *m/e* 181 (**200**), and *m/e* 170 (**201**) from the anhydromacrosalhine methine unit.

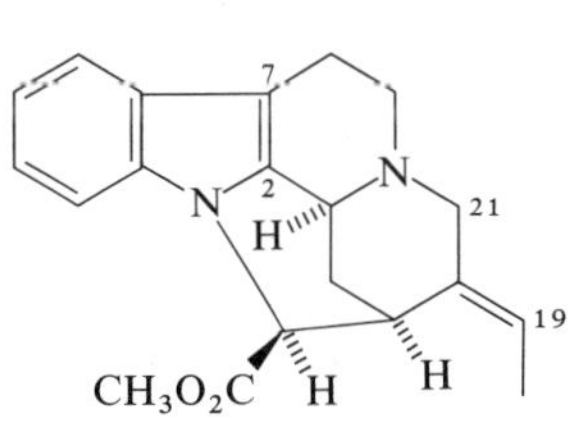

195 Pleiocarpamine

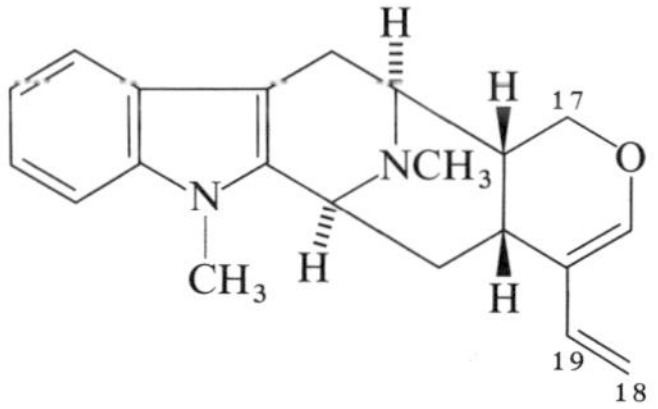

196 Anhydromacrosalhine methine

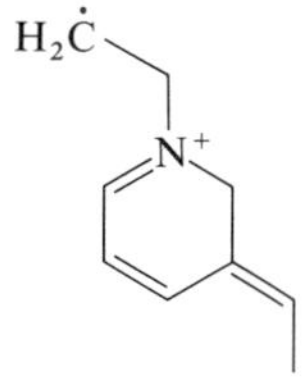

197 *m/e* 135

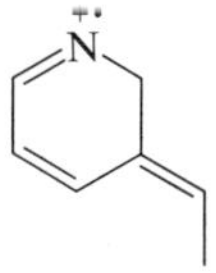

198 *m/e* 107

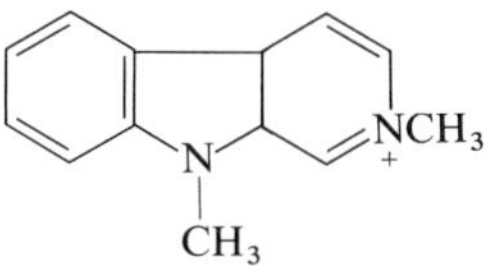

189 *m/e* 197

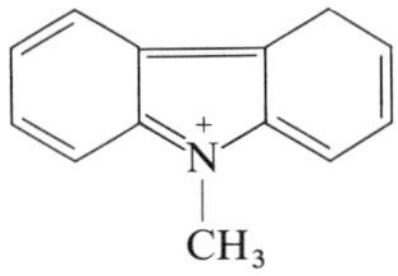

199 *m/e* 182

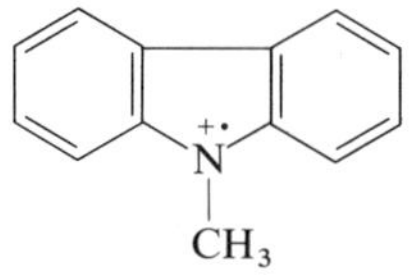

200 *m/e* 181

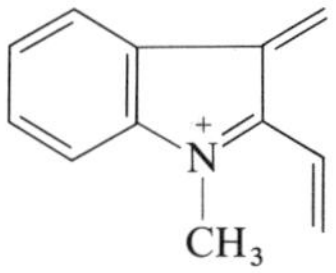

201 *m/e* 170

In the mass spectra of both dihydromacrocarpamine and tetrahydromacrocarpamine the base peak appeared at *m/e* 322, suggesting the facile cleavage of a bond adjacent to a carbon-bearing nitrogen. That is, the C-18 of **196** should be linked to C-2 of **195**. At low (12 eV) electron voltage, the parent compound still showed the ions at *m/e* 322 and 320; therefore, they are due to electron impact and not thermal fragmentation. The Swiss group rationalized the formation of these ions as shown in Scheme 9 (*135*).

The ^{13}C-NMR spectrum of macrocarpamine recorded at 90 MHz showed each of the 41 carbons resolved, although very few of these were in fact assigned.

The ^{1}H-NMR spectrum of macrocarpamine was studied by comparison with those of 2,7-dihydropleiocarpamine (**193**) and anhydromacrosalhine

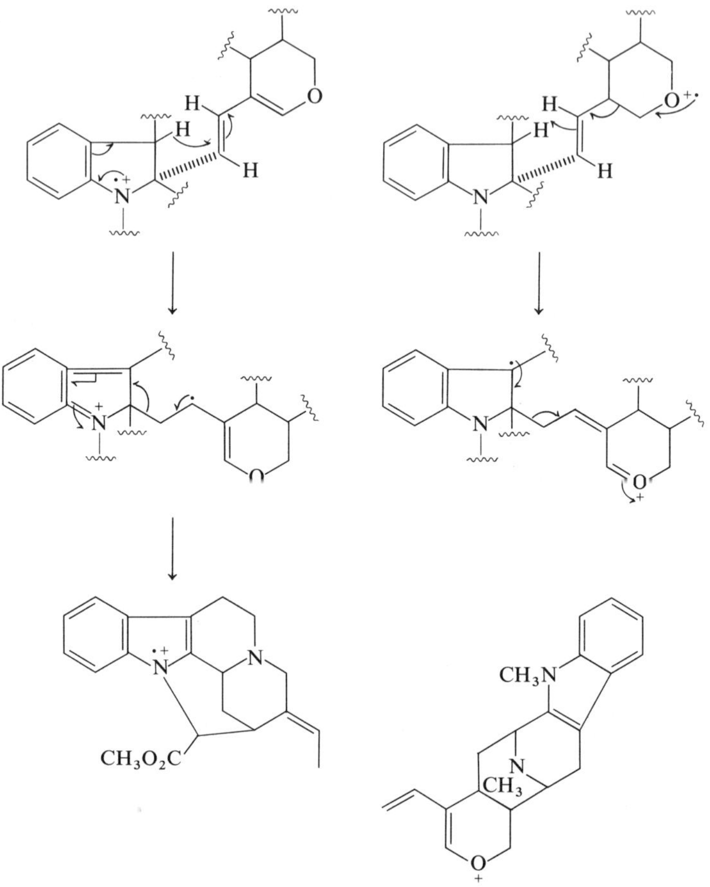

m/e 322 m/e 320

SCHEME 9. *Formation of ions m/e 320 and 322 in the mass spectrum of macrocarpamine.*

methine (**196**). In the spectrum of **196**, the 18, 19, and 21 protons appeared at 4.55 ppm ($J = 11$ Hz) and 4.40 ppm ($J = 17$ Hz), 6.00 ppm ($J = 11, 17$ Hz), and 6.46 ppm. In the spectrum of macrocarpamine, the C-21′ proton was observed as a singlet at 6.27 ppm with doublets ($J = 16$ Hz) at 5.44 and 4.58 ppm for the trans-related C-19′ and C-18′ protons, respectively.

The region 2.46–3.28 ppm was a particularly complex one, containing nine protons; three of these were readily assigned from the spectrum of **196** as being due to the C-6′ β-H (2.54 ppm, $J = 16$ Hz), C-5′ H (3.12 ppm, $J = 7$ Hz) and C-6′ α-H (3.34 ppm, $J = 7, 16$ Hz). The corresponding signals in macrocarpamine appeared at 2.46, 3.05, and 3.28 ppm. A doublet of doublets ($J = 4$ Hz) at 3.15 ppm in macrocarpamine was assigned to the C-15 H and the assignment established by double-resonance studies which showed coupling with the C-16 H at 4.18 ppm and the C-21 α-H at 2.97–2.89 ppm. Note that the C-21 β-H proton is substantially deshielded to appear as a broad doublet ($J = 13$ Hz) at 4.32 ppm.

The signal at 2.60 ppm, a doublet of doublets ($J = 7$ and 11 Hz), is a crucial signal in the structure determination of macrocarpamine because it is assigned to H-7 in the pleiocarpamine unit, the coupling being with the C-6 methylene protons. This establishes the structure of macrocarpamine (**202**) as being derived from a 2,7-dihydropleiocarpamine unit and an anhydro-macrosalhine methine unit linked between C-2 of the former and C-18′ of the latter. The chemical shift assignments for macrocarpamine are shown on structure **202** (*135*).

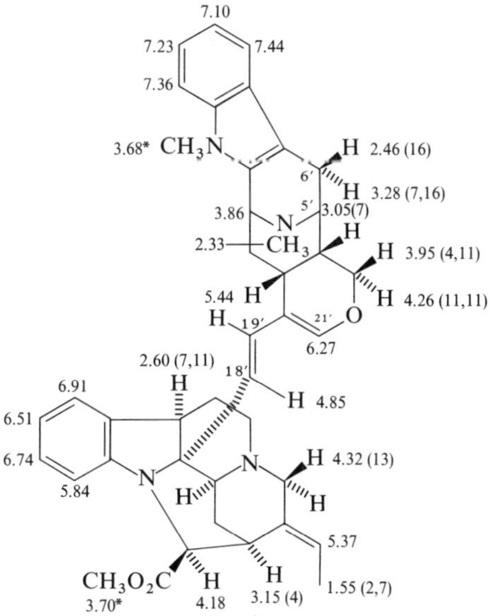

202 Macrocarpamine (asterisks indicate assignments may be reversed)

C. SYNTHESIS AND BIOGENESIS OF THE BISINDOLE *ALSTONIA* ALKALOIDS

As already indicated in this chapter, the key question for many of these dimeric alkaloids is their biogenesis, and this is a particularly important question when the two monomeric halves of the dimer are present in the same plant part in reasonable quantities.

For the alkaloids villalstonine (**184**), macralstonine (**186**), and alstonisidine (**185**), this question has been examined *in vitro* by Le Quesne and co-workers (*136–139*), who studied their formation from the monomeric units.

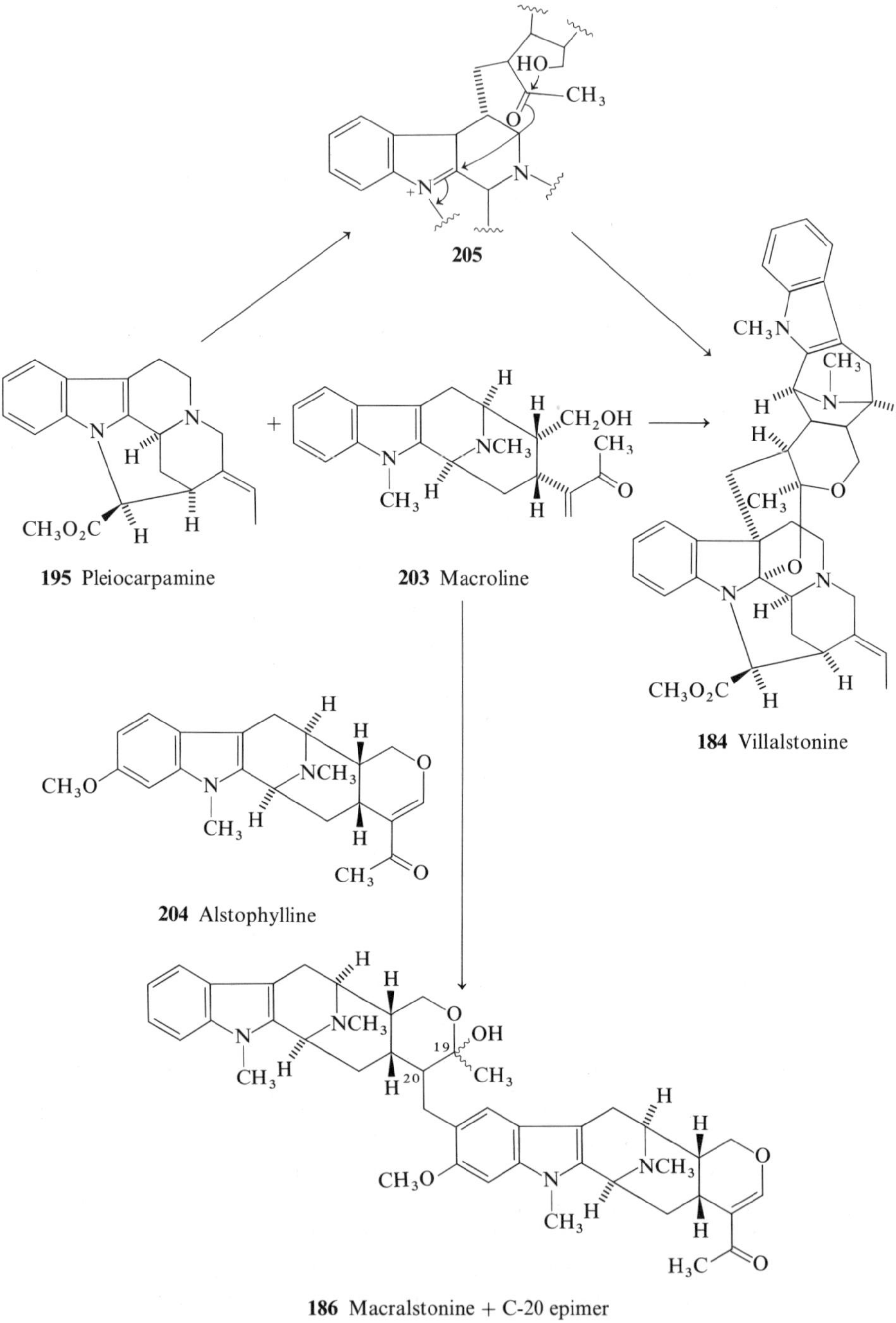

205

195 Pleiocarpamine

203 Macroline

184 Villalstonine

204 Alstophylline

186 Macralstonine + C-20 epimer

The biomimetic stereospecific synthesis of villalstonine (**184**) was achieved in 38% yield by allowing a mixture of pleiocarpamine (**195**) and macroline (**203**) to stand in 0.2 *N* aqueous HCl solution at 20° for 18 hr (*136*, *140*). The reaction is thought to proceed by way of the intermediate **205**, which can then undergo additional cyclization, as shown (*136*). Although this is clearly a facile process, it remains to be determined whether this is the true biosynthetic pathway, for macroline (**203**) itself has not yet been obtained from natural sources, but only from the degradation of villamine.

A similar reaction of macroline (**203**) and alstophylline (**204**) in 0.2 *N* HCl at 20° for 120 hr gave macralstonine in 40% yield (*138*). In this case, however, macralstonine (**186**) was not the sole product, but a mixture of the two C-20 (and C-19?) epimers was produced. In particular, the product showed doubling of the *O*-methyl signals (3.84 and 3.90* ppm), the aromatic *N*-methyl group (3.57 and 3.55 ppm), and one of the aliphatic *N*-methyl groups (2.38* and 2.42 ppm). On acetylation, a single product was obtained that was identical with *O*-acetylmacralstonine.

Before turning to the biomimetic synthesis of alstonisidine, some model reactions should be mentioned. In their biogenetic proposal for the formation of alstonisidine (**185**), the Michigan group had proposed a reaction between macroline (**203**) and quebrachidine (**206**) involving initial electrophilic substitution by the macroline enone to the quebrachidine indoline nitrogen followed by ring closure to afford the amino acetal group (*141*). This concept was evaluated with indoline and methyl vinyl ketone.

Heating a mixture of indoline (**207**) and methyl vinyl ketone at 90° for 6 hr and then 60° for 12 hr gave 6-methyl-lilolidine (**208**) in up to 43% yield. However, when 4-methoxy-2-tosyloxybutane (**209**) was treated with indoline and the product treated with phosphorus pentoxide–phosphoric acid at 200°, 4-methyl-lilolidine (**210**) was obtained in 30% yield (*139*). These results implied that if macroline is a true intermediate, it should condense with an orientation opposite to that predicted for the biosynthetic process.

Treatment of macroline (**203**) with quebrachidine (**206**) at 20° with 0.2 *N* aqueous HCl gave a labile compound, molecular weight 690, formulated as **211**. Further reaction with boron trifluoride etherate at 0° for 6 hr gave alstonisidine (**212**) identical with the natural product (*137*, *139*). No reaction took place when a mixture of macroline (**203**) and quebrachidine (**206**) was treated with boron trifluoride etherate.

Based on the model experiments, alstonisidine should therefore have the revised structure **212** and not **185**, the structure deduced originally (*141*) for this compound. The partial synthesis also establishes the stereochemistry of alstonisidine, except for two centers, the new ring junction.

* Chemical shift in natural compound.

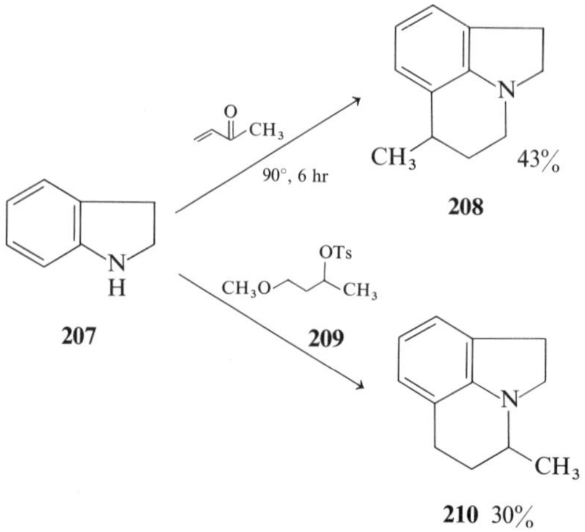

$CH_2=CHCOCH_3$

90°, 6 hr

207

208 43%

$CH_3OCH_2CH_2CH(OTs)CH_3$ **209**

210 30%

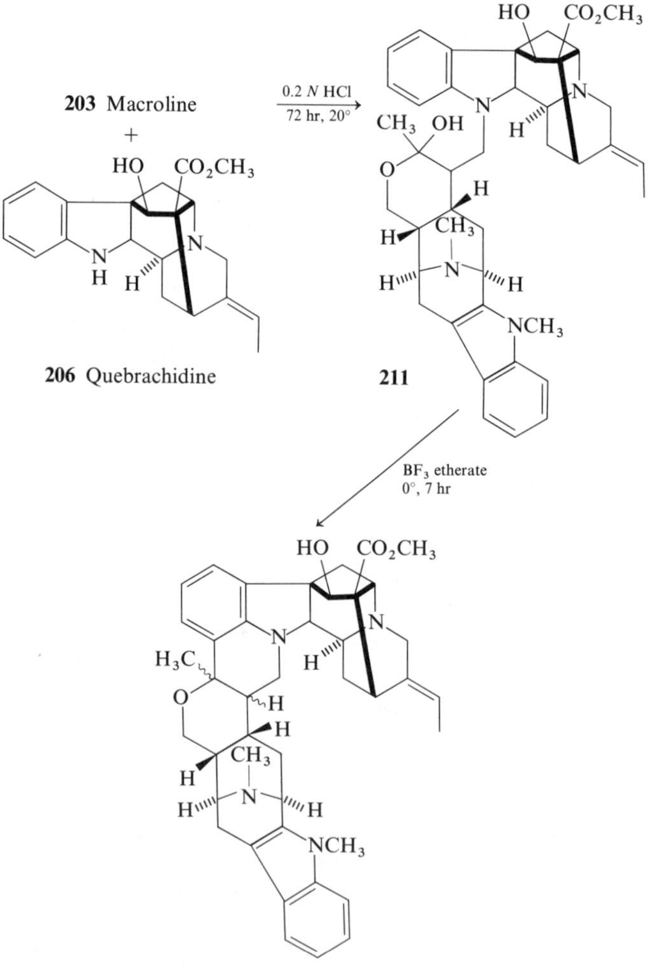

203 Macroline

\+

206 Quebrachidine

0.2 *N* HCl
72 hr, 20°

211

BF_3 etherate
0°, 7 hr

212 Alstonisidine (revised)

IX. Pleiocarpamine–Vincorine Type

PLEIOCORINE (**214**)

An interesting demonstration of the power of ^{13}C-NMR in structure elucidation is the analysis, by a group at Gif, of pleiocorine (*142*).

Pleiocorine, isolated from *Alstonia deplanchei* Van Heurck et Muell. Arg. (Apocynaceae), showed no melting point and a UV spectrum (λ_{max} 244, 297, and 344 nm) of an unknown type. No NH or OH absorptions were indicated in the IR spectrum, but two different ether functionalties were apparent. A molecular ion was observed at *m/e* 674 and analyzed for $C_{41}H_{46}N_4O_5$, but no further significant ions (except for $M^+ - CO_2CH_3$) were found.

The NMR spectrum (at 240 MHz) confirmed the presence of two carbomethoxy functions (3.65 and 3.72 ppm) and also indicated two ethylidene side chains, an *N*-methyl group, and six aromatic protons. Two of the aromatic protons were singlets, indicating their para relationship on a 10,11-disubstituted nucleus. A one-proton doublet ($J = 4$ Hz) at 4.66 ppm is characteristic of the C-16 proton in pleiocarpamine derivatives; therefore, a ^{13}C-NMR comparison with related alkaloids was made.

From the UV spectrum a 2,7-substituted pleiocarpamine derivative should be present in pleiocorine, and consequently it became necessary initially to assign the ^{13}C-NMR signals in such a compound. A readily available alkaloid which contains this moiety is villalstonine (**184**), but as this also has a "macroline" moiety, the assignments of this fragment were also necessary. Again, an alkaloid containing this unit, macralstonidine (**194**), was available. The second unit of **194** is *N*-methylsarpagine, and by comparison with sarpagine itself the chemical shifts of the "macroline" carbons could be deduced. Subtraction of these data from those of villalstonine then gave values for pleiocorine (Table V) and confirmed both the presence of this unit in the alkaloid and the substitution at C-7 by a carbon and at C-2 by an oxygen. Clearly, however, the nature of the substituting group was not the same as in villalstonine.

The remaining signals of pleiocorine (Table V), now apparent by subtraction, revealed a very low-field (97.5 ppm) quaternary carbon. Because all the aromatic and olefinic carbons could be accounted for (a necessarily even number), this lone carbon must be highly substituted by deshielding groups. Substitution by oxygen is not possible, for all oxygens are accounted for; consequently, an N—C—N group was proposed. One alkaloid isolated from *A. deplanchei* having this group was vincorine (**213**). Comparison (see Table V) of the ^{13}C-NMR spectral data with those of the remaining carbons of pleiocorine indicated that this was, indeed, the second unit of the dimer.

The data demonstrate that, as well as being substituted at C-10′ by oxygen, the vincorine part of pleiocorine is also substituted at C-11′ (previously

TABLE V

^{13}C-NMR DATA OF PLEIOCORINE (**214**) AND PLEIOCRALINE (**216**)[a,b]

Carbon	Villalstonine (**184**)	Pleiocorine (**214**)	Pleiocraline (**216**)	Carbon	Vincorine part of pleiocorine (**214**)	Vincorine (**213**)	Substituted akuammiline part of pleiocraline (**216**)	N_a-Methyldeacetyl-deformyl-1,2-dihydro-akuammiline (**215**)
C-2	92.2	103.2	104.3	C-2′	97.5	97.9	80.3	79.1
C-3	51.5	51.3	51.8	C-3′	40.6	40.6	53.0	52.8
C-5	53.1	52.0	52.1[d]	C-5′	55.0[f]	56.1[g]	55.0	54.7
C-6	28.6[a]	24.6	24.7	C-6′	20.2	20.4	31.5	31.1
C-7	44.2	54.0	54.1	C-7′	56.9	57.3	43.2	43.0
C-8	132.9	134.4	134.7	C-8′	134.8	138.2	140.3	*
C-9	120.9	121.6	122.8	C-9′	106.1	105.5	104.8	120.5
C-10	118.1[b]	119.2[c]	118.5[e]	C-10′	151.1	152.3	153.4	119.0
C-11	126.5	126.3	126.9	C-11′	127.4	111.7[h]	127.8	126.7
C-12	109.3	108.9	109.5	C-12′	100.1	112.1[h]	104.3	109.0
C-13	147.0	144.4	144.9	C-13′	143.6	143.6	148.2	*
C-14	28.9[a]	28.1	27.9	C-14′	26.3	26.3	34.2	33.9
C-15	31.9	32.2	32.2	C-15′	34.7	34.8	34.5	34.4
C-16	57.8	58.1	58.3	C-16′	50.9	50.7	47.5	47.3
C-17	170.8	169.3	169.9	C-17′	173.1	173.5	172.9	172.2
C-18	12.2	12.3	12.3	C-18′	13.4	13.6	12.9	13.0
C-19	118.4[b]	119.5[c]	119.9[e]	C-19′	122.5	123.2	120.1	118.7
C-20	136.4	136.1	135.2	C-20′	138.8	138.2	140.3	*
C-21	47.5	48.2	48.2[d]	C-21′	58.1[f]	58.1[g]	50.8	50.6
CO_2CH_3	51.9	50.6	51.0	CO_2CH_3	51.6	51.7	51.3	51.3
				N_a-CH_3	28.1	28.3	35.2	33.9

[a] Superscripts *a–h* indicate assignments may be reversed.
[b] Asterisk indicates quaternary carbon signals not observed.

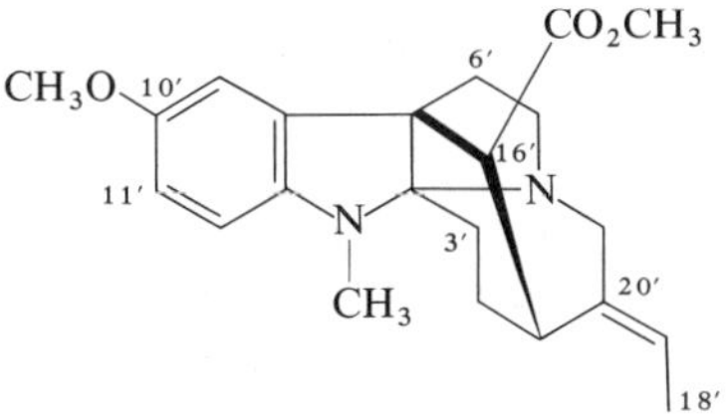

213 Vincorine

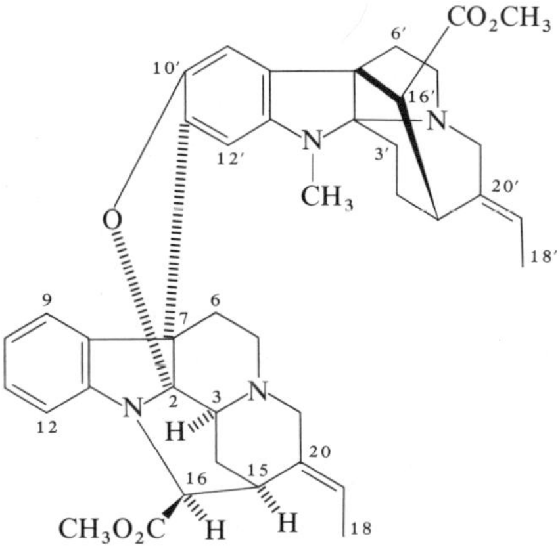

214 Pleiocorine

indicated by the ^{1}H-NMR spectrum). Because C-2 of the pleiocarpamine unit is of necessity substituted by both N and O, there is only one possible orientation of the two halves; pleiocorine has structure **214** (*142*).

X. Pleiocarpamine–Akuammiline Type

Pleiocraline (216)

The structure of a second dimeric alkaloid from *Alstonia deplanchei*, pleiocraline, has recently been elucidated by the group at Gif (*143*) using ^{13}C-NMR spectroscopy. The molecular formula, $C_{41}H_{46}N_4O_5$, was deduced by high-resolution mass spectrometry. Although the alkaloid formed colorless plates from MeOH, no melting point was observed below 300°. The UV spectrum showed λ_{max} 244, 295, and 344 nm, analogous to that pleiocorine; its IR spectrum, while lacking NH or OH absorption, indicated the presence of saturated ester (1725 cm^{-1}) and dihydroindole (1606 cm^{-1}) moieties. A 240-MHz NMR spectrum of pleiocraline demonstrated the presence of two carbomethoxy groups, with a singlet at 3.70 ppm integrating for six protons, a single *N*-methyl group at 2.65 ppm, and two ethylidene groups. The latter displayed three-proton doublets at 1.54 and 1.58 ppm and one-proton quartets at 5.33 and 5.42 ppm. Only six aromatic protons were observed; the nature of two of these, singlets at 6.35 and 6.60 ppm, indicated the presence of a 10, 11-disubstituted nucleus. As with pleiocorine, a doublet ($J = 4$ Hz) at 4.68 ppm taken with the other available information indicated the probable presence of a 2,7-dihydropleiocarpamine unit (**193**).

The presence of unit **193** was firmly established by the ^{13}C-NMR spectrum of pleiocraline. As shown in Table V, excellent agreement was obtained for the signals of pleiocorine (**214**) and pleiocraline, particularly with respect to unit **193**. Of some consequence was the establishment of the orientation of the substitution on **193**, which was found to be identical to that in pleiocorine in that C-2, with a shift of 104.3 ppm, was clearly shown to be substituted by both oxygen and nitrogen.

Comparison with pleiocorine (**214**) also indicated that the second unit of pleiocraline was not the same as that present in **214**. The chemical shift of the aromatic carbons of this unit establishes the nature of attachment between the two aromatic units. The aliphatic unit, however, was quite different, carbons 2′, 3′, 6′, 7′, 8′, 14′, and 21′ being particularly shifted. The C-2′ carbon was now shifted back to 80.3 ppm and appeared as a doublet in the SFORD spectrum, and the C-3′ carbon was now also a doublet, shifted downfield to 53.0 ppm. These data establish a C-3′—N bond and a proton at C-2′. This skeleton is that of akuammiline in which C-16′ is attached to C-7′. The stereochemistry at C-2′ was determined to be β by comparison with N_a-methyldeacetyldeformyl-1,2-dihydroakuammiline (**215**) in which C-2 was observed at 79.1 ppm. A compound having the 2α-H configuration exhibited this resonance at 70.6 ppm. The structure of pleiocraline can, therefore, be represented by **216**, in which the stereochemistry at C-2 and C-7 was deduced by analogy with pleiocorine (**214**) (*143*).

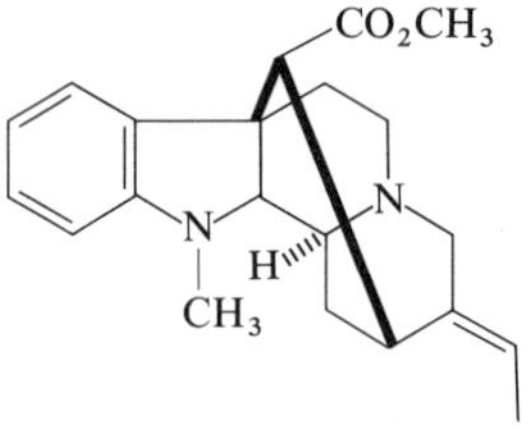

215

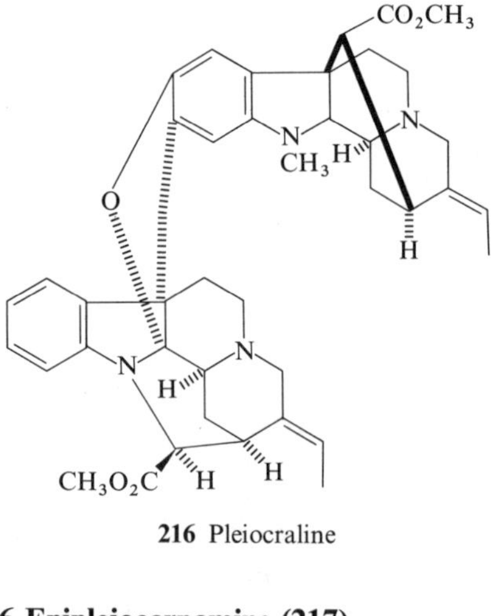

216 Pleiocraline

XI. Synthesis of 16-Epipleiocarpamine (217)

Pleiocarpamine (**195**) is a novel alkaloid, first isolated from *Pleiocarpa mutica* Benth., having C-16 linked to the indole nitrogen. It is quite widespread in the Apocynaceae and occurs both free and as part of several dimeric indole alkaloids, as we have seen.

In 1976 Sakai and Shinma (*144*) reported the synthesis of 16-epipleiocarpamine (**217**) from geissoschizine methyl ether (**218**) along biogenetic lines. This was the first chemical correlation of pleiocarpamine (**217**) with an indole of known absolute configuration. (See Scheme 10.)

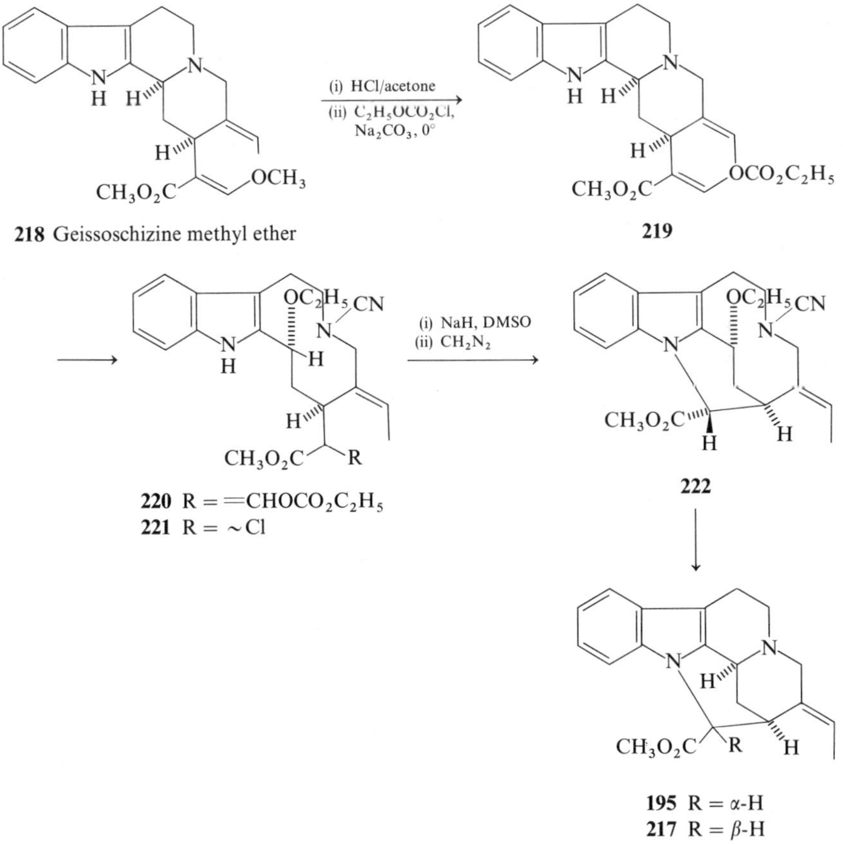

(i) HCl/acetone
(ii) $C_2H_5OCO_2Cl$, Na_2CO_3, 0°

218 Geissoschizine methyl ether

219

(i) NaH, DMSO
(ii) CH_2N_2

220 R = =$CHOCO_2C_2H_5$
221 R = ~Cl

222

195 R = α-H
217 R = β-H

SCHEME 10. *Correlation of geissoschizine methyl ether* (**218**) *and 16-epipleiocarpamine* (**217**).

Demethylation of **218** followed by reaction with ethyl chlorocarbonate in the presence of sodium carbonate gave **219**. Cleavage under von Braun conditions in 15% ethanol–chloroform gave the 3*R*-ethoxy derivative **220** as the main product in 44% overall yield from geissoschizine. Oxidation with *tert*-butyl hypochlorite at −78° following base hydrolysis gave the chloro ester **221**. Closure between N_a and C-16 was accomplished with sodium hydride in DMSO at 80° followed by methylation of any partially hydrolyzed carboxylic acid to give **222**. Closure to **217** was achieved in 22% yield by heating with aqueous acetic acid–ammonium acetate (*144*).

XII. Pseudoakuammigine–Eburnea Type

UMBELLAMINE (**223**)

From the root bark of *Hunteria umbellata* (K. Schum.) Hall. F., Schmid and co-workers obtained an interesting bisindole alkaloid, umbellamine (**223**) (*145*).

Umbellamine, $[\alpha]_D$ $-217°$, was isolated as pale yellow needles, but decomposed at about 250° prior to melting. At 380° in the mass spectrometer two high-mass ions at *m/e* 660 and 674 were observed, the latter peak becoming more intense during sample analysis. In addition, fragment ions characteristic of eburnamenine (**224**) were observed, and this thermal reaction was used preparatively to obtain in 40% yield (+)-eburnamenine (**224**) from umbellamine. Therefore, this alkaloid must comprise one-half the molecule, and it must be linked at some point to the second unit. The ion at *m/e* 660 analyzed well for $C_{41}H_{48}N_4O_4$.

The IR spectrum showed an ester group (ν_{max} 1733 cm^{-1}), and this was confirmed by the NMR spectrum which showed two three-proton singlets at 2.78 and 2.83 ppm corresponding to one *O*-methyl and one *N*-methyl group. As well as the methyl triplet of the eburnamenine moiety at 0.86 ppm, decoupling experiments demonstrated the presence of an ethylidene group having a methyl doublet at 1.37 ppm.

A shift in the UV spectrum of umbellamine from 295 to 309 nm on the addition of alkali indicated the probable presence of a phenolic group in the noneburnamenine "half," and the nature of this group was established by the formation of both *O*-acetyl and methyl ether derivatives.

Vigorous lithium aluminum hydride (LAH) reduction of umbellamine methyl ether (**225**) gave a diol (**226**), $C_{41}H_{52}N_4O_3$, which gave diacetyl derivative **227**, but $LiAlD_4$ reduction of **225** introduced only three deuterium atoms. Mild LAH reduction of **225** afforded an alcohol (**228**), $C_{41}H_{50}N_4O_3$, which yielded only a monoacetyl derivative, **229**. These data indicate that umbellamine contains a single carbomethoxy group and that the fourth oxygen is present in the form of a carbinolamine ether moiety.

The mass spectrum of diol **226** showed peaks at *m/e* 252, 237, and 197, typical of the dihydroeburnamenyl moiety, and a set of peaks at *m/e* 196, 178, and 166, characteristic for the aliphatic moiety of pseudoakuammigol (**230**) (*145*). The latter peaks, of which *m/e* 196 is suggested to have structure **231**, were all shifted by 2 mu in the deuterodiol derivative, indicating the carbinol carbinolamine ether to be attached at C-2′ and not C-3′, C-21′, or C-5′. In addition, the high-molecular-weight region of the spectrum analyzed well for a pseudoakuammigol moiety attached to eburnamenine (**224**). (In this analysis 30 mass units were deducted to allow for the presence of a methoxy group in the pseudoakuammigol moiety.)

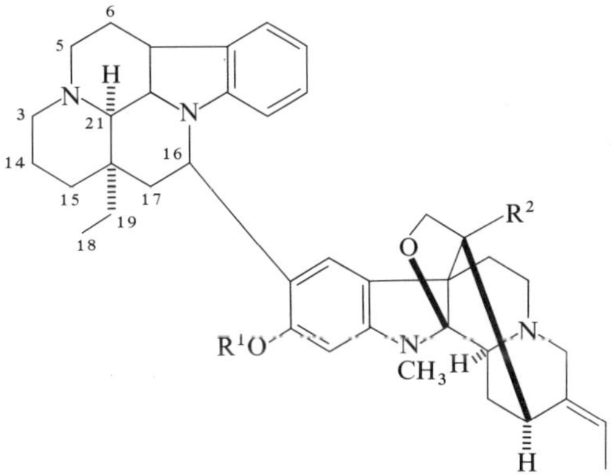

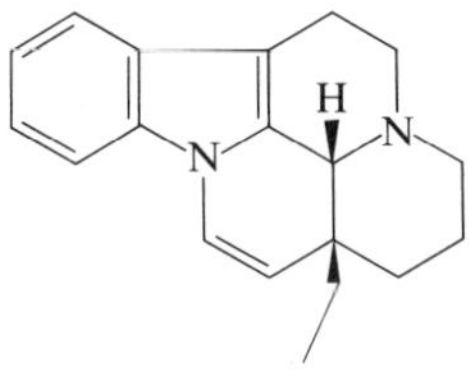

224 (+)-Eburnamenine

223 $R^1 = H, R^2 = CO_2CH_3$ Umbellamine
225 $R^1 = CH_3, R^2 = CO_2CH_3$
228 $R^1 = CH_3, R^2 = CH_2OH$
229 $R^1 = CH_3, R^2 = CH_2OCOCH_3$

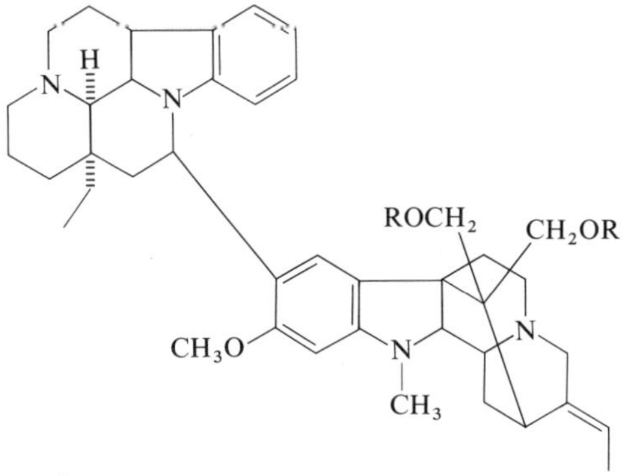

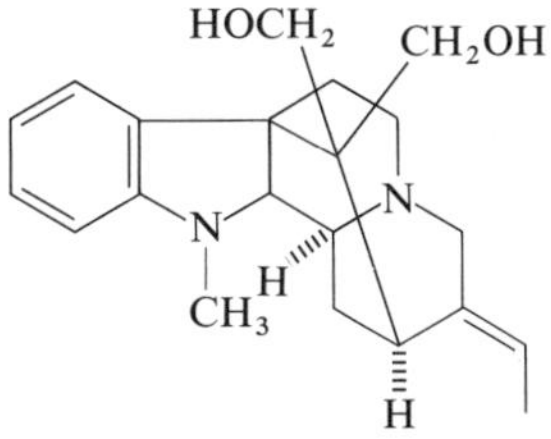

226 R = H
227 $R = COCH_3$

230 Pseudoakuammigol

The ion at *m/e* 265 in the mass spectrum of **226** is suggested to have structure **232**, and therefore indicates that the point of attachment to the dihydroeburnamenyl unit is through C-16. The other half of the molecule appears as the corresponding fragment *m/e* 383, formulated as **233**. This, and the ions at 648 ($M^+ -1$), 617 ($M^+ -31$), 617 ($M^+ -CH_2OH$), 573, 466, and 452 ($M^+ -196$), indicate that C-16 is attached to an aromatic carbon of the **230** unit.

In the aromatic region of the NMR spectrum of umbellamine, two singlets were observed at 6.52 ppm and 6.10 ppm for the C-9′ and C-12′ protons, respectively. The remaining aromatic protons appeared at 7.40 (H-9), 7.00 (H-11), 6.82 (H-10), and 6.52 (H-12).

Two points remain to be discussed: the location of the phenolic group at C-10′ or C-11′ and the stereochemistry of C-16. Reductive cleavage of umbellamine with Zn or Sn and concentrated HCl gave an isodihydrostrictamine derivative (**234**) in low yield. The UV spectrum of this product

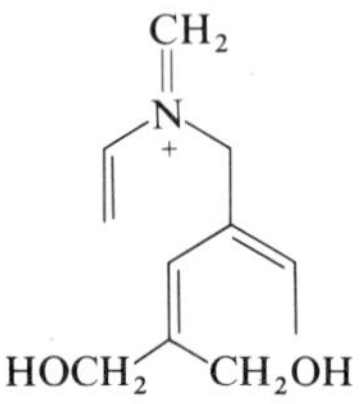

231 *m/e* 196 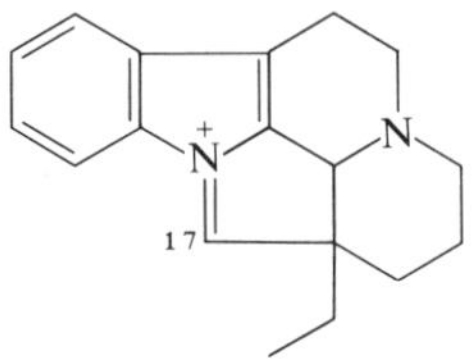 **232** *m/e* 265 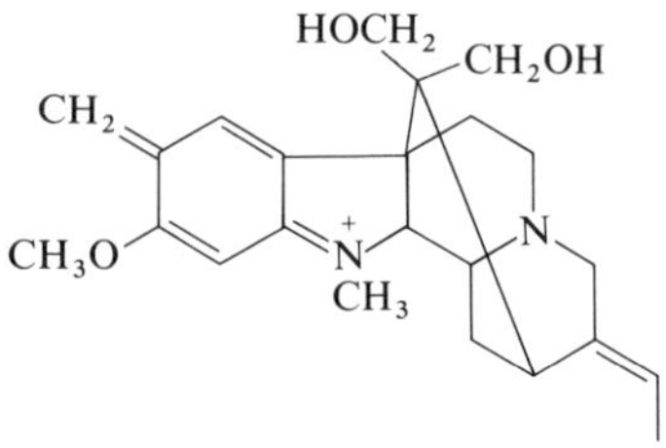 **233** *m/e* 383

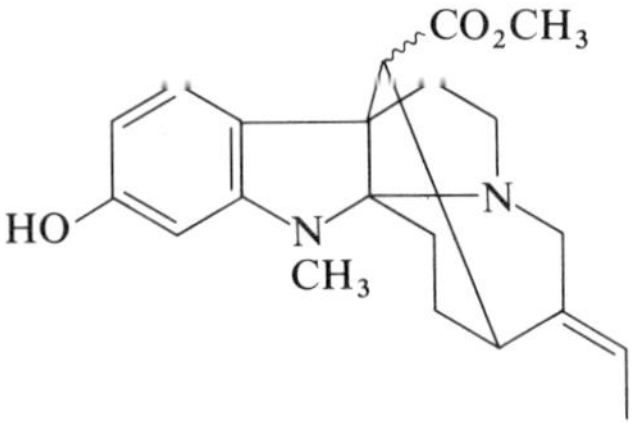

234

corresponded more closely with that of an 11-hydroxy- than with a 10-hydroxyindoline; therefore, structure **223** was suggested for umbellamine (*145*).

The evidence for the stereochemistry at C-16 is less clear. The C-16 proton appears as a complex multiplet at 5.90 ppm when the spectrum is taken in pyridine, but this gives little stereochemical information. One important clue may be the consistent shielding of the carbomethoxy group of umbellamine by about 1 ppm which must be due to the anisotropic effect of an aromatic nucleus in the dihydroeburnamenine moiety. Molecular models indicate that such a situation is most likely when the substituent at C-16 is *β*, but this suggestion remains to be proven.

XIII. *Strychnos–Strychnos* Type

The dimeric *Strychnos–Strychnos* alkaloids have been discussed in previous volumes of this treatise (*1*, *5*, *11*), by Gorman *et al.* (*13*), and by others (*22–26*). This work will not be discussed here; instead, emphasis will be placed on more recent progress.

A. Alkaloids of *Strychnos dolichothyrsa*

The liana *Strychnos dolichothyrsa* Gilg ex Onochie et Hepper displayed muscle relaxant activity, and from the stem bark Verpoorte and Baerheim Svendsen have obtained several new bisindole alkaloids in the dimeric *Strychnos* series (*146*, *147*).

One of the major alkaloids was identified as bisnordihydrotoxiferine (**235**), and a second minor alkaloid was deduced to be bisnor-C-curarine (*147*). Two other compounds which showed chromogenic reactions and UV spectra similar to **235** were obtained from more polar fractions. One of these showed an M^+ ion in the mass spectrum at m/e 568 and a large fragment ion at $M^+ - 16$, suggesting that it was an *N*-oxide. Indeed, when bisnordihydrotoxiferine (**235**) was treated with 5% hydrogen peroxide solution, each of the natural compounds was produced, indicating that they are bisnordihydrotoxiferine *N*-oxide and bisnordihydrotoxiferine *N,N′*-dioxide. Reduction of the natural products with sulfurous acid afforded **235**.

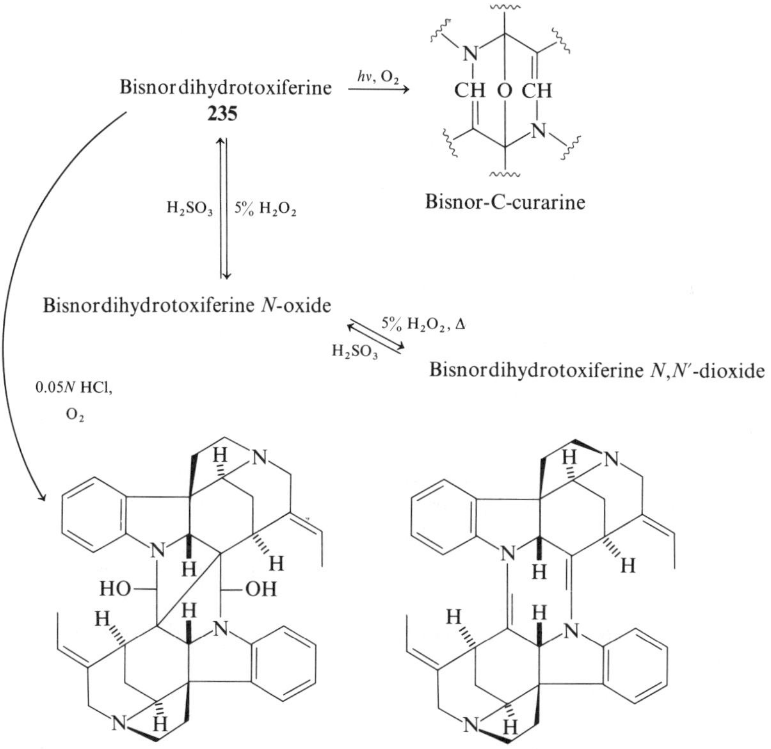

Bisnordihydrotoxiferine **235** $\xrightarrow{h\nu,\ O_2}$ Bisnor-C-curarine

235 ⇌ Bisnordihydrotoxiferine *N*-oxide (H_2SO_3 / 5% H_2O_2)

Bisnordihydrotoxiferine *N*-oxide ⇌ Bisnordihydrotoxiferine *N,N′*-dioxide (5% H_2O_2, Δ / H_2SO_3)

235 $\xrightarrow{0.05N\ HCl,\ O_2}$ **236**

236 Bisnor-C-alkaloid D

235 Bisnordihydrotoxiferine

One other bisindole alkaloid was identified as bisnor-C-alkaloid D (**236**), a product of the treatment of **235** with 0.05 *N* HCl (*146*). No NMR spectral data were reported for the new isolates.

B. Caracurine V *N*-Oxide (238) and Caracurine V *N*,*N*′-Dioxide (239)

The stem bark of *Strychnos dolichothyrsa* has also afforded (*147*) caracurine V (**237**) and its *N*-oxide (**238**) and *N*,*N*′-dioxide (**239**).

The last two were obtained by preparative TLC of column fractions after elution of **237**, and were also available from Wieland–Gumlich aldehyde (**240**). Thus, heating **240** at 120° for 8 hr in the presence of pivalic acid according to the method of Battersby and Hodson (*148*) gave a mixture (proportions not reported) containing **237**, **238**, and **239**. Identity with the

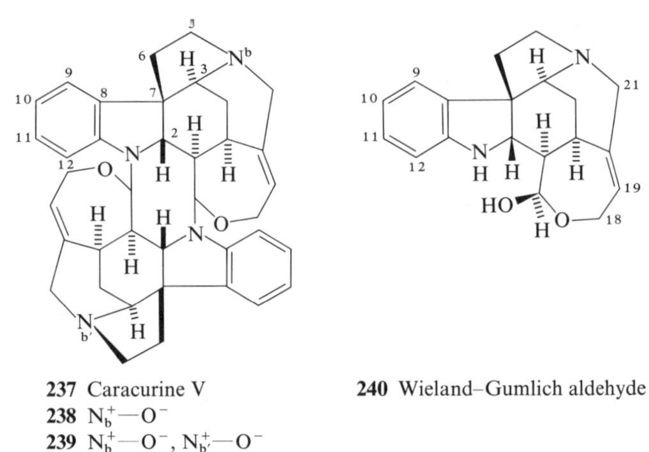

237 Caracurine V
238 $N_b^+—O^-$
239 $N_b^+—O^-$, $N_{b'}^+—O^-$

240 Wieland–Gumlich aldehyde

natural materials was established by comparison of TLC, and UV, and IR spectral properties.

Oxidation of **237** with 5% hydrogen peroxide solution at room temperature gave a 1:1 mixture of **238** and **239**, but heating on a water bath yielded only **239**. Reduction of **238** with 5% sulfurous acid solution at room temperature for 10 min gave one major product identified as caracurine V (**237**) (*147*).

The ^{13}C-NMR spectra of these compounds were also examined, and the data are shown in Table VI. Comparison was also made with strychnine and its *N*-oxide. Thus, on N-oxidation, C-3, C-5, and C-21 of strychnine were shifted downfield by 22.6, 17.5, and 17.8 ppm, respectively. In caracurine V the values for the same carbons on N-oxidation were 22.7, 17.4, and 17.2 ppm, respectively. Two carbons—C-8 and C-20—were shifted upfield on N-oxidation. For caracurine V (**237**) these shifts were 2.7 and

TABLE VI
^{13}C-NMR SPECTRA OF CARACURINE V AND RELATED ALKALOIDS

	Chemical Shift δ (ppm)			
	Caracurine V	Caracurine V *N*-oxide (**238**)		Caracurine V *N,N'*-dioxide
Carbon	(**237**)	a	b	(**239**)
C-2	56.8	56.9	55.8	55.8
C-3	59.9	59.9	82.6	82.6
C-5	51.2	51.3	68.7	68.7
C-6	40.8	40.7	37.2	37.2
C-7	55.5	55.8	56.9	56.9
C-8	133.6	133.3	130.6	130.7
C-9	119.5	120.3	120.7	121.2
C-10	121.7	121.8	121.8	121.9
C-11	128.2	128.6	129.6	129.6
C-12	110.2	110.7	111.5	111.6
C-13	152.3	152.1	152.0	151.6
C-14	26.1	25.9	24.5	24.4
C-15	34.0	33.8	32.5	32.4
C-16	52.5	52.4	52.1	52.0
C-17	98.9	98.9	98.5	98.6
C-18	66.5	66.4	65.9	65.9
C-19	127.3	128.1	133.8	133.8
C-20	141.2	140.7	136.0	136.2
C-21	53.3	53.3	70.5	70.5

4.7 ppm, and these paralleled strychnine for which the shifts were 3.3 and 5.2 ppm, respectively. The only other carbon to be substantially (> 1.5 ppm) shifted was C-19, for which downfield shifts of 5.6 and 6.8 ppm were observed for caracurine V and strychnine, respectively (*147*).

In the mouse screen grip test the *N*-oxides were considerably less active as muscle relaxants than caracurine V (**237**), and also showed less toxicity (*147*).

C. SYNTHESIS OF THE CHROMOPHORE OF THE CALEBASH CURARE ALKALOIDS

In a series of papers (*149–151*), Fritz and co-workers have described the synthesis of several model compounds having the chromophore of C-curarine (**157**).

Condensation of **241** and **242** in ether afforded the dioxadiazaadamantane derivative **243** which in methanolic HCl equilibrated with the dehydration product **244**. This compound has the same central ring system as C-curarine (**157**); indeed, both the color reaction with the ceric reagent and the UV spectrum were very similar (*152*).

Subsequently, this type of reaction was extended to the series with compounds containing a ring analogous to the E ring of *Strychnos* alkaloids (*149*) and to compounds having ethylamine side chains (*150*). Examples of the products are shown in **245** and **246**. This work, however, left the stereochemistry at C-9 and C-18a unresolved.

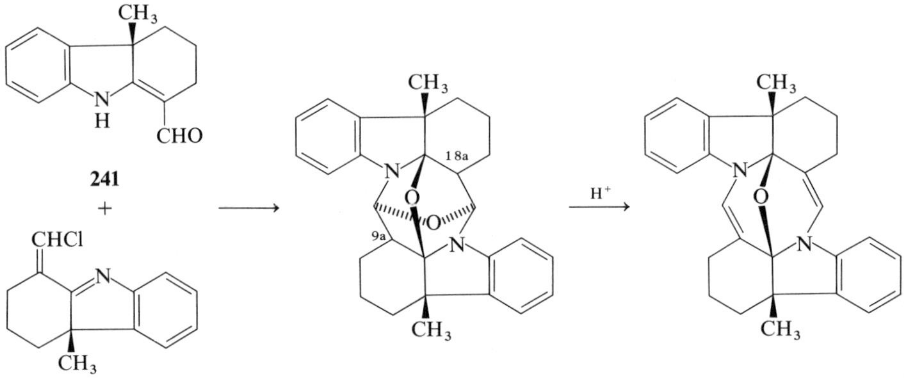

241 + **242** → **243** $\xrightarrow{H^+}$ **244**

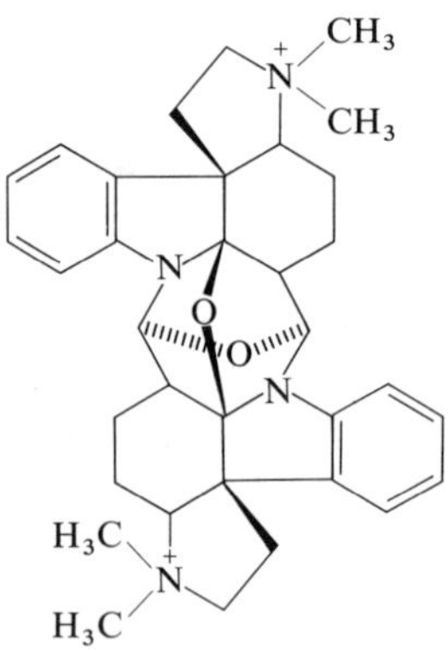

245

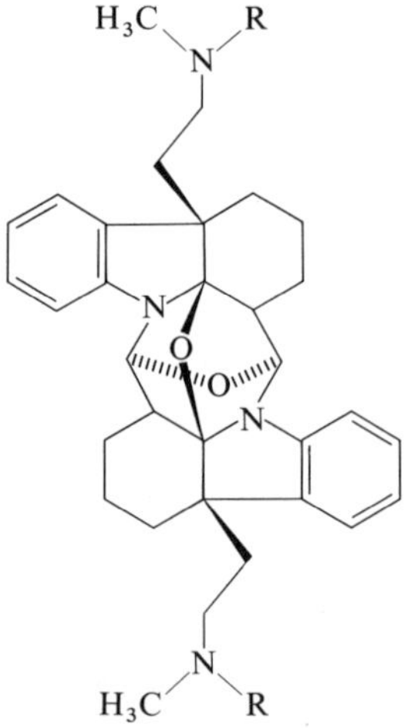

246 R = CH_3 or $CO_2C_2H_5$

Treatment of **243** with concentrated H_2SO_4 gave the bis salt **247**, and rehydration with base then yielded compound **248**, stereoisomeric at C-9 and C-18 with **243** (*151*). In the NMR spectrum of **243** the aminoacetal

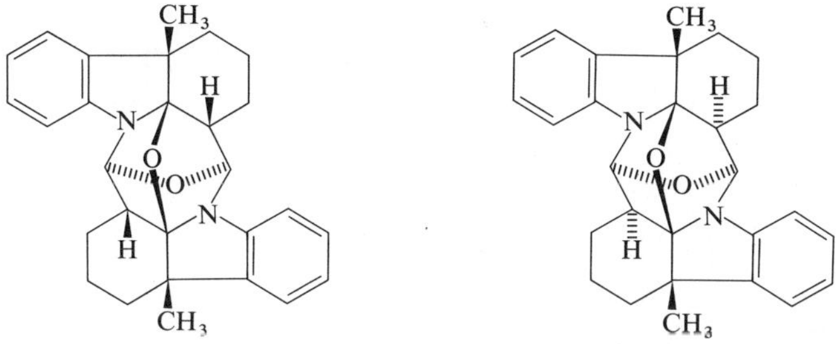

243 **248**

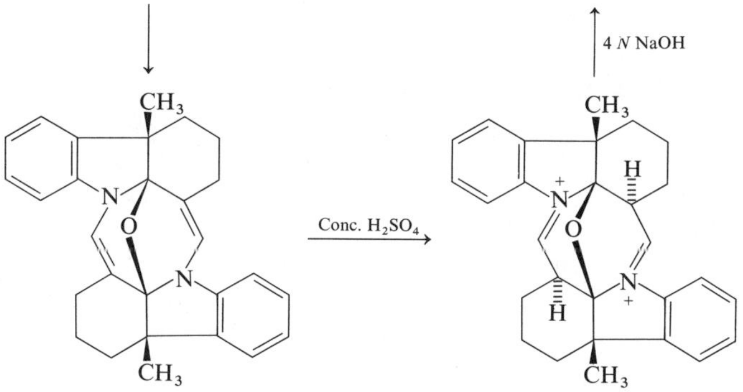

4 *N* NaOH

Conc. H_2SO_4

247

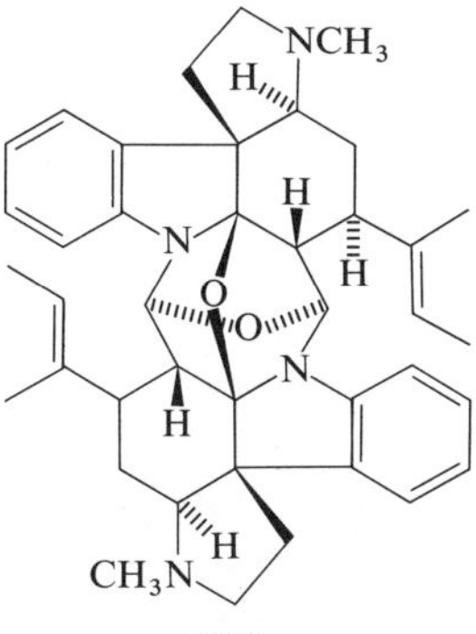

249

protons C-9 and C-18α appeared as a singlet at 5.37 ppm. In **248**, however, they were observed as a two-proton doublet ($J = 3$ Hz) at 5.24 ppm. These findings indicate the relationship between the two compounds.

Of more significance, however, as far as the natural dimers are concerned, was a study of the corresponding compounds in the optically active series with respect to their ORD and CD curves. Indeed, both ORD and CD spectra of these compounds were distinctly different; in particular, the optically active **243** showed characteristic positive maxima at 248 nm (CD) and 252 nm (ORD) (*151*). These data were used to determine the structure of an indoline base (**249**) obtained some years previously by Schmid and co-workers (*153*).

XIV. Secodines, Presecamines, and Secamines

The genus *Rhazya* in the family Apocynaceae is comprised of the two species *R. stricta* Decsne., native to northwest India, and *R. orientalis* A. DC., native to northwest Turkey. It has been studied for its indole alkaloids by several groups (*154*). The genus has provided three groups of closely related, biogenetically interesting alkaloids. Two of these are true dimeric indole alkaloids; the third is the monomeric unit which produces the dimers.

The secamines were the first compounds in this series to be isolated (from *R. stricta*) (*155*), and the lowest molecular weight compound to be obtained exhibited a molecular ion at 676 mu and was named secamine. Two other related compounds having molecular ions at *m/e* 678 and 680 were named dihydro- and tetrahydrosecamine, respectively. Each of these compounds was obtained as mixtures of differing complexity; this point will be referred to subsequently.

All these compounds showed purely indolic UV spectra, indicating the additional unsaturation to be distant from the indole chromophore. For convenience, most of the structure work was carried out with tetrahydrosecamine.

The IR spectrum indicated NH, saturated ester, and disubstituted benzene systems, and the molecular ion at *m/e* 680 was overshadowed by an extremely intense base peak at *m/e* 126. No class of indole alkaloids had previously been found to contain such an intense ion. High-resolution mass measurement, biogenetic considerations, and work with model compounds, e.g., **250**, indicated that the structure of this ion was **251**.

One of the tetrahydrosecamines showed two carbomethoxy singlets at 3.76 and 3.63 ppm, eight aromatic protons, a very broad NH, a broad one-proton singlet at 4.3 ppm, and a triplet at 1.0 ppm integrating for six protons, indicating the presence of two ethyl groups. The signal at 4.3 ppm was assigned to an α-indolylacetic ester proton because it was exchanged with MeONa–MeOD at 100°.

The ^{1}H-NMR spectrum of secamine was quite similar to that of tetrahydrosecamine except that two broad olefinic signals at 5.5 and 5.36 ppm were observed. Biogenetic considerations indicated that the double bonds in the piperideine rings were at the 15,20-position, and this was confirmed by Hofmann degradation.

Reduction of tetrahydrosecamine with LAH gave a diol (M^+ *m/e* 624) indicating reduction of two carbomethoxy functions, and showing losses of 18 and 30 mu. Therefore, the carbomethoxy groups are on tetra- and trisubstituted carbon atoms. Alkaline hydrolysis of tetrahydrosecamine and heating in 1 *N* HCl gave a didecarbomethoxytetrahydrosecamine along with other products, and reductive didecarbomethoxylation with stannous chloride gave **252**, whose structure was confirmed by synthesis. At this point, biogenetic speculation permitted two alternative structures, **253** and **254**, to be proposed for tetrahydrosecamine.

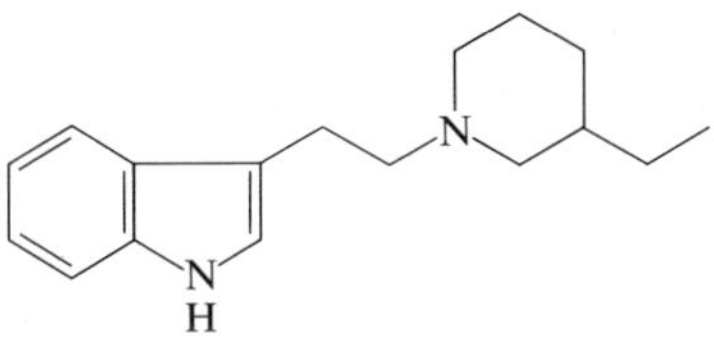

250

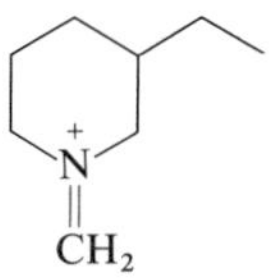

251 *m/e* 126

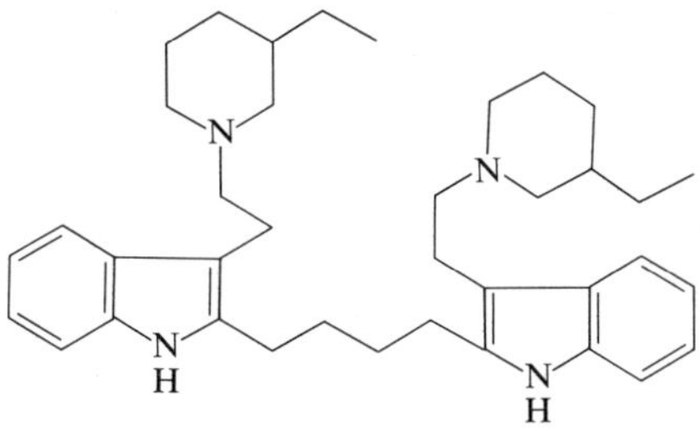

252

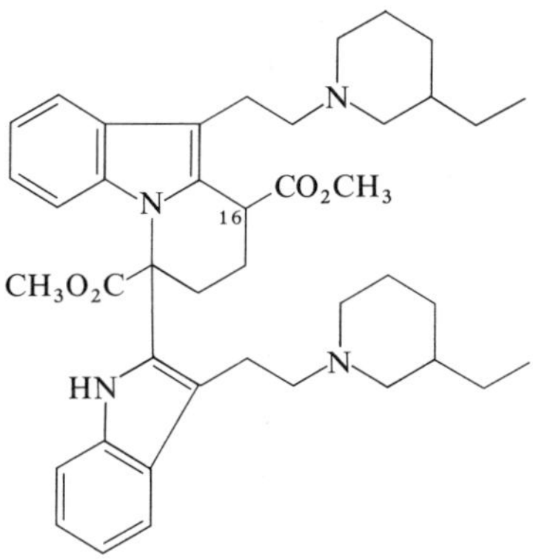

253

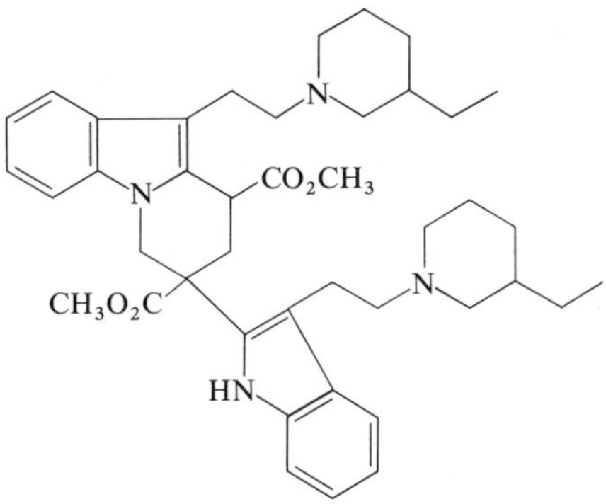

254

Several of the fractions of tetrahydrosecamine, although apparently homogeneous by TLC, exhibited complex carbomethoxy regions in the ^{1}H-NMR spectrum. One fraction for example showed singlets at 3.78, 3.77, 3.74, and 3.65 ppm in the ratio 2:3:2:3, but in other samples this ratio varied. This complexity was thought to be due to the presence of two components, one having carbomethoxyl signals at 3.78 and 3.74 ppm ("narrow" type) and the other showing signals at 3.77 and 3.65 ppm ("wide" type).

A key observation was that treatment of a mixture of tetrahydrosecamines with methoxide–methanol at room temperature overnight gave an almost pure wide tetrahydrosecamine. The conclusion was made, therefore, that the wide and narrow tetrahydrosecamines were diastereoisomers at C-16 and not skeletal alternatives.

Tetrahydrosecamine has also been isolated from *R. orientalis* (*156*), *Amsonia elliptica* (*157*, *158*) and *A. tabernaemontana* (*159*), although its diastereomeric purity has not always been given.

Subsequent to the isolation of the secamines, Smith and co-workers (*160*, *161*) examined the alkaloid fractions of *R. orientalis* and *R. stricta* by mass spectrometry for fractions which exhibited the *m/e* 126 ion (**251**), and as a result two additional series of alkaloids were found in each plant.

One of the purified fractions from *R. orientalis* displayed a UV spectrum containing both an indole and a β-anilinoacrylic ester chromophore, (**255**) and although apparently homogeneous by TLC, it was thought that possibly the complex UV spectrum was due to a mixture. However, at low probe temperatures, essentially only two ions were observed in the mass spectrum, at *m/e* 340 ($C_{21}H_{28}NO_2$) and 126. A thermal process occurring on the probe of the mass spectrometer to give a volatile, monomeric species explains this observation (*161*).

In a previous review of *Aspidosperma* alkaloids (*162*) it was emphasized that the retro-Diels–Alder fragmentation pathway of β-anilinoacrylate esters is highly diagnostic. In this instance, it was surmised that because of the simplicity of the mass spectrum, one unit must constitute both halves of the dimer, and in addition, this acrylic ester unit must be capable of Diels–Alder dimerization.

Biogenetically, the structure considered most likely for the acrylic ester was **256**, and because one of the indole units remains unreacted in the starting material (from the UV spectrum), it is the acrylic ester portion of this unit which is the ethylenic product in the reverse Diels–Alder reaction. Two gross structures, **257** and **258**, were suggested for the parent compound on this basis (*161*).

The IR spectrum of the fraction showed NH absorptions at 3420 and 3360 cm^{-1} and a complex carbonyl region showing absorptions at 1730, 1675, and 1610 cm^{-1}. The latter are typical for a β-anilinoacrylate unit.

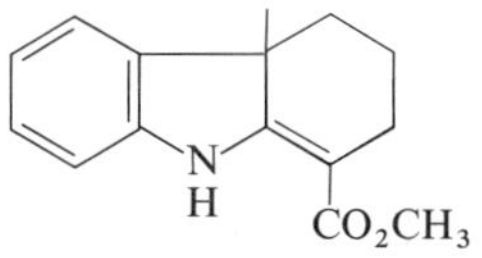

255

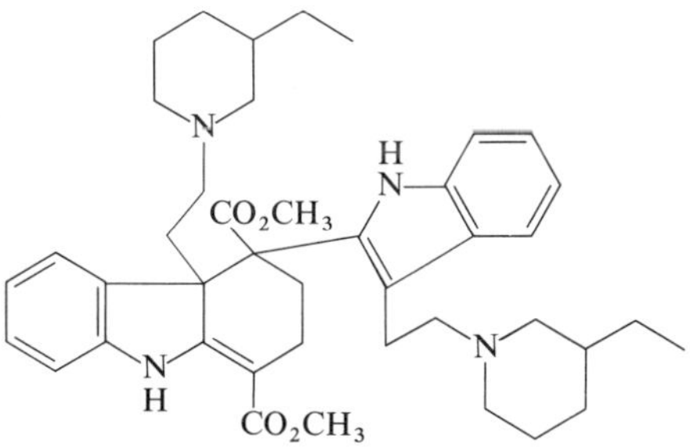

257 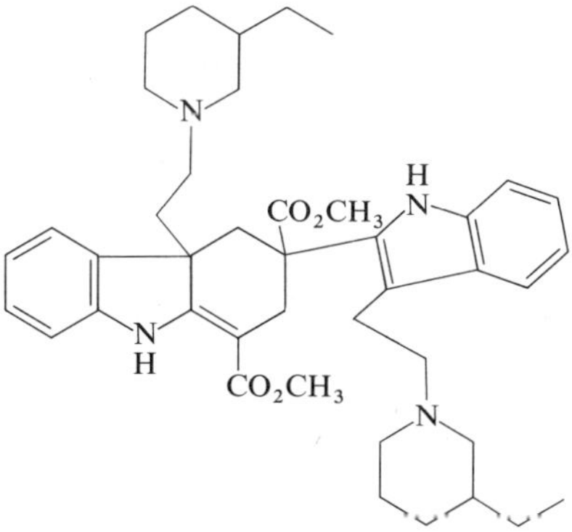 **258**

The ^{1}H-NMR spectrum confirmed the presence of two exchangeable NH protons and two carbomethoxy groups at 3.56 and 3.76 ppm, the former being attributed to the acrylate ester. Except for the observation of eight aromatic protons, little additional information could be gained from the ^{1}H-NMR spectrum.

Several fractions from *R. stricta* contain the indole plus β-anilinoacrylate chromophore, and from these three alkaloids were isolated (*161*). One of these was identical to the alkaloid from *R. orientalis*, a second showed *m/e* 338 and 124, and the third alkaloid displayed *m/e* 340, 338, 126, and 124.

The ^{1}H-NMR spectrum of the second alkaloid showed two olefinic protons at 5.46 and 5.36 ppm. The double bonds were deduced to be at C-15 and C-20 in each of the two piperideine rings.

The alkaloid, MW 680, was treated with $NaBH_4$–formic acid in an attempt to reduce the β-anilinoacrylate unit to the indoline. The UV spectrum of the product was purely indolic, however, and no reduction had taken place because the product showed M^+ *m/e* 680.

The same alkaloid was treated with formic acid, and after 4 min at room temperature a purely indolic UV spectrum was obtained. A similar product was also obtained by treatment with 2 *N* dilute HCl after 15 min. Two properties of this product were of crucial importance at this point. The acid-catalyzed product was shown to be identical to tetrahydrosecamine by TLC and mass spectral fragmentation. The tetrahydropresecamine, as the starting material was named, had rearranged quantitatively to tetrahydrosecamine!

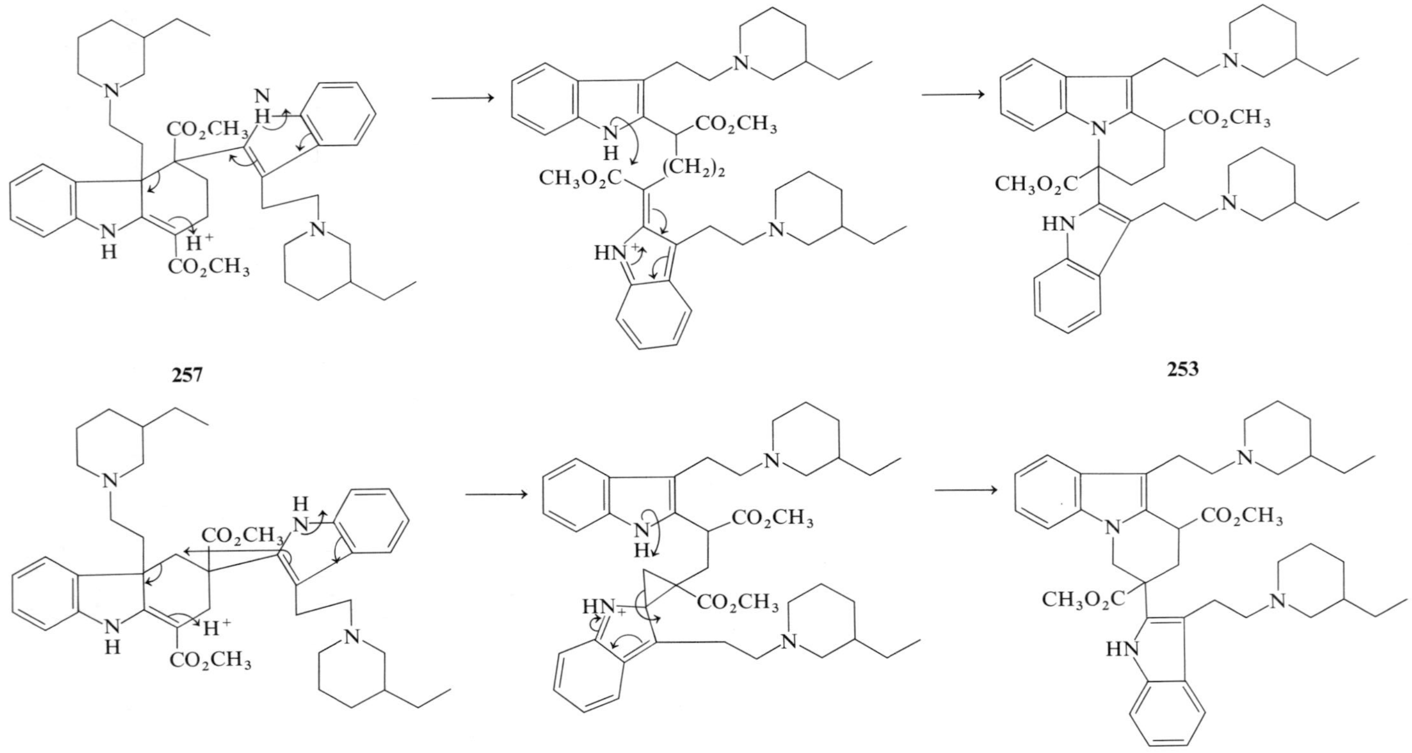

257 253 258 254

SCHEME 11. *Acid rearrangement of the alternative presecamine structures.*

Unfortunately, the mechanisms that explain this rearrangement (Scheme 11) do not distinguish between the two possible skeleta. Indeed, from one of the presecamines, each secamine skeleton can be obtained!

Evidence confirming the rearrangement to the tetrahydrosecamine skeleton came from an examination of the NMR spectrum of the product. Somewhat surprisingly, the crude product was found to be the pure "narrow" tetrahydrosecamine with singlets at 3.75 and 3.72 ppm. Therefore, the rearrangement is kinetically controlled, and from previous work it gives rise to the least thermodynamically stable tetrahydrosecamine.

At this juncture in the investigation, trapping of the monomeric species produced by the thermal decomposition of the presecamines was carried out. Once again, because of the quantities available, tetrahydropresecamine was used.

The material trapped on a cold finger after sublimation of 0.5 mg of tetrahydropresecamine showed a unique UV absorbance, exhibiting λ_{max} (Et_2O) at 278 (sh) and 312 nm. Several reactions were used to establish structure **247** for this compound, and some of these are shown in Scheme 12.

Catalytic reduction gave 15,16,17,20-tetrahydrosecodine (**259**), a compound also isolated from *Rhazya stricta* (*160*), and LAH reduction afforded alcohol **260** which could also be prepared from **259**. Additional evidence for the acrylic ester unit was provided from reaction with diazomethane to give the pyrazoline **261**.

On a larger scale, sufficient acrylic ester was obtained for NMR analysis. Two doublets ($J = 1.2$ Hz) at 6.46 and 5.99 ppm and the carbomethoxy group at 3.93 ppm confirmed the acrylic ester unit, and the remainder of the NMR spectrum was in agreement with the proposed structure, 15,20-dihydrosecodine (**256**).

In the condensed phase at 0° the acrylic ester underwent a Diels–Alder reaction to give tetrahydropresecamine as a mixture of isomers. The NMR spectrum of the mixture displayed carbomethoxy singlets at 3.79, 3.76, 3.69, and 3.56 ppm in the ratio 2:9:2:9, separable by TLC. The major product (now designated "major" tetrahydropresecamine) was identical to the isolated tetrahydropresecamine both from its UV and NMR spectra and by its conversion to the "narrow" tetrahydrosecamine.

The minor product ("minor" tetrahydropresecamine) displayed a similar but characteristically different UV spectrum. Two other distinct differences that were noted are the carbomethoxy signals in the NMR spectrum and the rate of the rearrangement to the secamine skeleton (*163*).

The skeleton of the "minor" tetrahydropresecamine became clear when the tetrahydrosecamine derived from it by acid rearrangement gave the *same* narrow tetrahydrosecamine as did rearrangement of the "major" tetrahydropresecamine. This "minor" tetrahydropresecamine was therefore deduced to

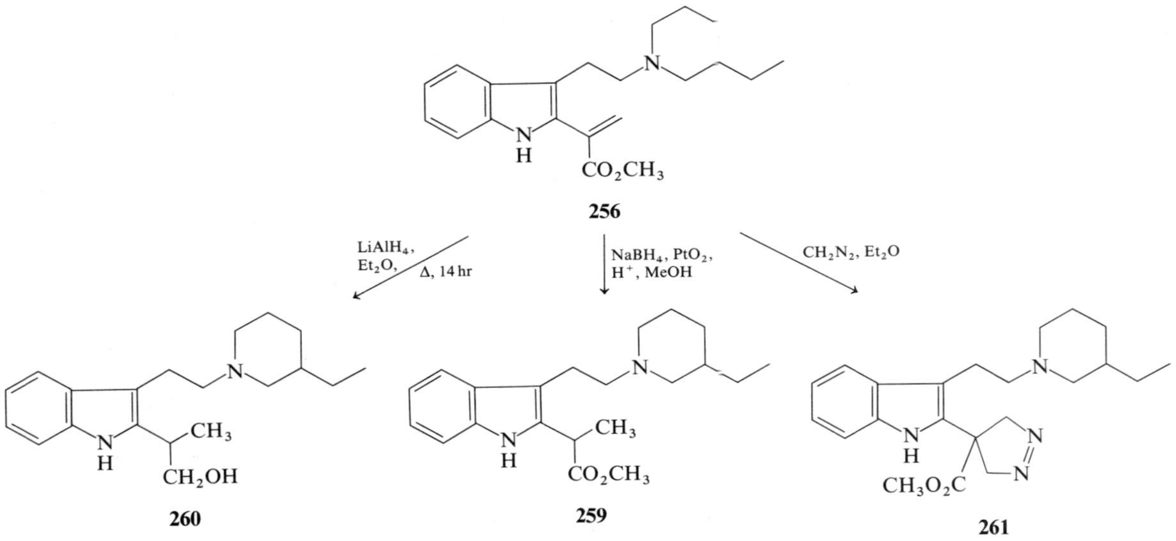

CO_2CH_3
256
$LiAlH_4$,
Et_2O,
Δ, 14 hr
$NaBH_4$, PtO_2,
H^+, MeOH
CH_2N_2, Et_2O
CH_3
CH_2OH
260
CH_3
CO_2CH_3
259
CH_3O_2C
261

SCHEME 12.

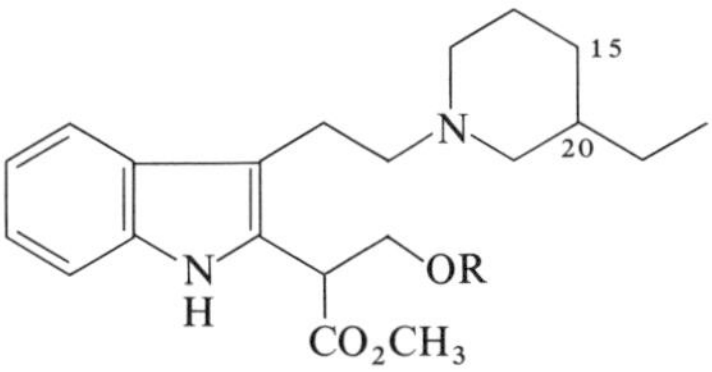

262 R = H
263 R = $COCH_3$
264 R = H, $\Delta^{15,20}$

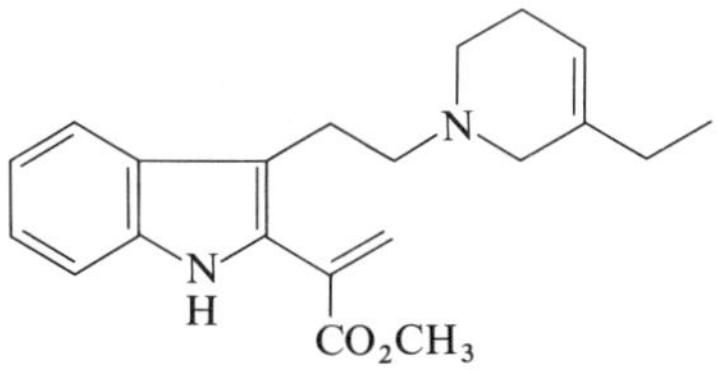

265 Secodine

be a stereoisomer at C-16′ of the natural tetrahydropresecamine from *R. orientalis*, and not a compound in a different skeletal series. Since this original isolation, tetrahydropresecamine has also been isolated from *Amsonia tabernaemontana* (*159*) and *Pandaca minutiflora* (*164*). In the latter case at least, the isolated isomer was the "major" one.

A distinction between the two possible skeleta for the presecamines and the secamines was eventually made by synthesis of the series of compounds in the skatole series. Before this could be achieved, however, one further reaction in the sequence should be discussed.

At about this time, alcohol ester **262** was isolated from *R. orientalis* (*160*), and because so little material was obtained its structure was proved by total synthesis. On acetylation this alcohol ester did not give the expected acetate **263**, which would have an indolic UV spectrum, but rather a product with an indole plus β-anilinoacrylic ester chromophore. This product was pure tetrahydropresecamine. Thus, in one step, acetylation, elimination of acetic acid, and Diels–Alder dimerization had occurred (*160*). The presecamines themselves were prepared in analogous fashion from the alcohol ester **264** by way of secodine (**265**).

Therefore, this elimination reaction together with the dimerization and acid rearrangement processes constitute a total synthesis of the secamine skeleton. These reactions were crucial in deducing the nature of the central ring substituents.

The compounds in the series derived from skatole (3-methylindole) were synthesized from the alcohol ester **266** (*163*). The acetylation–elimination

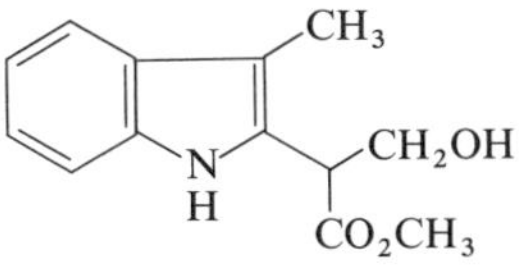

266

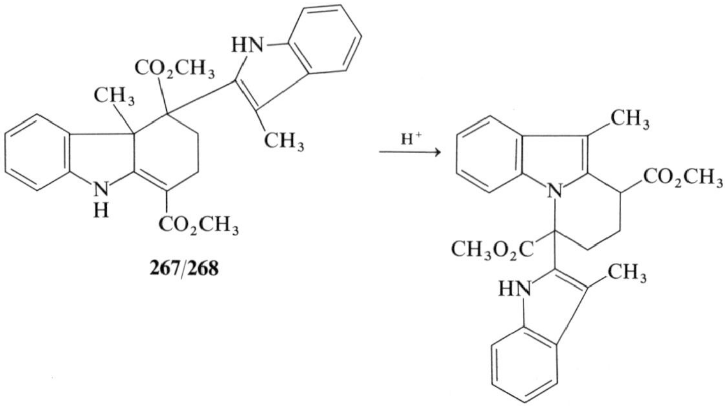

HN CO_2CH_3 CH_3 CH_3 N H CO_2CH_3

267/268

H^+ →

CH_3 CO_2CH_3 N CH_3O_2C HN CH_3

269/270

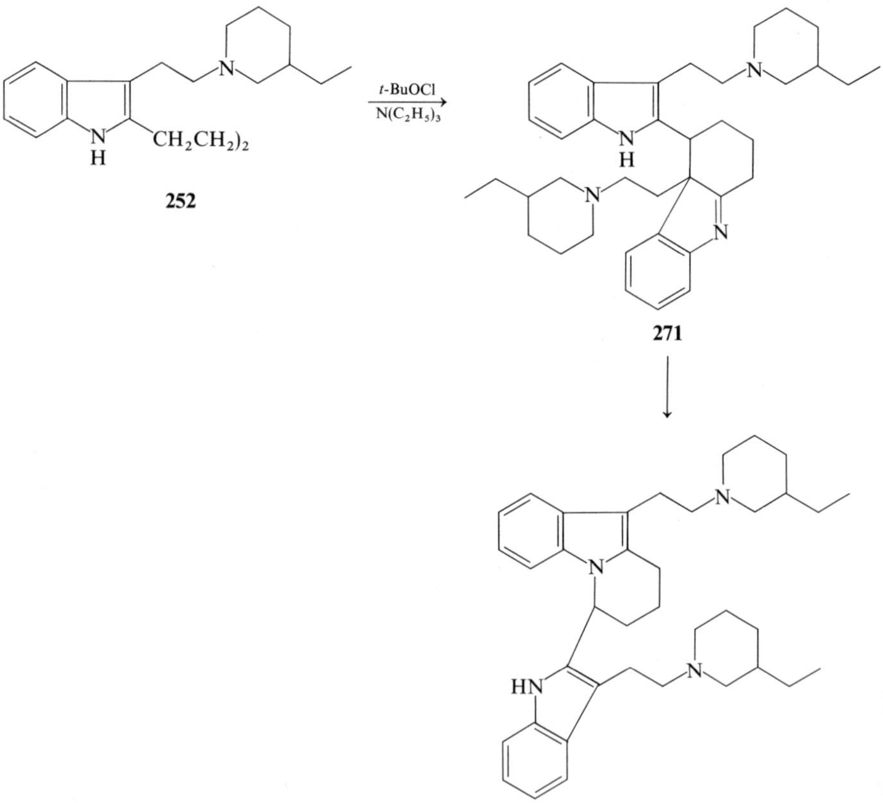

N N H $CH_2CH_2)_2$

252

t-BuOCl, $N(C_2H_5)_3$ →

N N H N N

271

↓

N N HN N

272

reaction was carried out with Ac_2O–py in the presence of triethylamine, to give a mixture of the skatolylpresecamine isomers **267** and **268**. These could be rearranged to the skatolylsecamine isomers **269** and **270**. The NMR spectra of these two isomers confirmed the substitution by the groups on the central alicyclic ring (*163*).

Sakai and co-workers (*157*, *158*) have reported confirmatory evidence of the central ring arrangement. The key reaction is the oxidation of **252** with *tert*-butyl hypochlorite and triethylamine to give indolenine **271**, which has the presecamine skeleton. This compound was rapidly rearranged with acid to the secamine skeleton in the form of didecarbomethoxytetrahydrosecamine (**272**), identical to the product resulting from degradation of natural tetrahydrosecamine.

Although the acid rearrangement of natural tetrahydropresecamine proceeded stereospecifically, the stereochemistry of the substituents on the nucleus was not known.

In the synthetic work which was developed to prove the structure of the natural secamines and presecamines, two other groups of compounds—the

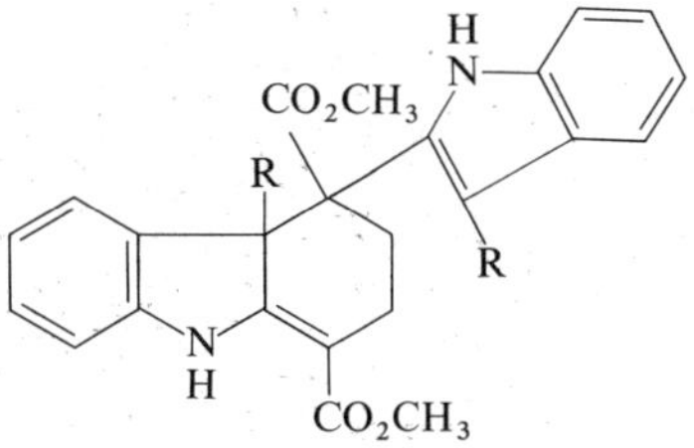

267/268 R = —CH_3
274 R = —$CH_2CH_2N(CH_3)_2$
257 R = —CH_2CH_2N(3-ethylpiperidin-1-yl)

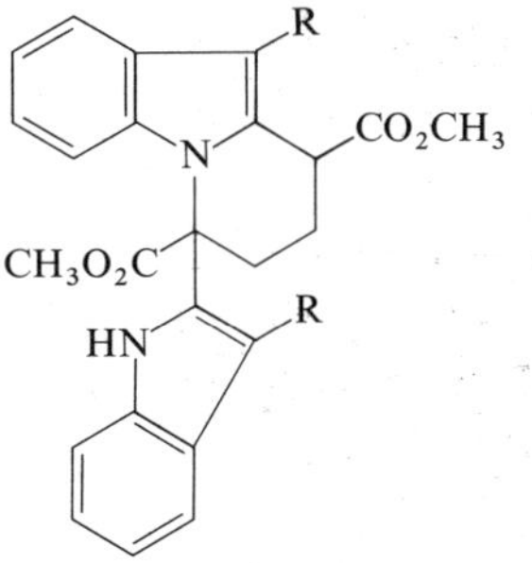

269/270 R = —CH_3
273 R = —$CH_2CH_2N(CH_3)_2$
253 R = —CH_2CH_2N(3-ethylpiperidin-1-yl)

N,N-dimethyltryptamine **273/274** [R = $CH_2CH_2N(CH_3)_2$] and skatolyl **270/268** (R = CH_3) secamines and presecamines—were also obtained. Although each diastereoisomer in each series was available and distinguishable by chromatographic and spectroscopic analysis, stereochemical conclusions which were made at that time (*165*) were only tentative. The corresponding isomer in each series, however, did behave chemically in a similar manner to its analogs in the other two series. Thus, acid rearrangement of either presecamine diastereoisomer consistently gave the least thermodynamically stable "narrow" secamine diastereoisomer, which epimerized in base to the "wide" secamine diastereoisomer. Because of this finding it was believed that

TABLE VII

CORRELATION OF THE PRESECAMINES AND SECAMINES

	Tetrahydropresecamines		Dimethylaminoethylpresecamines		Skatolylpresecamines	
	Major	Minor	Minor	Minor	Major	Minor
Relative amounts formed in dimerization of secodine monomer	82%	18%	70%	30%	70%	30%
Ester CH_3 (ppm)	3.76, 3.56	3.79, 3.69	3.76, 3.56	3.80, 3.71	3.76, 3.57	3.84, 3.66
	Tetrahydrosecamines		**Dimethylaminoethylsecamines**		**Skatolylsecamines**	
	Narrow	Wide	Narrow	Wide	Narrow	Wide
Relative amounts formed in acid rearrangement	>95%	<5%	95–98%	2–5%	60%	40%
Relative amounts in NaOMe equilibrium	<5%	>95%	7%	93%	25%	75%
Ester CH_3 (ppm)	3.75, 3.72	3.76, 3.63	3.77, 3.73	3.77, 3.63	3.76, 3.72, 3.62	

deduction of the stereochemistry in one series would permit correlation with each of the other series. Some of the data which evince the validity of these correlations are shown in Table VII.

The diastereoisomeric secamines differ by a cis or trans arrangement of two ester functions on a cyclohexene ring. In principle, therefore, these isomers could be distinguished by formation of a cyclic system only from the cis isomer (*166*).

Anhydride formation was attempted initially but failed as a result of mixture formation. The diastereoisomeric skatolylsecamine diols **275** and **276**, produced by LAH reduction of **269**/**270**, were more amenable, but even traces of acid caused elimination of water to **277**; consequently, the possibility of S_N2 reactions was explored. Attempts to cyclize the C-22 monotosylate **278** or **279** were unsuccessful, again because of elimination reactions (λ_{max} ~310 nm of a vinylindole). Similar problems were also encountered with the ditosylates **280** and **281**. However, treatment of the C-22′ monotosylate **282** with sodium hydride in hexamethylphosphoramide gave the desired ether (**283**) in high yield.

The C-22′ tosylate **284** from the other diastereoisomer was not available directly and was prepared from diacetate **285** by selective removal of the C-22′ acetyl group with triethylamine in methanol, tosylation, and removal of the acetate group with methoxide. No C-16 epimerization occurred under these conditions.

The *trans*-C-22′ tosylate **284** on treatment with sodium hydride–hexamethylphosphoramide gave a stable indolic product, which on the basis of chemical and spectral data was assigned structure **286**.

Reinvestigation of the reaction of the *cis*-C-22′ tosylate **282** revealed that the reaction was more complex than had at first been supposed. An initial product, formed in high yield, was assigned structure **287**, diastereomeric with the compound obtained in the trans series. Allowing the reaction to continue afforded the desired cis cyclic ether **283** isolated previously.

From these data it was established (*166*) that the thermodynamically more stable "wide" skatolylsecamine has trans stereochemistry, and consequently that the "narrow" isomer which is favored in the acid rearrangement of the presecamines has cis stereochemistry. By correlation, "wide" tetrahydrosecamine has structure **288** and "narrow" tetrahydrosecamine structure **289**.

The last major stereochemical problem concerns the relative configurations at C-7 and C-16′ in the presecamine series. Once again, although all the stereoisomers were available, early stereochemical conclusions (*165*) were based only on the interpretation of NMR data.

Refluxing the skatolylpresecamines **267** and **268** in dioxan in the presence of air gave, stereospecifically, one diastereoisomer of an oxidation product

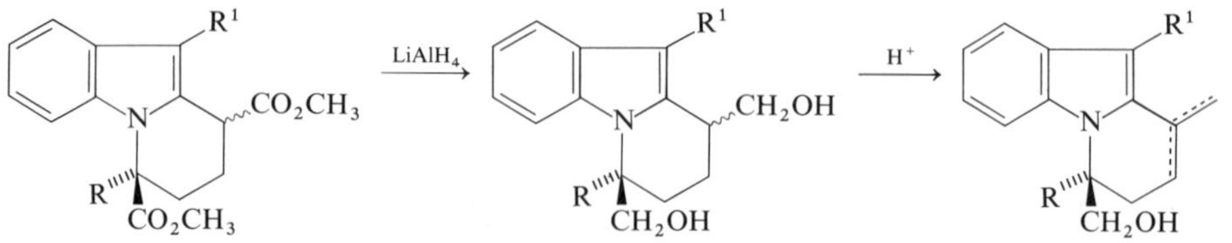

R^1 CO_2CH_3 N R CO_2CH_3 **269/270** $\xrightarrow{LiAlH_4}$ R^1 CH_2OH N R CH_2OH **275/276** $\xrightarrow{H^+}$ R^1 N R CH_2OH **277**

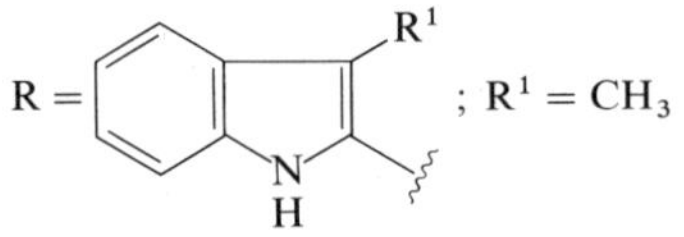

R = (3-methylindol-2-yl, R^1, N H) ; $R^1 = CH_3$

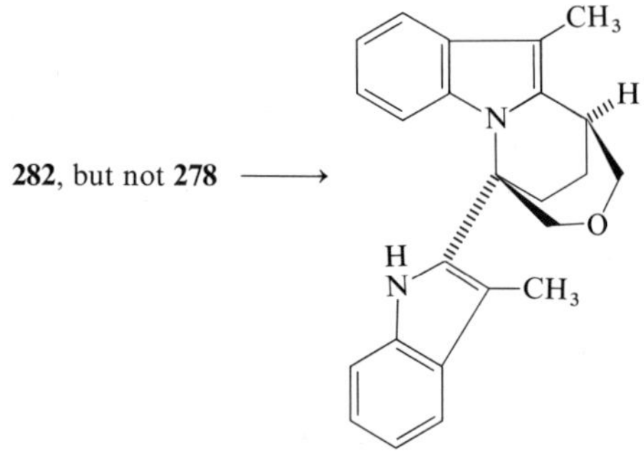

282, but not **278** ⟶ CH_3 H N O H N CH_3 **283**

CH_3 N 16 H CH_2OR^1 22 16′ CH_2OR^2 22′ H N CH_3

CH_3 N H CH_2OR^1 R^2OCH_2 H N H_3C

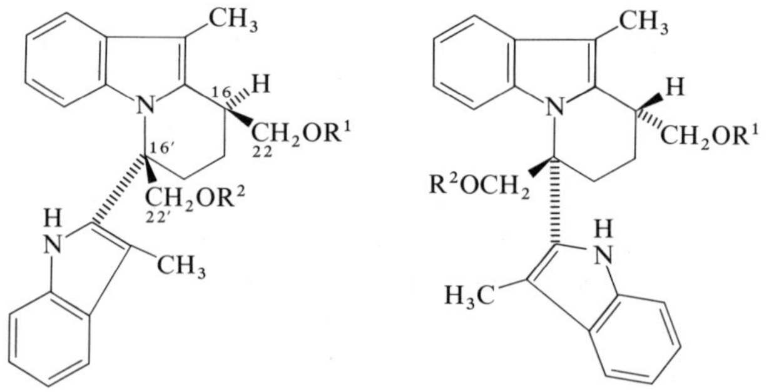

cis ("narrow") *trans* ("wide")

	R^1	R^2	
275	H	H	**276**
278	Ts	H	**279**
280	Ts	Ts	**281**
282	H	Ts	**284**

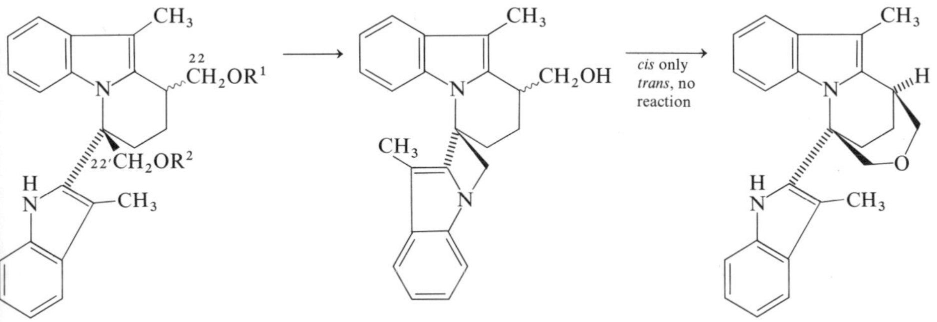

cis	R^1	R^2	*trans*
275	H	H	**276**
	Ac	Ac	**285**
282	H	Ts	**284**

286 *trans*
287 *cis*

cis only
trans, no reaction

283

Ac_2O–py → no reaction

NaH, HMPT, CH_3I → mainly monomethylation

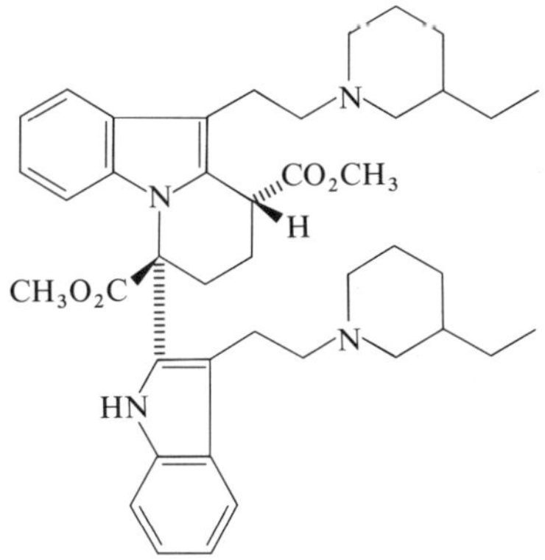

288 "Wide" tetrahydrosecamine

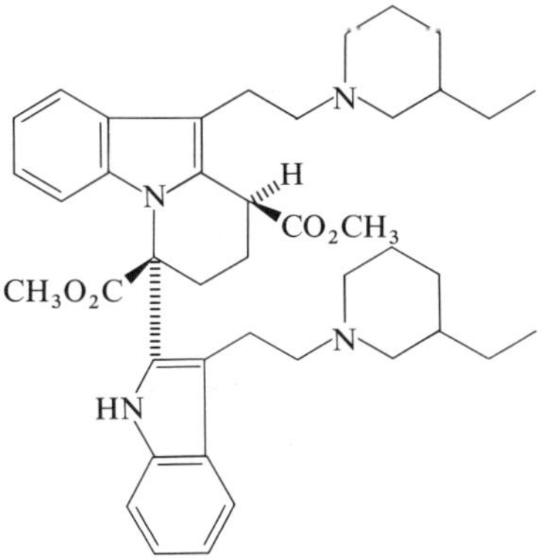

289 "Narrow" tetrahydrosecamine

from each isomer. In addition, the minor isomer was more susceptible to autoxidation under these conditions.

Each isomer showed a molecular ion at m/e 446 analyzing for $C_{26}H_{26}N_2O_5$ and showed UV spectra corresponding to the addition of indole and indolenine chromophores. The products were therefore formulated as **290** and **291**. Such oxidations are well known in the *Aspidosperma* alkaloid series which contain a β-anilinoacrylate unit (*162*).

Acid-catalyzed rearrangement of either oxidation product (**290** or **291**) gave the same unsaturated skatolylsecamine **292**, the structure for which was established spectroscopically (*166*).

267 "Minor" isomer

268 "Major" isomer

O_2

O_2

290

291

292

293 Natural tetrahydropresecamine

Molecular models indicated that the least hindered side of the molecule was that on which the angular methyl group was located. Since autooxidation was not only regioselective but stereoselective, these oxidation products should have the relative stereochemistries shown in **290** and **291**.

A distinction between these two structures was made on the basis of their mass spectra. Only in the oxidation product **290** derived from the minor skatolylpresecamine was an ion observed at *m/e* 400 (M* 258.7) for the direct elimination of dimethyl ether, a reaction possible only when the carbomethoxy groups are cis to each other. The other isomer (**291**), on the other hand, preferentially lost methanol between the C-16 carbomethoxy group and the C-16′ α-hydroxy group.

On this basis, therefore, the minor isomer of the skatolylpresecamines was assigned structure **267** and the major isomer structure **268**. By analogy, natural tetrahydropresecamine has structure **293**.

The major isomer from the dimerization process, which is also the only isomer obtained naturally, is the most thermodynamically stable. This stability may be due to the more substantial steric requirements for the indolic system vs. the carbomethoxy group. In this respect it is noteworthy that as the size of the indole 3-substituent increases from $-CH_3$ to $-CH_2CH_2N(CH_3)_2$ and finally ethylpiperidinoethyl, the yield of the "minor" isomer in the dimerization process correspondingly decreases (see Table VII).

The acid-catalyzed rearrangement of the presecamines to the secamines was the first clue to their structure. On initial inspection it appeared to be a simple reaction sequence of two isomers A and B going to a third product C. By UV and TLC no intermediates were detected, the only discernible difference being that the "minor" isomer consistently took longer to rearrange than did the "major" isomer (*165*). It was only when HPLC was applied to the reaction that the real complexity revealed itself.

Analysis of the reaction of the "major" skatolylpresecamine (**268**) indicated that it was transformed directly to the "narrow" skatolylsecamine (**269**). However, analysis of the reaction of the "minor" isomer (**267**) indicated that it was first rapidly transformed into the "major" isomer (**268**), which was *then* transformed slowly to a mixture containing the same "narrow" skatolylsecamine **269**. Thus, in 2 *N* HCl–CH_3OH at 40° after 20 min, the reaction mixture contained virtually no "minor" isomer but about 70% of the "major" isomer and about 30% "narrow" skatolylsecamine.

An analogous situation was revealed in the *N,N*-dimethyltryptamine presecamines, where the "minor" isomer was converted to the "major" isomer prior to rearrangement to the "narrow" secamine diastereoisomer.

These results were interpreted by the Manchester group in terms of the mechanism shown in Scheme 13 (*166*).

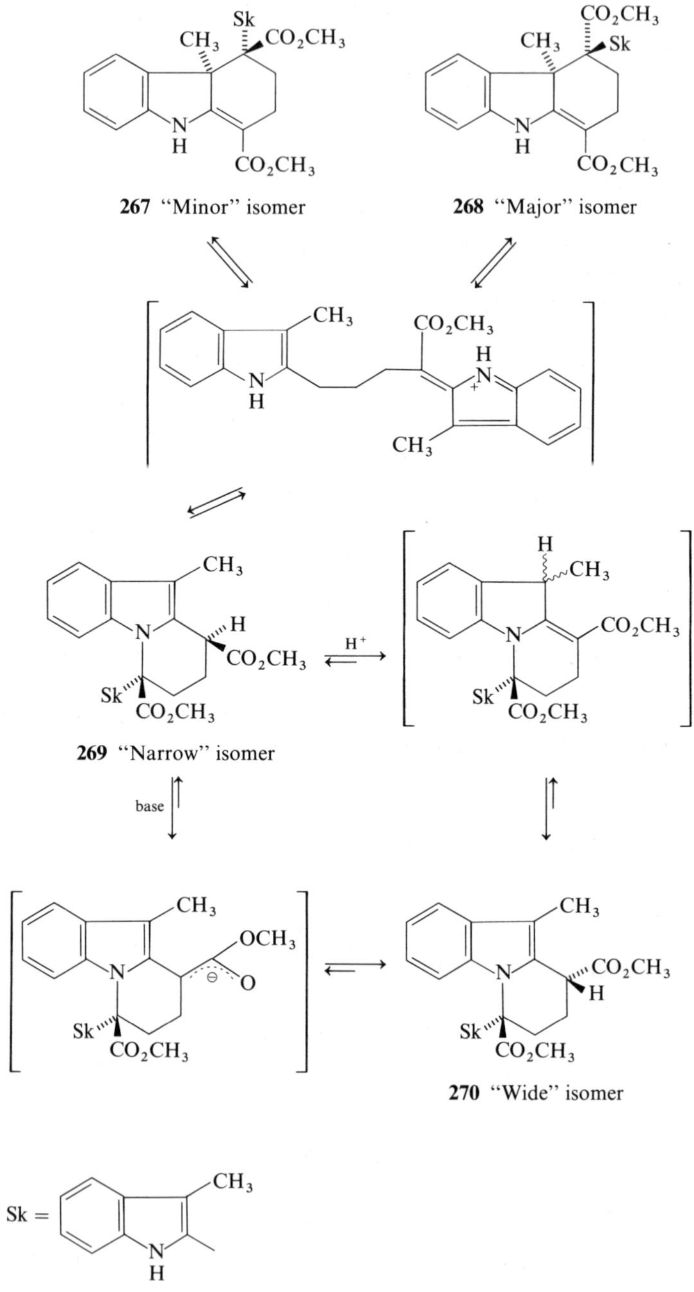

267 "Minor" isomer

268 "Major" isomer

269 "Narrow" isomer

270 "Wide" isomer

SCHEME 13. *Presecamine–secamine interconversions in the skatolyl series.*

XV. Iboga–Canthinone Type

BONAFOUSINE (296)

From the leaves of *Bonafousia tetrastachya* (Humb., Bon. & Kunth) Mgf. (Apocynaceae) Ahond and co-workers obtained, as the major alkaloid, the novel dimeric alkaloid bonafousine, mp 199°–200° (*167*).

The molecular formula $C_{35}H_{40}N_4O_3$ was deduced by high-resolution mass spectrometry, and the oxygen functions were traced to a carbomethoxy group (ν_{max} 1715 cm^{-1}) and a hydroxy group (ν_{max} 3600 cm^{-1}). In alkali, the UV spectrum (λ_{max} 228, 286, 294(sh), and 300 nm) was shifted to λ_{max} 228, 286, 294, and 320 nm in which the band at 286 nm increased in intensity. These data suggested the presence of indole and 11-hydroxyindole moieties. From the ^{1}H-NMR spectrum it was deduced that one of the aromatic nuclei was 11,12-disubstituted, because two ortho-coupled protons were observed at 7.09 and 7.51 ppm. The linkage from the hydroxyindole unit, therefore, must be to an aliphatic portion of the indole moiety.

The only other readily recognizable peaks in the ^{1}H-NMR spectrum were the triplet of an ethyl group at 0.84 ppm and a singlet carbomethoxyl at 3.54 ppm. Analysis of the mass spectrum of bonafousine indicated that the alkaloid fragmented into two "halves" of mass 353 and 211. The fragmentation of the former unit was of the iboga type, but the latter type could not be determined precisely although it was clearly a carboline (*m*/*e* 184, 170, 169). Two possibilities (**294** and **295**) were considered based on biogenetic considerations, but the ^{13}C-NMR spectrum, although substantiating these

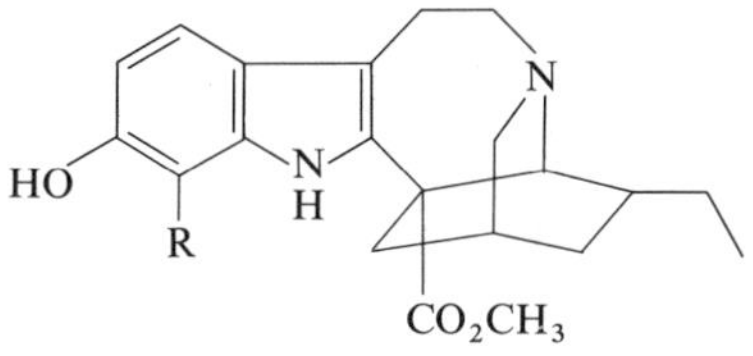

HO N H R N CO_2CH_3

294 R = N NH

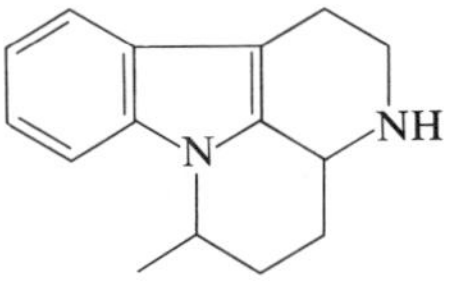

or

295 N H N

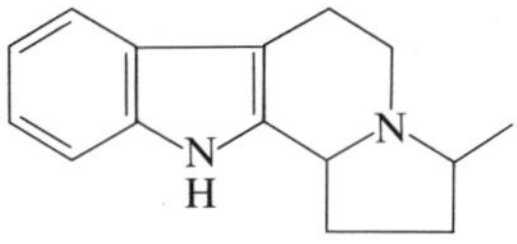

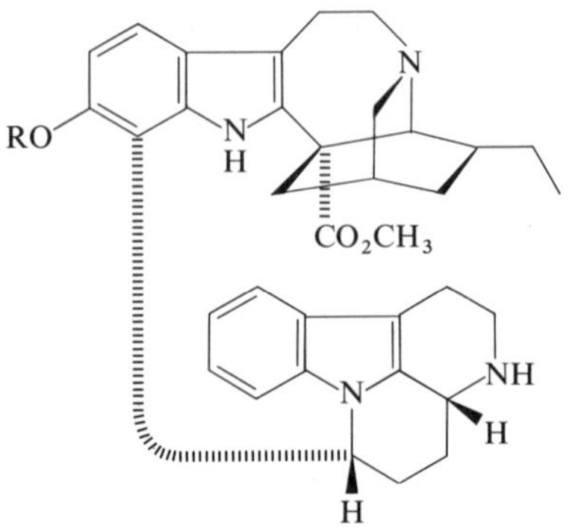

296 R = H Bonafousine
297 R = CH_3

assignments, did not distinguish between them. The differentiation was made possible by single-crystal X-ray crystallography of bonafusine dihydrochloride trihydrate, which indicated structure **296** for the parent alkaloid. The circular dichroism of 11-*O*-methylbonafousine (**297**) established that the ibogane unit was of the absolute stereochemistry indicated (*167*). Bonafousine is the first alkaloid to be isolated containing a perhydrocanthinone moiety.

XVI. Iboga–Vobasine Type

Previous mention has been made in this series (*9*) of the iboga–vobasine bisindole alkaloids such as voacamine, voacamidine, voacorine, 19′-epivoacorine, conoduramine, conodurine, and gabunine. In the intervening years, several new alkaloids in the series have been isolated.

A. *N*-Demethylvoacamine (298)

The third alkaloid isolated by Achenbach and Schaller (*122*) from the root bark of *Tabernaemontana accedens* was a simple derivative of voacamine, namely, *N*-demethylvoacamine (**298**).

The UV and IR spectra of **298** and of voacamine (**299**) were similar, but in the mass spectrum the molecular ion of the former compound appeared at 14 mu less than in **299**. In the NMR spectrum of **299** an *N*-methyl singlet was observed at 2.69 ppm which was absent in the spectrum of **298**. The NH

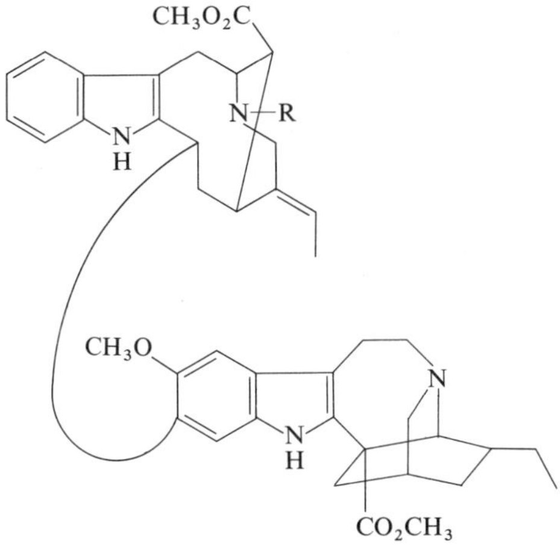

298 R = H *N*-Demethyl voacamine
299 R = CH_3 Voacamine

group was confirmed by acetylation with acetic anhydride–pyridine and N-methylation, which gave voacamine (**299**) identical with the natural product (*122*).

B. 19-Epivoacorine (302)

In 1973, Poisson and co-workers (*168*) reported the isolation from the bark of *Tabernaemontana brachyantha* Stapf. of both voacorine (**300**) and a new isomer of voacorine, mp 260° (dec).

The UV spectrum of the alkaloid was purely indolic, and the IR spectrum indicated the presence of both NH and OH groups. From the mass spectrum a molecular weight of 720 was deduced, and this ion readily gave a $M^+ - H_2O$ fragment. Comparison of the ^{1}H-NMR spectra of voacorine (**300**) and the new compound (Table VIII) indicated that the only major difference was the chemical shift of the C-19 olefinic proton, which was now somewhat deshielded (to 5.35 ppm).

The NMR spectrum of 19′-epivoacorine (**301**) (*169*) differed from that of voacorine in the chemical shift of the C-18′ protons but showed essentially the same chemical shifts for the remaining major peaks. The new compound could not be an isomer at positions 16, 3, or 20′ or at the location of the methoxy group. The only remaining reasonable possibility is of an isomerization of the 19,20-bond; consequently, the new compound would be 19-epivoacorine (**302**).

TABLE VIII

COMPARISON OF THE PRINCIPAL ^{1}H-NMR DATA OF VOACORINE (**300**), 19′-EPIVOACORINE (**301**), AND THE ALKALOID FROM *TABERNAEMONTANA BRACHYANTHA*[a]

	Voacorine	19′-Epivoacorine	Alkaloid from *T. brachyantha*
$\overset{19'}{CH_3}CH(OH)$	1.07	1.22	1.0
$\overset{19}{CH_3}CH{=}C$	1.67	1.66	1.67
$\overset{16}{-}CO_2CH_3$	2.48	2.47	2.55
$\overset{16'}{-}CO_2CH_3$	3.68	3.63	3.74
$-NCH_3$	2.60	2.58	2.64
$ArOCH_3$	3.99	3.98	4.00
$CH_3\overset{19}{CH}{=}C$	5.15	5.15	5.35

[a] From Patel *et al.* (*168*).

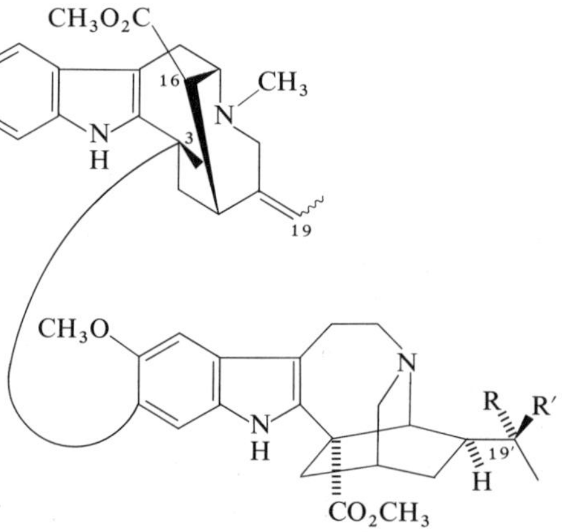

300 R = H, R′ = OH Voacorine
301 R = OH, R′ = H 19′-Epivoacorine
302 R = H, R′ = OH 19-Epivoacorine

C. TABERNAMINE (303)

Ethanol extracts of *Tabernaemontana johnstonii* Pichon (Apocynaceae) exhibited antitumor and cytotoxic activities. One of the constituents responsible for the activity was found to be tabernamine (**303**), which was active in the P-388 lymphocytic leukemia system in cell culture (*170*).

The molecular ion was observed at *m/e* 616 with a base peak at *m/e* 122. From the UV spectrum, the structure was deduced to be a bisindole without aromatic substitution. Only one carbonyl absorption (1720 cm^{-1}) was observed, and the ^1H-NMR spectrum indicated this group to be a methyl ester (2.56 or 2.50 ppm) shielded by close proximity to an indole nucleus. Such shielding is quite typical of a vobasine (**172**) type compound. The second three-proton singlet (2.50 or 2.56 ppm) was assigned to an *N*-methyl group.

Signals at 1.68 ppm (a three-proton doublet) and 5.38 ppm (a one-proton multiplet) indicated the presence of a vinyl methyl group with the corresponding carbons in the other half of the molecule being an ethyl group (three-proton triplet at 0.90 ppm).

Cleavage of tabernamine under acidic conditions (*171*) afforded ibogamine (**304**) as shown by TLC, HPLC, and mass spectral comparisons. To confirm the nature of the two indolic units, vobasinol (**172**) was condensed with ibogamine (**304**) to afford tabernamine, identical to the natural material. The point of linkage of the vobasine unit to the aromatic portion was deduced from the ^1H-NMR spectrum, which showed a doublet ($J = 8$ Hz) for the

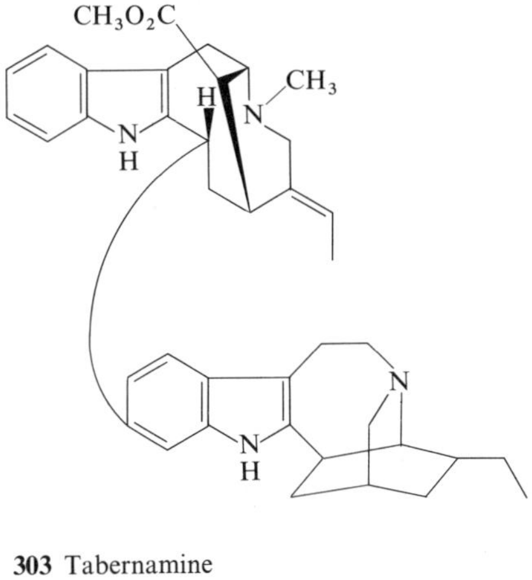

303 Tabernamine

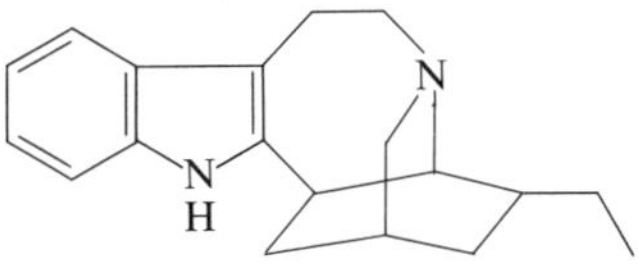

304 Ibogamine

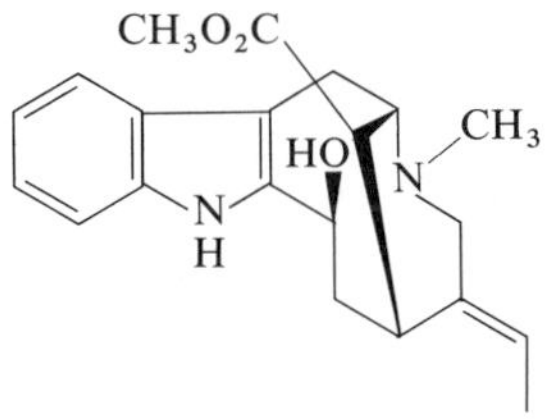

172 Vobasinol

9′ proton. The linkage is therefore between C-11′ of the iboga unit and C-3 of the vobasine unit. The stereochemistry was inferred from an attack taking place from the least hindered (α) face of vobasinol (**172**). Consequently, tabernamine was assigned structure **303** (*170*).

D. 19,20-Epoxyconoduramine (305)

A second new dimeric alkaloid isolated from *Tabernaemontana johnstonii* was 19,20-epoxyconoduramine (**305**) (*172*), whose structure relies on spectroscopic data. The UV spectrum showed a typical purely indolic chromophore, the IR spectrum a carbomethoxy group (ν_{max} 1730 cm^{-1}), and the mass spectrum indicated the compound to be a bisindole alkaloid of the conoduramine type.

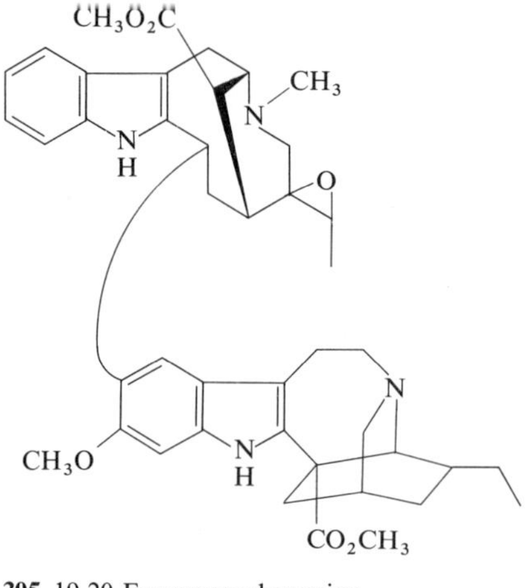

305 19,20-Epoxyconoduramine

The molecular ion was observed at *m/e* 720, 16 mu higher than conoduramine (**306**), indicating the presence of an additional oxygen somewhere in the molecule. Analysis of the mass spectrum indicated that peaks for the iboga unit were observed as usual at *m/e* 122, 136, and 148, but peaks from the alicyclic unit of the vobasine portion were shifted to *m/e* 196, 198, and 210. The two ions at *m/e* 676 ($M^+ - 44$) and *m/e* 704 were analyzed in terms of loss of C_2H_4O from a 19,20-epoxide and direct loss of oxygen from an epoxide.

The NMR spectrum of the isolate showed three-proton singlets at 2.42, 2.60, and 3.66 ppm, virtually the same positions as in conoduramine and assigned to the vobasinyl carbomethoxy group, the N_b-methyl group, and the iboga carbomethoxy group, respectively. The aromatic region of the spectrum was quite similar to that of conoduramine (**306**), indicating that the vobasine unit is linked to the 12′ position of the isovoacangine (**307**) moiety. The key signal in the NMR spectrum, however, was a three-proton doublet at 1.34 ppm which could be assigned to the C-18 methyl group with a 19,20-epoxy group.

On this basis the structure 19,20-epoxyconoduramine (**305**) was assigned to this compound (*172*).

E. Gabunamine (308)

Kingston and co-workers have described another new dimeric indole alkaloid from the stem bark of *Tabernaemontana johnstonii* (*172*).

Gabunamine showed in the IR spectrum the presence of a secondary amine (ν_{max} 3430 cm^{-1}) and a carbomethoxy group (ν_{max} 1715 cm^{-1}). In the NMR spectrum a carbomethoxy group was observed at 3.70 ppm and a highly shielded three-proton singlet at 2.48 ppm was assigned to a carbomethoxy group in a vobasine moiety. An ethylidene group (doublet at 1.70 ppm), an ethyl group (triplet at 0.90 ppm), and an aromatic methoxy group (singlet at 3.96 ppm) were also observed. The mass spectrum of gabunamine showed a molecular ion at *m/e* 704, major fragment ions at *m/e* 352, 338, 336, 225, 194, 183, 182, 181, 180, and 172, and a base peak at *m/e* 166. These data suggested that gabunamine was a vobasine–iboga alkaloid having a secondary amino group. Indeed, methylation of gabunamine gave conoduramine (**306**), and treatment with 10% HCl in methanol gave isovoacangine (**307**). Gabunamine therefore has structure **308** and is *N*-demethylconoduramine (*172*).

Synthesis of gabunamine (**308**) was accomplished by refluxing isovocangine (**307**) under nitrogen with 1.5% methanolic HCl and adding portions of perivinol (**309**). Two compounds, gabunamine (**308**) and gabunine (**310**), were isolated from the reaction mixture, each identical with the corresponding natural product (*172*).

F. Cytotoxicity of the Alkaloids of *Tabernaemontana johnstonii*

Tabernaemontana johnstonii also yielded the dimeric alkaloids conodurine (**311**), conoduramine (**306**), pericyclivine, and gabunine (**310**), in addition to the new alkaloids gabunamine (**308**) tabernamine (**303**), and 19,20-epoxyconoduramine (**305**). Of the alkaloids examined for cytotoxic activity gabunamine (**308**), gabunamine (**310**), and tabernamine (**303**) were marginally active in the P-388 lymphocytic leukemia system in cell culture (*172*).

Kingston has compared the cytotoxicities of several bisindole alkaloids in this series with a view to establishing the structure–activity relationships

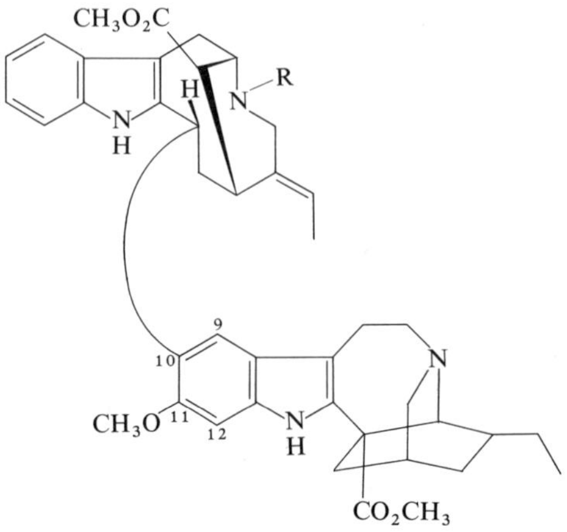

307 Isovoacangine

306 R = CH_3 Conoduramine
308 R = H Gabunamine

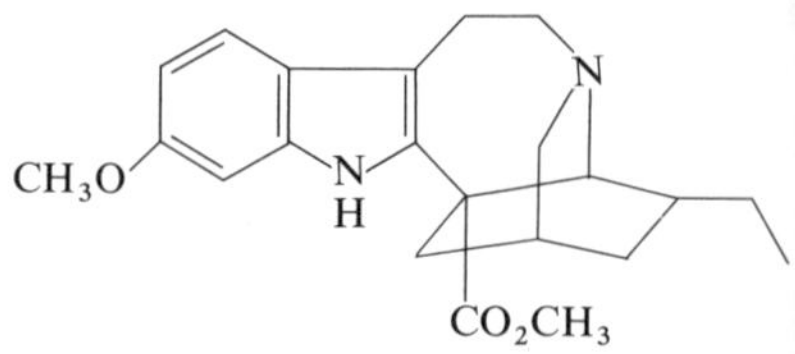

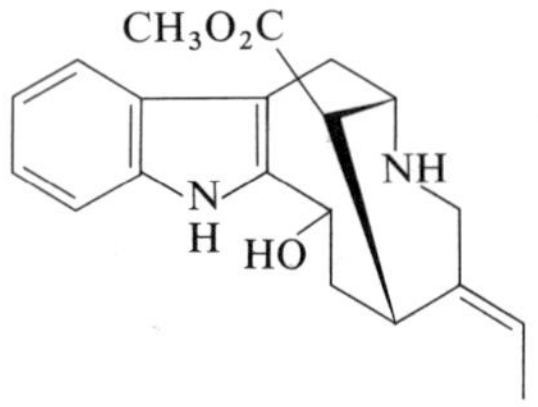

309 Perivinol

307 Isovoacangine + **309** Perivinol ⟶ **308** Gabunamine +

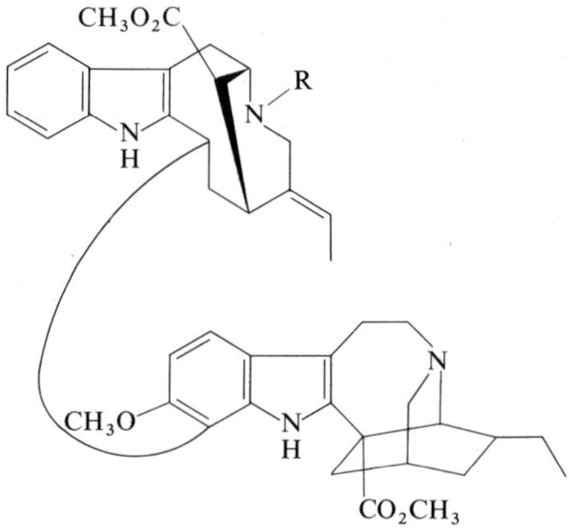

310 R = H Gabunine
311 R = CH_3 Conodurine

(*173*). Of the monomeric alkaloids, isovoacangine (**307**) and perivine (**312**) did not show cytotoxic activity but voacangine (**313**) could be regarded as almost active (ED_{50} 6.8 μg/ml).

In the bisindole series the presence of a vobasinyl *N*-methyl was found to be detrimental to cytotoxicity. Thus, although gabunamine (**308**) and gabunine (**310**) were cytotoxic, the corresponding *N*-methyl derivatives conoduramine (**306**) and conodurine (**311**) were inactive (*173*).

The position of attachment of the vobasane unit to the iboga unit was also found critical for cytotoxic activity. Thus, voacamidine (**314**), conoduramine (**306**), and conodurine (**311**) were not active, but 11′-substituted

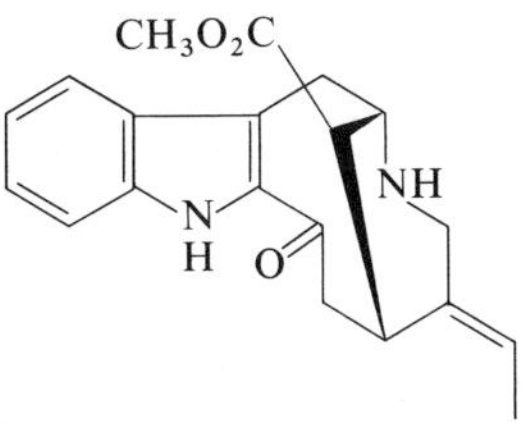

312 Perivine

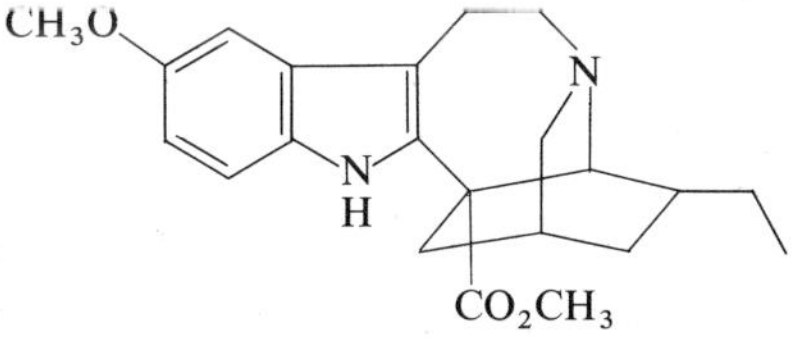

313 Voacangine

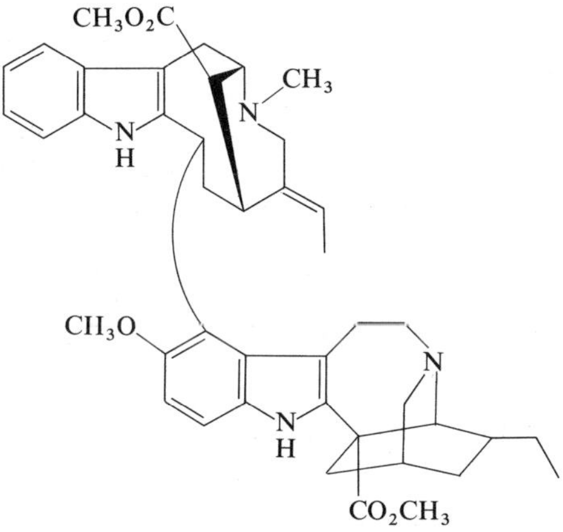

314 Voacamidine

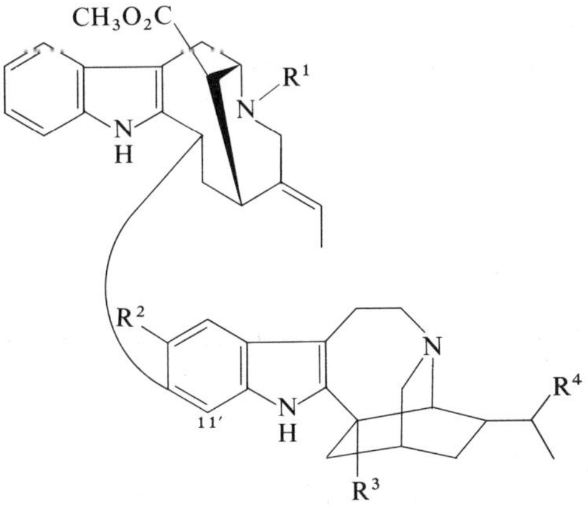

R^1	R^2	R^3	R^4	
H	OCH_3	CO_2CH_3	H	**315** *N*-Demethylvoacamine
H	H	H	H	**316** *N*-Demethyltabernamine
CH_3	OCH_3	CO_2CH_3	H	**299** Voacamine
CH_3	OCH_3	CO_2CH_3	OH	**301** Epivoacorine
CH_3	H	H	H	**303** Tabernamine

compounds such as voacamine (**299**), epivoacorine (**301**), and tabernamine (**303**) were all cytotoxic in spite of the fact that the iboga units are quite different.

To examine whether these two effects act synergistically, the alkaloids *N*-demethylvoacamine (**315**) and *N*-demethyltabernamine (**316**) were synthesized.

TABLE IX
CYTOTOXICITY OF BISINDOLE ALKALOIDS IN THE IBOGA–VOBASINE SERIES[a]

Compound	ED_{50} (μg/ml) P-388 test system	Compound	ED_{50} (μg/ml) P-388 test system
Gabunamine (**308**)	1.3	Voacamine (**299**)	2.6
Conoduramine (**306**)	20	Epivoacorine (**301**)	1.7
Gabunine (**310**)	3.2	Tabernamine (**303**)	2.1
Conodurine (**311**)	26	*N*-Demethylvoacamine (**315**)	0.39
Voacamidine (**314**)	14	*N*-Demethyltabernamine (**316**)	0.44

[a] From Kingston (*173*).

Condensation of perivinol (**309**) with voacangine (**313**) in 1.5% HCl in refluxing methanol for 12 hr gave *N*-demethylvoacamine (**315**). A similar condensation of perivinol (**309**) and ibogamine (**304**) gave *N*-demethyltabernamine (**316**).

A synergistic effect was indeed observed, and both **315** and **316** showed ED_{50} values of about 0.40 μg/ml (*173*). Some of the cytotoxicity data for these alkaloids are shown in Table IX.

G. 20′-βH-16-DECARBOMETHOXYDIHYDROVOACAMINE (319) AND 20′-αH-16-DECARBOMETHOXYDIHYDROVOACAMINE (322)

Among a number of alkaloids obtained from the bark of *Ervatamia orientalis*, Knox and Slobbe have obtained two new bisindole alkaloids in the voacamine series (*174*).

No melting points were observed for either dimer. Their UV spectra indicated purely indolic nuclei, and from their IR spectra indolic NH, ester carbonyl and ortho-disubstituted benzene functionalities were deduced. The C-3 methine signals in these compounds were observed at 4.93 and 5.03 ppm, and two methoxyls were observed at 3.91 and 2.57 ppm in one compound and at 3.94 and 2.44 ppm in the second compound. In each case the more downfield signal for an aromatic methoxy was somewhat broadened and clearly the second methoxy of a carbomethoxy group was quite shielded, probably by proximity to the indole nucleus. One *N*-methyl group in each compound was also observed, as well as two para-related aromatic protons at 6.50 ppm.

The mass spectra of each compound indicated a molecular ion at *m/e* 648, with a weak intermolecular methylene transfer ion at *m/e* 662. These ions were much weaker than those in compounds in the voacamine series which have a carbomethoxy group in the iboga unit. Significant peaks around *m/e* 466 and 452 were analogous to those resulting from loss of the piperidine ring system of the 3,4-secosarpagine unit (*175*). Typical ions for the iboga system were observed at *m/e* 225, 149, 136, 135, and 122.

On this basis the compounds were suggested to be derived from the combination of an ibogaine with a dregamine unit or with a tabernaemontanine unit. Acid cleavage of one of the dimers gave ibogaine (**317**) in low yield, but the other half of the molecule could not be obtained. No other information distinguishing the two structures could be obtained from spectroscopic data at this time, and the structures were confirmed and individual assignments made by partial synthesis.

Condensation of dregaminol (**318**) and ibogaine (**317**) by heating in dilute methanolic HCl solution gave 20′-*βH*-16-decarbomethoxydihydrovoacamine (**319**) in moderate yield, which was identical to the more polar natural compound.

Sodium borohydride reduction of tabernaemontanine (**320**) gave tabernaemontaninol (**321**) which was condensed with ibogaine (**317**) under the acidic conditions described above to give 20′-*αH*-16-decarbomethoxy-20′-epidihydrovoacamine (**322**) in low yield. This compound was identical to the less polar natural compound (*174*).

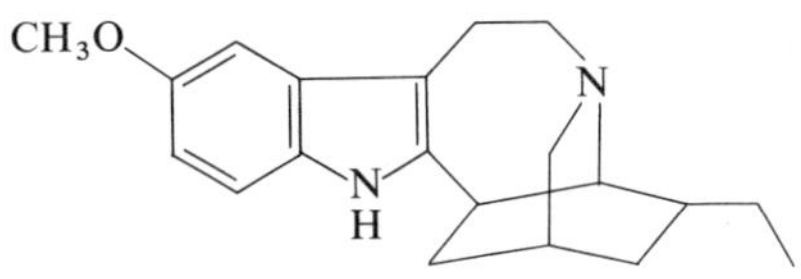

317 Ibogaine

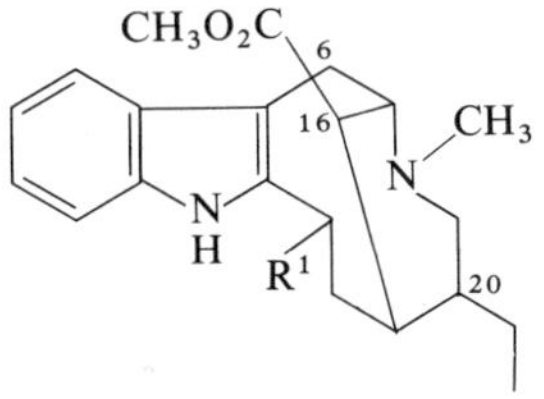

318 R^1 = *β*-OH, 20*β*-H
320 R^1 = =O, 20*α*-H
321 R^1 = *β*-OH, 20*α*-H
324 R^1 = OH
327 R^1 = =O, 20*β*-H

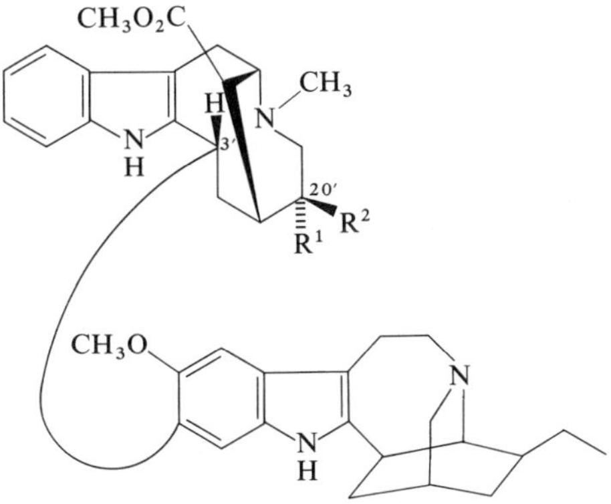

319 $R^1 = C_2H_5$, R^2 = H
322 R^1 = H, $R^2 = C_2H_5$

The slightly broadened nature of the aromatic methoxyl group was thought to be due to restricted rotation about the bond linking the two monomer units.

H. TABERNAELEGANTINES AND RELATED ALKALOIDS

Examination of the root bark of *Tabernaemontana elegans* Stapf. (Apocynaceae) afforded seven new bisindole alkaloids (*176*), in addition to the known bisindole conoduramine (**306**). These alkaloids, tabernaelegantines A–D, tabernaelegantinines A–B, and tabernaelegantinidine, were isolated from the pH 2.6 extract and have some quite interesting features (*176*, *177*).

Tabernaelegantines A–D were found to have the same molecular formula, $C_{43}H_{54}N_4O_5$, and displayed indole plus alkoxyindole chromophores. Their IR spectra were also quite similar and indicated the presence of NH and carbomethoxy functionalities. Overall, the mass spectra were also very similar, although differences in intensity of certain peaks were observed. Of diagnostic significance were the ions at *m/e* 196, 182 (**323**), and 124, which are found in 19,20-dihydrovobasinols (**324**), and at *m/e* 136 and 122 (**325**), which are typical of voacangine (**313**).

The data suggested that these compounds were comprised of 19,20-dihydrovobasine and voacangine units, and this suggestion was confirmed when dihydrovoacamine (**326**) gave a similar mass spectrum to the tabernaelegantines. In addition, the ^{1}H-NMR spectra of these compounds displayed the characteristic three-proton singlets of the aromatic methoxy ($\sim$3.9 ppm), voacangine carbomethoxy ($\sim$3.7 ppm), vobasine N—CH_3 ($\sim$2.6 ppm), and shielded vobasine carbomethoxy ($\sim$2.5 ppm) groups, and a one-proton multiplet at about 5 ppm for H-3. Each compound displayed six aromatic protons, but tabernaelegantines A and C showed the presence of ortho-related protons whereas compounds B and D, like voacamine (**299**), showed a singlet for para-related protons.

Renner and Fritz (*178*) had found previously that voacamidine (**314**) on treatment with 2 *N* HCl at reflux gave voacamine (**299**). When tabernaelegantines A and C were treated with concentrated HCl refluxing in methanol, tabernaelegantines B and D were obtained, respectively. Some isovoacangine (**307**) was obtained in each of these reactions. Thus, if it is assumed that H-3 is β in these compounds (*171*), tabernaelegantines A and C are isomers at C-20, as are compounds B and D. These two alternative stereochemistries are present in the alkaloids dregamine (**327**) and tabernaemontanine (**320**), where the C-20 stereochemistry has been established by X-ray analysis (*179*).

Although it is possible to distinguish **327** and **320** by ^{1}H-NMR spectroscopy, in the bisindole compounds the 1–2 ppm region is too complex for

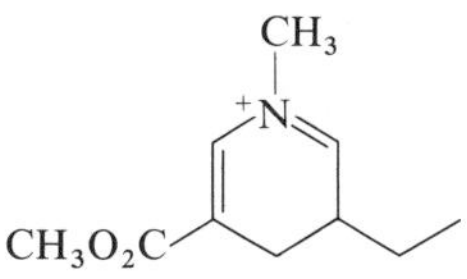

323 *m/e* 182

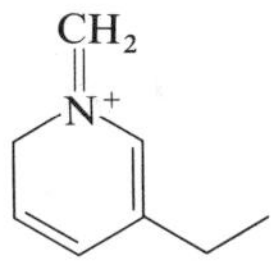

325 *m/e* 122

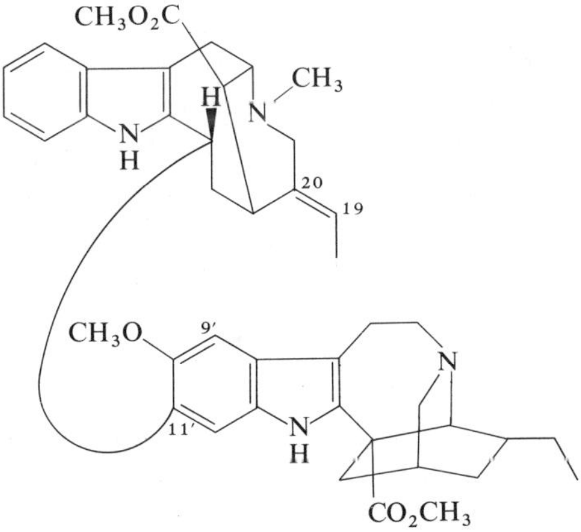

299 11′-linkage
314 9′-linkage
326 11′-linkage, 19,20-dihydro-

definitive study. Consequently, attention was turned to the ^{13}C-NMR spectra of these derivatives, and this indeed proved useful. The compounds examined initially were dregaminol (**318**) and tabernaemontaninol (**321**). Of particular significance are the chemical shift differences of carbons 6, 14, 16, 18, 19, 20, and 21 (Table X). Most important are the shifts of the C-16

TABLE X
SELECTED ^{13}C-NMR DATA OF *TABERNAEMONTANA* ALKALOIDS[a]

			Tabernaelegantine			
Carbon	Dregaminol (**318**)	Tabernaemontaninol (**321**)	A (**328**)	B (**329**)	C (**330**)	D (**331**)
C-6	19.3	17.7	17.9	17.6	19.5	19.9
C-14	30.6	38.2	36.9	37.1	29.2	31.5
C-16	50.3	44.2	44.0	44.1	49.9	49.8
C-18	11.6	42.8	12.9	13.0	11.4	10.7
C-19	23.6	25.6	25.7	25.8	23.5	22.9
C-20	43.8	42.9	43.1	43.2	43.9	41.6
C-21	49.3	47.1	47.0	47.2	49.5	48.4

[a] From Bombardelli *et al.* (*177*).

and C-14 resonances. If the C-20 ethyl group is axial (as in **321**), C-16 experiences a marked steric interaction with C-19 and is consequently deshielded. Alternatively, if the C-20 group is α-oriented (as in **318**), it is C-14 which interacts and this is reflected in the deshielding of this resonance in **318**.

Before this information could be used in the tabernaelegantines, it was necessary to assign the carbon spectrum of the other half of the molecule, isovoacangine (**307**). For reasons of sample availability, it was actually voacangine (**313**) which was examined. Comparison of the aliphatic carbon resonances of voacangine with those of the tabernaelegantines substantiated the structural assignment. These data permitted assignment of the aliphatic carbons in the vobasine half of the alkaloids. By examining the chemical shifts of C-14 and C-16, tabernaelegantines A and B could be assigned the C-20 β stereochemistry of the ethyl side chain, and tabernaelegantines C and D the C-20 α stereochemistry.

It therefore remained only to provide additional evidence for the position of linkage of one aromatic nucleus to the other. In all the tabernaelegantines, a resonance at ~152 ppm was assigned to the carbon bearing the methoxy group, and because isovoacangine (**307**) was obtained in the acid rearrangement reactions, this carbon must be C-11 of **307**. The frequencies of carbons 2′, 7′, 9′, 11′, and 13′ changed little in the tabernaelegantines, but carbons 8′, 10′, and 12′ showed characteristic shifts depending on the substitution as well as differences in multiplicity. The two alternative sites of substitution, C-10′ and C-12′, were easily distinguished on this basis. Thus tabernaelegantines B and D, which are C-10′,C-11′-substituted, displayed a singlet for C-10′ at 127 ppm and a doublet for C-12′ at ~93 ppm. On the other hand, tabernaelegantines A and C, which are C-11′,C-12′-substituted, displayed a doublet for C-10′ at 105 ppm and a singlet for C-12′ at ~115 ppm. These assignments are permitted because of the consistency of the aromatic resonances of the vobasine half of these molecules. In sum, tabernaelegantines A–D can be represented by structures **328**–**331**, respectively (*177*).

The isomeric tabernaelegantinines A and B, having the molecular formula $C_{46}H_{58}N_4O_6$, were also found from their spectral data to be members of the dihydrovoacamine series. Indeed, the aromatic regions of the carbon spectra of these compounds were essentially identical with those of **328** and **329**, respectively.

Three additional carbon atoms were observed in the carbon spectra of these compounds, and comparison of mass and carbon spectra showed that these were located in the isovoacangine unit. This unit was deduced to be CH_2COCH_3 based on the ^{13}C- and ^{1}H-NMR spectra, and its location was deduced by comparison of the aliphatic regions of the isovoacangine unit of **328**–**331** with those of the tabernaelegantinines.

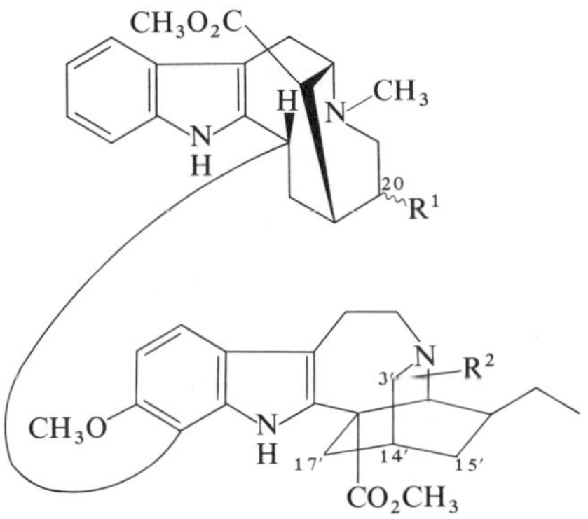

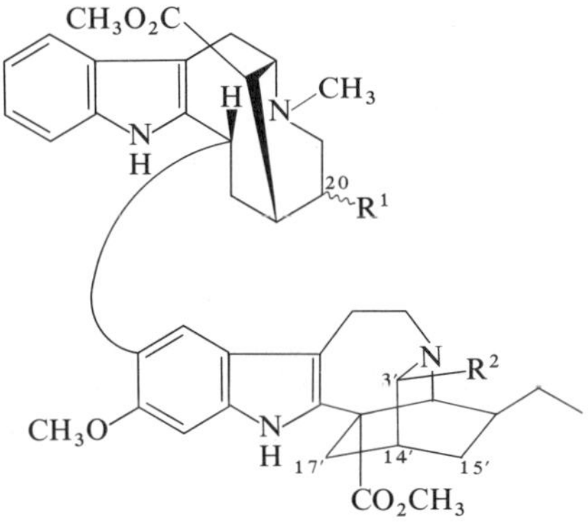

328 $R^1 = \beta\text{-}C_2H_5$, $R^2 = H$
330 $R^1 = \alpha\text{-}C_2H_5$, $R^2 = H$
332 $R^1 = \beta\text{-}C_2H_5$, $R^2 = —CH_2COCH_3$

329 $R^1 = \beta\text{-}C_2H_5$, $R^2 = H$
331 $R^1 = \alpha\text{-}C_2H_5$, $R^2 = H$
333 $R^1 = \beta\text{-}C_2H_5$, $R^2 = —CH_2COCH_3$

The C-6′, C-17′, and C-20′ resonances were relatively unaffected, but C-14′ was shifted to 31 ppm and C-15′ to ~27 ppm from 32 ppm in these compounds. In addition, C-3′ was shifted to 55 ppm from 51.5 ppm and C-5′ from 53 ppm to 51.3 ppm. The C-3′/C-5′ assignments here are clearly critical to the structure, but shifts of the resonances of carbons 14′ and 15′ (and not 6′ or 7′) would indicate both substitution at C-3′ and a stereochemistry as indicated.

The stereochemistry of the C-20 ethyl group and the linkage of the aromatic moieties (both β) were analyzed as discussed previously for the tabernaelegantines; therefore, tabernaelegantinines A and B were assigned structures **332** and **333**, respectively (*177*).

One other point that should be mentioned relates to transmethylation in these compounds. Although quite variable in proportion to the M^+, the transmethylation ion was only observed for **329**, **331**, and **333** and not for **328**, **330**, and **332**. Molecular models indicated that indeed only compounds with the C-3–C-10′ linkage would permit transfer of a methyl of a carbomethoxyl to the basic nitrogen of the other unit.

I. 3′-Oxoconodurine (335) and 3′-(2-Oxopropyl)conodurine (334)

An alkaloid closely related to the tabernaelegantinines has been isolated by Kingston and co-workers from the roots of *Tabernaemontana holstii* K. Schum (*180*). The alkaloid, 3′-(2-oxopropyl)conodurine (**334**), and a second possible artifact, 3′-oxoconodurine (**335**), were obtained by normal isolation techniques. They once again demonstrate how fine the line is between natural and isolated artifactual alkaloids.

3′-Oxoconodurine (**335**) displayed a typically indolic UV spectrum; its IR spectrum, in addition to showing the ester carbonyl absorption at 1730 cm^{-1}, gave evidence for a second carbonyl function at 1670 cm^{-1}. With a molecular ion at *m/e* 714, 14 mu higher than conodurine (**311**), and fragment ions at *m/e* 122, 180, 181, and 194 for the vobasan group, it was clear that the carbonyl function should be located in the iboga moiety. Confirmatory evidence for such an assignment came from the higher mass region of the spectrum, where an abundant ion was found at *m/e* 523 (**336**). Typically, in a bisindole alkaloid of the voacamine type, an ion is observed at *m/e* 509 which has been attributed (*175*) structure **337**.

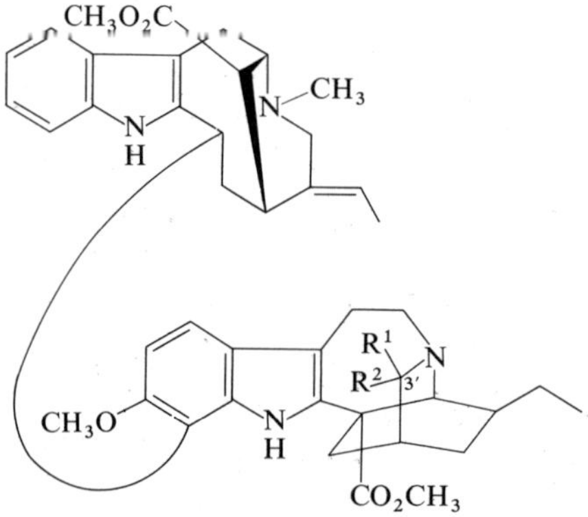

311 $R^1 = R^2 = H$ Conodurine
335 $R^1, R^2 = {=}O$ 3′-Oxoconodurine
334 $R^1 = -CH_2COCH_3$, $R^2 = H$ 3′-(2-Oxopropyl)conodurine

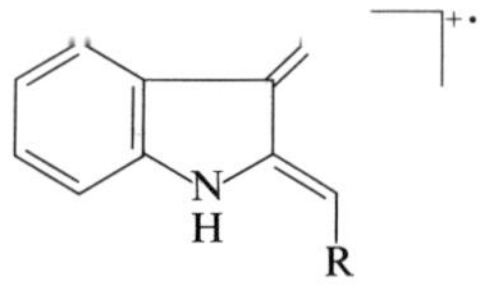

337 *m/e* 509 R = voacangine or isovoacangine

336 *m/e* 523 R = oxovoacangine or oxoisovoacangine

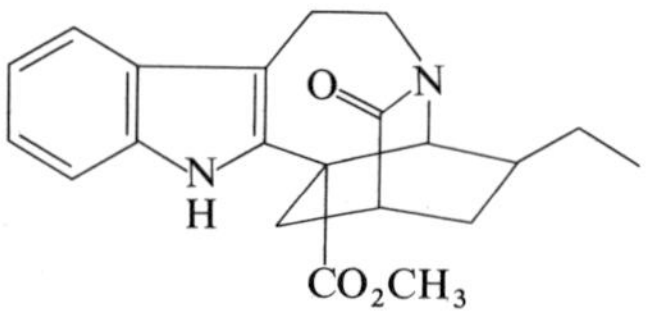

338

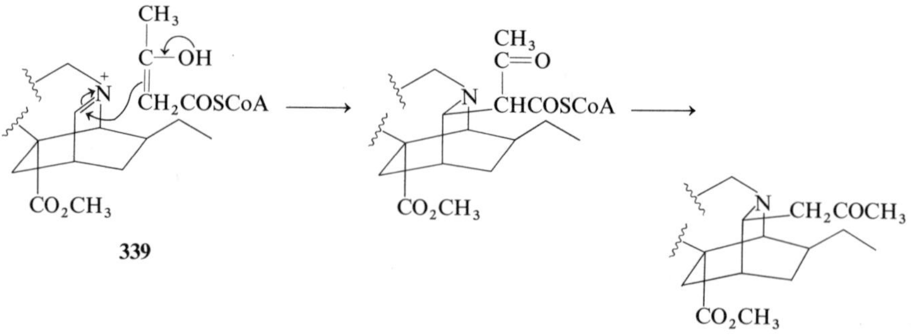

339

The location of the carbonyl group at C-3′ was revealed by the NMR spectrum. Signals typical of a conodurine-type alkaloid were observed with an aromatic AB system at 6.90 and 7.24 ppm and a carbomethoxy singlet at 3.62 ppm. The voacamidine (**314**) type substitution could be excluded, because the latter signal is shielded to about 3.08 ppm in this series. A complex two-proton multiplet at 4.30 ppm was attributed to H-5 and H-2, as had been the case for 3-oxovoacangine (**338**). On this basis, the isolate was assigned the structure 3′-oxoconodurine (**335**) (*180*).

The mass spectrum of the second isolate indicated the presence of an additional 56 mu, which could be explained in terms of a $—CH_2COCH_3$ grouping. After losses from the molecular ion of 57 and 58 mu, the remaining part of the mass spectrum was quite similar to that of conodurine (**311**). The NMR spectrum showed a three-proton singlet at 2.16 ppm and a two-proton doublet at 3.04 ppm, suggesting the presence of the unit $—CHCH_2COCH_3$. Again, analysis of the aromatic portion of the spectrum, revealing in particular a doublet at 7.02 ppm and a singlet for a carbomethoxy group at 3.84 ppm, indicated the alkaloid to be in the conodurine series.

Location of the three-carbon fragment was made from the mass spectrum, which indicated that it should be attached to the iboga moiety at either C-3′ or C-5. The former position was favored on the basis of known oxidations which occur more readily at C-3′. The isolate was therefore suggested to have the structure 3′-(2-oxopropyl)conodurine (**334**) (*180*).

Although these alkaloids (**332**–**334**) are thought to be artifacts formed by reaction with trace quantities of acetone during the extraction process, it is also conceivable that they are natural products. One could imagine, for example, attack of acetoacetate on an iminium species such as **339** to occur with little difficulty.

XVII. Cleavamine–Vobasine Type

A. CAPUVOSINE (343)

Capuvosine was isolated from the leaves and stem bark of *Capuronetta elegans* Mgf. in the Apocynaceae (*181*). High-resolution mass spectrometry established the molecular ion to be at *m/e* 634, with a molecular formula $C_{40}H_{50}N_4O_3$. The UV spectrum was purely indolic (λ_{max} 236, 287, and 295 nm), and the presence of a hydroxy group was established by acetylation.

Both the mass and ^{1}H-NMR spectra indicated that capuvosine might contain a vobasinol (**172**) unit. The nature of the second unit was determined by acid hydrolysis. One of the products obtained was identified as (+)-capuronine (**340**), a compound also isolated from *C. elegans* by the same group (*181*).

To determine the structure of capuvosine, a methanolic solution of vobasinol (**172**) and capuronine acetate (**341**) was heated in the presence of HCl under nitrogen to afford a mixture containing capuvosine acetate (**342**). On this basis capuvosine was assigned structure **343**, although there was no evidence presented that the linkage was between C-3′ and C-11 (*181*).

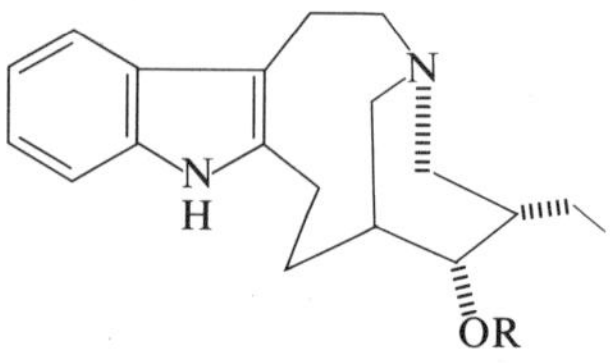

340 R = H Capuronine
341 R = $COCH_3$

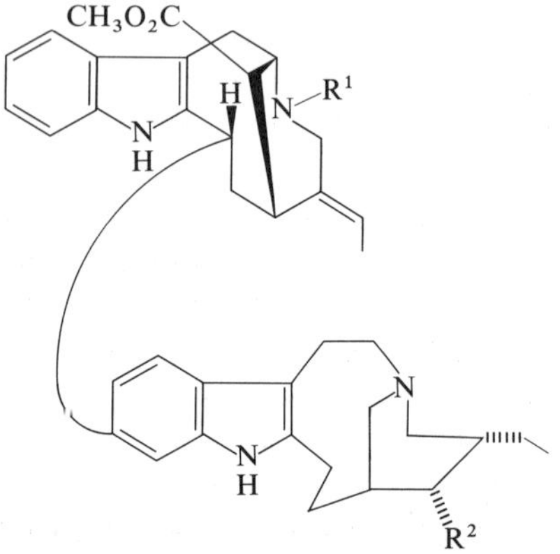

342 $R^1 = CH_3$, $R^2 = OCOCH_3$
343 $R^1 = CH_3$, $R^2 = OH$ Capuvosine
344 $R^1 = CH_3$, $R^2 = H$ Dehydroxycapuvosine
345 $R^1 = H$, $R^2 = OH$ Demethylcapuvosine

B. Dehydroxycapuvosine (344) and Demethylcapuvosine (345)

Husson and co-workers have continued their study of the alkaloids of *Capuronetta elegans* and have obtained three additional new bisindole alkaloids, two related to capuvosine (**343**) and the third, capuvosidine, the first member of a new series of pseudo-*Aspidosperma*–vobasine alkaloids (*182*).

Dehydroxycapuvosine was an amorphous alkaloid, $[\alpha]_D$ −98°, having a molecular ion at *m*/*e* 618 measuring for the formula $C_{40}H_{50}N_4O_2$. The UV spectrum indicated the presence of unsubstituted indolic nuclei, and the IR spectrum showed NH (3460 cm^{-1}) and ester (1720 cm^{-1}) functionalities. No hydroxyl absorption was observed. In the NMR spectrum of the alkaloid, a methyl triplet was observed at 0.87 ppm and a methyl doublet at 1.67 ppm for an ethylidene group. An *N*-methyl group was observed at 2.57 ppm and the carbomethoxyl singlet was typically shielded at 2.44 ppm. Seven aromatic protons were observed in the region 6.7–7.4 ppm, and a one-proton multiplet for the C-3′ proton was found at 4.55 ppm. Overall, the spectrum was very similar to that of capuvosine (**343**), indicating that the linkage should be between C-3 of a vobasine unit and C-11 of a cleavamine moiety.

The base peak in the mass spectrum appeared at *m/e* 138, 16 mu less than in capuvosine. This suggested that the C-15 hydroxy group of **343** was not present in the isolate; consequently, it was designated as dehydroxycapuvosine (**344**) (*182*). Additional evidence came from an examination of the ^{13}C-NMR spectrum of the alkaloid.

A second amorphous alkaloid, demethylcapuvosine, $[\alpha]_D$ $-85°$, gave a molecular ion at *m/e* 620 which analyzed for $C_{39}H_{48}N_4O_3$, a difference of CH_2 compared with capuvosine (**343**). The IR and UV spectra were very similar to those of capuvosine, indicating the presence of NH (3460 cm^{-1}), OH (3300 cm^{-1}), ester (1720 cm^{-1}), and indole (λ_{max} 279, 287, and 294 nm) functionalities. The NMR spectrum indicated the presence of ethylidene (three-proton doublet at 1.65 ppm, one-proton quartet at 5.08 ppm), ethyl (three-proton triplet at 0.85 ppm), and shielded carbomethoxy (singlet at 2.45 ppm) groups. The C-3′ proton was observed as a broad multiplet at 4.58 ppm. However, comparison with the spectrum of capuvosine (**343**) indicated that the *N*-methyl group was not present in the isolate which was, therefore, ascribed the structure demethylcapuvosine (**345**) (*182*). In support of this assignment, fragments in the mass spectrum which contained the vobasine moiety were all shifted by 14 mu, whereas the cleavamine fragments appeared at the same mass as in capuvosine (*182*).

XVIII. Pseudo-*Aspidosperma*–Vobasine Type

Capuvosidine (347)

Capuvosidine was obtained as an amorphous alkaloid, $[\alpha]_D$ $-7°$, by Husson and co-workers from *Capuronetta elegans* (*182*). A molecular ion was observed at *m/e* 616 ($C_{40}H_{48}N_4O_2$), and a vobasine unit was detected from the IR (ν_{max} 3420 and 1720 cm^{-1}) and UV (λ_{max} 295 nm) spectra. Confirmation of the presence of this unit came from the ^{1}H-NMR spectrum, which indicated an ethylidene group (three-proton doublet at 1.75 ppm), a shielded carbomethoxy group (singlet at 2.45 ppm), and an *N*-methyl group (singlet at 2.55 ppm). The second half of the molecule contained an ethyl side chain (triplet at 1.0 ppm) but no indolic NH or additional olefinic bonds. Because the UV spectrum indicated the presence of an indolenine (λ_{max} 280 nm), this unit should be in the *Aspidosperma* or pseudo-*Aspidosperma* series. Sodium borohydride reduction afforded a dihydro product giving fragment ions at *m/e* 124 (**346**) and *m/e* 190, characteristic of the pseudoaspidospermane nucleus. The linkage to the vobasine unit was deduced to be C-3′ from a complex proton signal at 4.65 ppm. By analogy with the co-occurring alkaloids, the other linkage was thought to be to C-10.

Capuvosidine was therefore assigned structure **347** (*182*), and this was subsequently confirmed by chemical correlation (*181*).

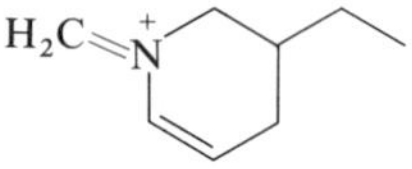

346 *m/e* 124

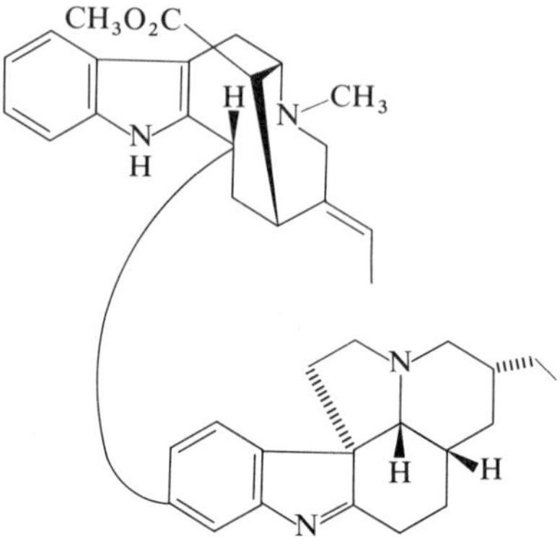

347 Capuvosidine

XIX. Iboga–Iboga Type

12,12′-BIS(11-HYDROXYCORONARIDINE) (348)

Ahond and co-workers (*183*) have reported the first member of a new group of alkaloids derived by the linkage of two iboga units through the aromatic rings. The alkaloid was obtained, along with several monomeric iboga alkaloids, from *Bonafousia tetrastachya* (Humb., Bon. & Kunth) Mgf. in the Apocynaceae. The amorphous alkaloid afforded a molecular ion at 706 mu and significant fragment ions at *m/e* 353 (M^{++}), 136, and 122. The last two peaks are characteristic of the iboga system. In the UV spectrum an almost typical hydroxyindole chromophore was observed which on addition of alkali exhibited the characteristic bathochromic shift of a phenol.

The equivalence of the two monomer units, suspected from the mass spectrum, was confirmed by the ^{1}H-NMR spectrum, which showed a singlet for a carbonyl methyl ester at 3.66 ppm, exchangeable NH and OH protons, and a pair of ortho-related protons at 6.90 and 7.41 ppm. Substitution is, therefore, at positions 11 and 12.

Comparison of the ^{13}C-NMR data of the alkaloid with those of related iboga alkaloids, particularly isovoacangine (**307**), indicated that the structure of the dimer was **348**. The characteristic shift induced on an aromatic carbon by an attached hydroxy group confirmed the location of substituents at positions 11 and 12, and from the upfield shift of the C-10 absorption and the downfield shift of the C-11 signal the hydroxy group was assigned to

C-11 in each half of the dimer. Therefore, the compound is a true dimeric alkaloid having structure **348** (*183*).

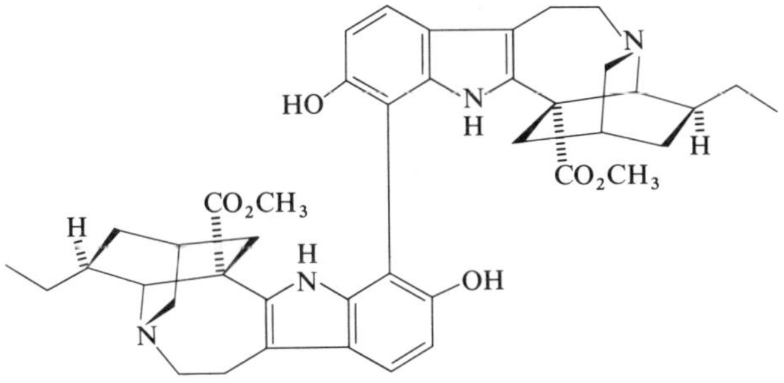

348

XX. *Aspidosperma–Pleiocarpa* Type

One of the rare, biogenetically interesting groups of bisindole alkaloids is the cluster of alkaloids derived from decarbomethoxyvindolinine (**349**) and pleiocarpamine (**195**). Pycnanthine and pleiomutinine are members of this group.

A. PYCNANTHINE (**358**) AND 14′,15′-DIHYDROPYCNANTHINE (**361**)

Pycnanthine was obtained from the root bark of *Pleiocarpa pycnantha* (K. Schum.) Stapf. var. *pycnantha* M. Pichon in reasonable yield (*184*) and pycnanthinine as a very minor base from the same plant (*185*). Pleiomutinine was isolated as minor alkaloid of *P. mutica* Benth. (*186*) and subsequently (*184*) shown to be 14′,15′-dihydropycnanthine.

An initial problem in the structure elucidation of pycnanthine was an interpretation of its complex UV spectrum, which displayed maxima at 267, 309, and 325 nm. The spectrum bore no relationship to the summation of any two of the common indole alkaloid UV chromophores, and it was only subsequently discovered that the complexity was due to the spatial interaction of two indoline chromophores. This chromophore, however, can now be regarded as characteristic of the alkaloid group.

No hydroxyl or amino groups were observed, but the IR spectrum did show the presence of two ester bands at 1754 and 1724 cm^{-1} characteristic of a pleiocarpamine-type alkaloid (*187*). The NMR spectrum indicated the presence of ethylidene, carbomethoxy, disubstituted olefin, and secondary methyl groups.

Hydrogenation of pycnanthine catalytically under acidic conditions gave 14′,15′-dihydropycnanthine, which was found to be identical to pleiomutinine. The ethylidene group remained in the reduction product and the UV spectrum was unchanged. Decoupling experiments indicated the presence of an allylamine unit in which the δ carbon was quaternary.

Treatment of pycnanthine with 11 *N* methanolic HCl at 120° gave formaldehyde, (+)-pleiocarpamine (**195**), and the upper half of the molecule. Similar cleavage of pleiomutinine also afforded **195** and formaldehyde.

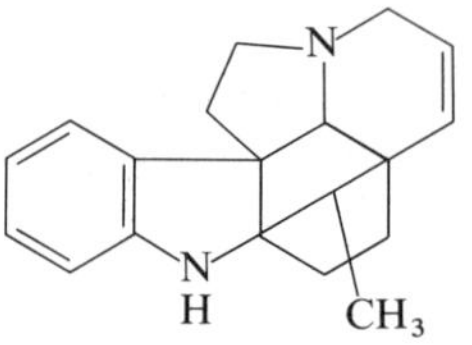

349 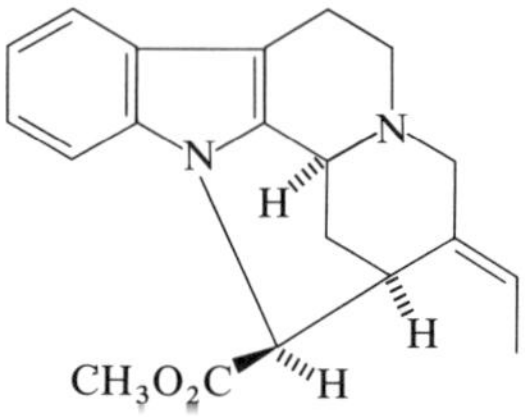 **195**

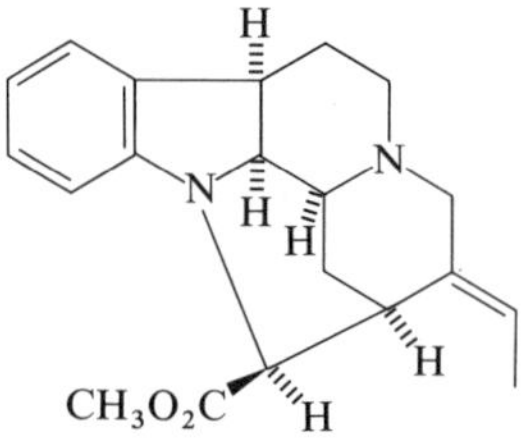

193 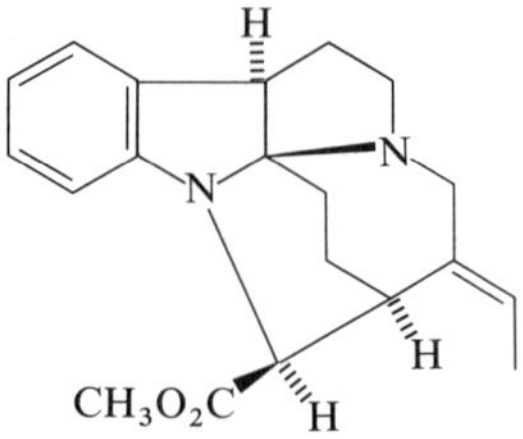 **350**

Cleavage of pycnanthine or 14′,15′-dihydropycnanthine with zinc and HCl gave a mixture of 2,7-dihydropleiocarpamine (**193**) and isodihydropleiocarpamine (**350**), characteristic products of Zn–HCl reduction of pleiocarpamine (**195**). From cleavage of pycnanthine, N_a-methyl-14,15-dehydrotuboxenine (**351**) and N_a-methyl-14,15-dehydroisotuboxenine (**352**) were also isolated, and isotuboxenine (**353**) and N_a-methylisotuboxenine (**354**) from 14′,15′-dihydropycnanthine. Only traces of tuboxenine (**355**) and N_a-methyltuboxenine (**356**) were obtained. The data clearly establish that the methylene group, which is the source of formaldehyde on acid hydrolysis, is also the group which gives the N_a-methyl under reducing conditions. The stereochemistry of the tuboxenine unit in dihydropycnanthine was deduced from the chemical shift of the methyl doublet, which in dihydropycnanthine was observed at 0.56 ppm and at 0.52 ppm in isotuboxenine (**353**).

The UV, NMR, and mass spectra of pycnanthine indicated that one half of the molecule was 2,7-dihydropleiocarpamine (**193**); of particular significance in the mass spectrum were two ions at *m/e* 135 (**197**) and 107 (**198**) not present in the spectrum of **195**.

351 R = CH_3
355 R = H, 14,15-dihydro
356 R = CH_3, 14,15-dihydro

352 R = CH_3
353 R = H, 14,15-dihydro
354 R = CH_3, 14,15-dihydro
359 R = H

197 *m/e* 135

198 *m/e* 107

357

358 Pycnanthine
361 14′,15′-Dihydropycnanthine (pleiomutinine)

The NMR spectrum of pycnanthine indicated the presence of only seven aromatic protons, four of which were located on the 2,7-dihydropleiocarpamine unit, and three adjacent protons, with the C-12′ proton missing, on the isotuboxenine unit.

At this point the partial structure **357** could be proposed in which the linkage of the *N*-methylene is to C-7 of the dihydropleiocarpamine unit, and in which C-2 of the latter is linked to C-12′ of the isotuboxenine unit; i.e., **358** is preferred for pycnanthine. One piece of evidence for this structure concerns the chemical shift of C-11, which appears about 1 ppm to lower field than expected. This was rationalized as deshielding by the dehydroisotuboxenine unit **359**.

A second piece of evidence concerns the projected biosynthesis of these compounds (Scheme 14) from an *N*-methyleneimmonium tuboxenine (**360**) and pleiocarpamine (**195**). The reversibility of the process was suggested for the facile cleavage (*114*). Thus, pycnanthine could be represented by **358** and pleiomutinine (14′,15′-dihydropycnanthine) by **361** (*184*).

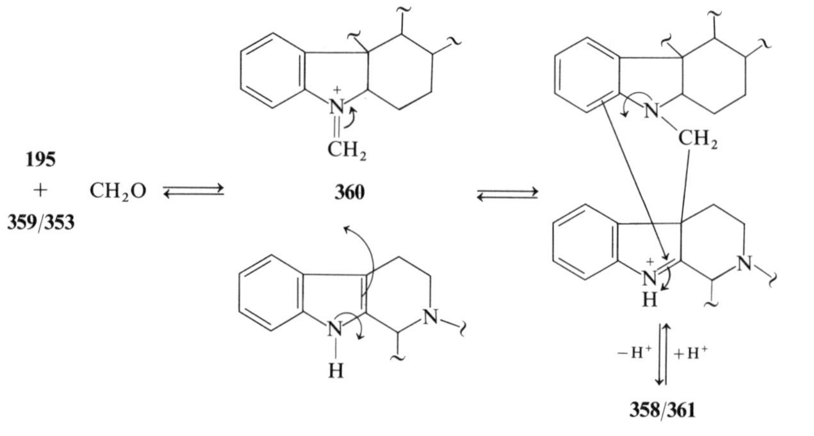

SCHEME 14.

One series of reactions remains to be explained, namely the products of hydrogenation under acidic and basic conditions.

Hydrogenation of pycnanthine (**358**) in ethyl acetate in the presence of potassium carbonate gave 3,14,15-tetrahydro-N_4-3-*chano*-pycnanthine (**362**). When the hydrogenation was conducted in methanol–acetic acid, three hydrogenation products were formed, namely, 14′,15′-dihydropycnanthine (**361**), **362**, and **363**, the N_4-methyl derivative of **362**. The mechanism of formation of **363** is unknown, but **362** apparently results from Emde cleavage of the allylamine unit.

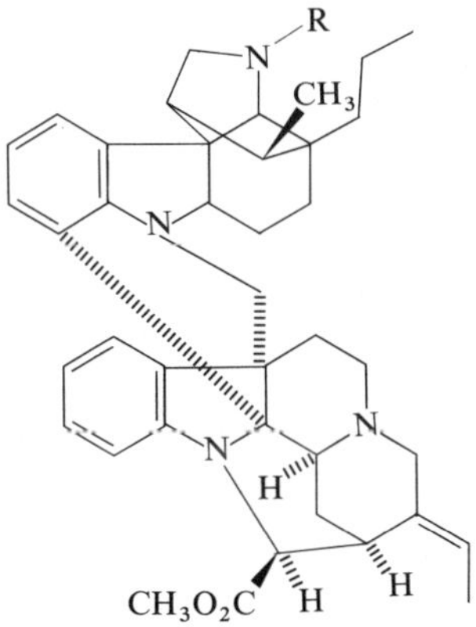

362 R = H
363 R = CH_3

B. 14′,15′-DIHYDROPYCNANTHINE (PLEIOMUTININE) (368): STRUCTURE REVISION

The Madagascan plant *Gonioma malagasy* Mgf. et P. Bt (Apocynaceae) has also afforded a dimeric alkaloid in this series (*188*). Its physical and spectral properties indicated that it was identical to 14′,15′-dihydropycnanthine, an alkaloid previously obtained from *Pleiocarpa mutica* (*142*) as pleiomutinine, and by catalytic reduction of pycnanthine (*184*).

Previous discussion of this alkaloid has indicated that it contains a decarbomethoxy-14′,15′-dihydrovindolinine unit and a 2,7-dihydropleiocarpamine moiety. The former unit was ascertained on the basis of a similar mass spectral fragmentation pattern. With the revision of the structure of vindolinine based on a ^{13}C-NMR analysis (*189*), it became clear that a re-examination of the dihydrovindolinine unit in 14′,15′-dihydropycnanthine would be necessary.

Once again, assignment of the carbon resonances of each of the individual units was required, and from study of alkaloids in the villalstonine/macralstonine series (Section VIII) assignments for the pleiocarpamine "half" of the molecule were available.

In the villalstonine series substitution at C-2 is by oxygen and leads to a pronounced (92.2 ppm) downfield shift for this resonance. In 14′,15′-dihydropycnanthine this carbon resonated at 67.6 ppm, indicating a connection with only one heteroatom. Except for a minor change in the C-7 frequency, the carbon resonances for the 2,7-dihydropleiocarpamine units in villalstonine (**184**) and 14′,15′-dihydropycnanthine were extremely close. The nature and stereochemistry of this unit were therefore confirmed. With these assignments in hand, subtraction gave the carbon resonances of the dihydrovindolinine unit, and there are several points to be established. First, the nature of the skeleton must be ascertained. Then, the nature of the linkage of the two units should be explored. Finally, the stereochemistry of the linkage at C-19′ in the dihydrovindolinine unit should be determined.

In the previous analysis of vindolinine (**364**) (*189*), a quaternary carbon resonance at 81.4 ppm had been assigned to C-2 and was crucial in revising the position of the linkage of C-19 from C-6 to C-2. 14′,15′-Dihydropycnanthine displayed such a resonance at 80.1 ppm, and together with the similar chemical shifts of C-6′ and C-7′, was indicative of the revised skeleton for the dihydrovindolinine unit.

The position of the linkage was readily determined, for in *N*-methylvindolinine (**365**) the NCH_3 was observed at 30.0 ppm, while in 14′,15′-dihydropycnanthine this resonance was deshielded to 41.4 ppm. The second point of attachment was clear from the shift of the C-12′ resonance, from 105.6 ppm in **365** to 120.6 ppm in 14′,15′-dihydropycnanthine. Because C-7 in villalstonine and 14′,15′-dihydropycnanthine were very close (44.2 ppm and 46.3 ppm, respectively), it was thought that C-12′ was linked to C-2 and not to C-7.

The stereochemistry at C-19′ was a more complex problem because no decarbomethoxyvindolinine or decarbomethoxydihydrovindolines were examined, and the spectrum only of 14′,15′-dihydro-19*R*-vindolinine (**366**) was determined. As the resonances of C-17′, C-18′, C-19′, C-20′, and C-21′ change on inversion at C-19′ owing to steric factors, assignment of stereochemistry is highly dependent on these chemical shift differences. The lack of appropriate model compounds precluded the use of C-17′, C-19′, and C-20′ in the assignment, but significant information was obtained both from the

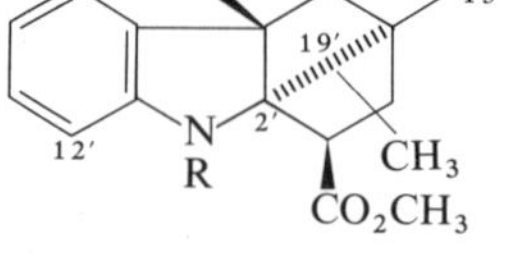

364 R = H
365 R = CH_3
366 R = H, 14′,15′-dihydro
367 R = H, 19′-epi

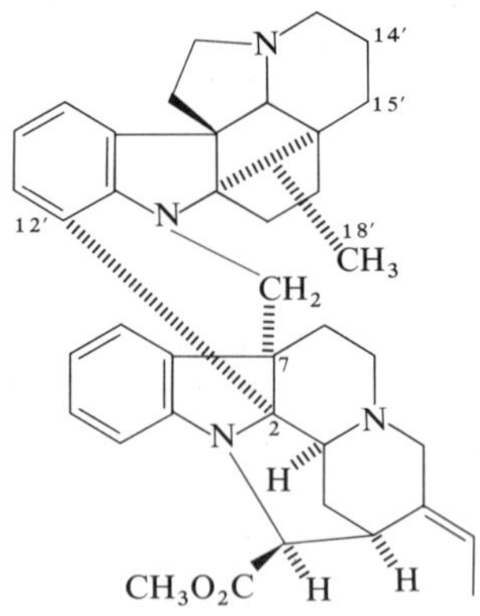

368 14′,15′-Dihydropycnanthine (revised)
369 Pycnanthine, $\Delta^{14',15'}$(revised)

^{1}H and ^{13}C resonances of the methyl group (C-18′) in 14′,15′-dihydropycnanthine and from the shift of the C-21′ resonance.

Reduction of the 14′,15′-bond in 19*R*-vindolinine (**364**) has essentially no effect on the chemical shift of C-18′ which appears at 7.4 ppm and little effect on C-21′ (~78.4 ppm). In 19*S*-vindolinine (**367**) these resonances appear at 10.1 ppm and 74.2 ppm, respectively. In 14′,15′-dihydropycnanthine these resonances were found at 11.0 and 74.5 ppm, respectively, and on this basis the 19*S*-stereochemistry was assigned to this center. This conclusion is in accord with the ^{1}H-NMR spectrum, where the 19*S*-methyl series displays a three-proton doublet at ~0.55 ppm, and the 19*R* series a doublet at ~0.9 ppm. 14′,15′-Dihydropycnanthine is therefore represented by the complete structure (**368**) (*188*). The structure of pycnanthine should then be revised to **369**.

C. PYCNANTHININE (372)

Gorman and Schmid also isolated a bisindole alkaloid, pycnanthinine, from *Pleiocarpa pycnantha* var. *pycnantha* (*185*). Pycnanthinine was found to be isomeric with 14′,15′-dihydropycnanthine having the molecular formula $C_{40}H_{46}N_4O_2$ by high-resolution mass spectrometry. The UV spectrum was identical to that of pycnanthine, and the IR spectrum confirmed a carbomethoxy group of a pleiocarpamine unit. No NH or OH groups were observed, and two ions at *m/e* 135 (**197**) and *m/e* 107 (**198**) indicated the presence of a 2,7-dihydropleiocarpamine unit. Acid hydrolysis gave (+)-pleiocarpamine (**195**). Also produced in the hydrolysis were formaldehyde and a base, $C_{19}H_{24}N_2$, mp 111.5°–113°, identified as 14,15-dehydroaspidospermidine (**370**). Cleavage under reducing conditions gave **193**, **350**, **370**, and its *N*-methyl derivative, **371**. The latter derivative establishes that N_a of the

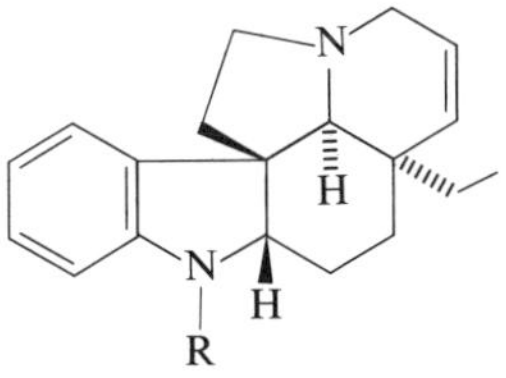

370 R = H
371 R = CH_3

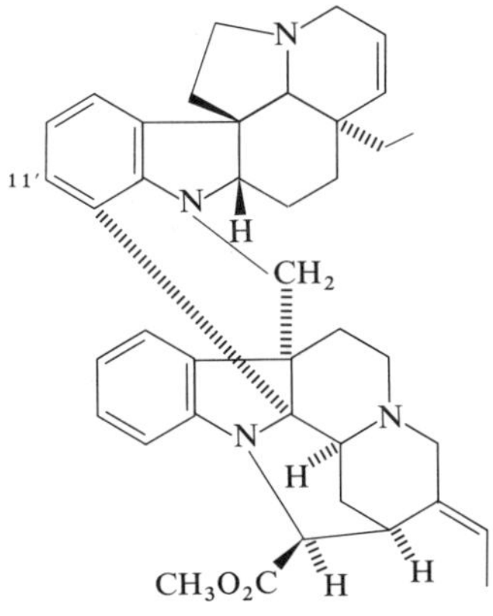

372 Pycnanthinine

Aspidosperma unit is the point of attachment of the methylene between the two units.

The NMR spectrum of pycnanthinine showed only seven aromatic protons with the C-12′ proton missing. Again, the C-11′ proton was deshielded, though not as markedly as in pycnanthine. This evidence together with the UV spectrum and hydrolysis products were taken as indicating structure **372** for pycnanthinine (*185*).

XXI. *Aspidosperma–Aspidosperma* Type

A. Bistabersonine (373)

Crioceras dipladeniiflorus (Stapf) K. Schum. has yielded a number of monomeric indole alkaloids in the *Aspidosperma* and eburnea series (*190*) and also two bisindole alkaloids. One of these was vobtusine (see Section XXII,E), and the second was a bistabersonine whose structure (**373**) was partially deduced.

Bistabersonine was obtained in amorphous form and, from the UV (ν_{max} 300 and 329 nm) and IR (1670 and 1610 cm^{-1}) spectral data was found to contain a β-anilinoacrylate unit. The principal features of the ^{1}H-NMR spectrum were a six-proton complex at 0.8 ppm for two terminal methyl groups, a six-proton singlet at 3.75 ppm for two carbomethoxy groups, and a complex multiplet at 7.10 ppm integrating for eight protons and indicating that the aromatic nuclei were unsubstituted. In addition, four olefinic protons were observed for H-14, H-14′, H-15, and H-15′.

The molecular ion in the mass spectrum appeared at *m/e* 670, corresponding to two tabersonine (molecular weight 336) units linked by a σ bond (−2H). Other major ions were observed at *m/e* 456, 442, 350, 335, 291, 263, 228, 214, 170, 168, 154, 135, 122, and 108. The observation of ions at *m/e* 228 and 442 (M^+ − 228) indicates that substitution in at least half of the molecule must be limited to carbons 3 or 21, and the intensity of the *m/e* 228 ion suggests that this may be true for both halves of the molecule.

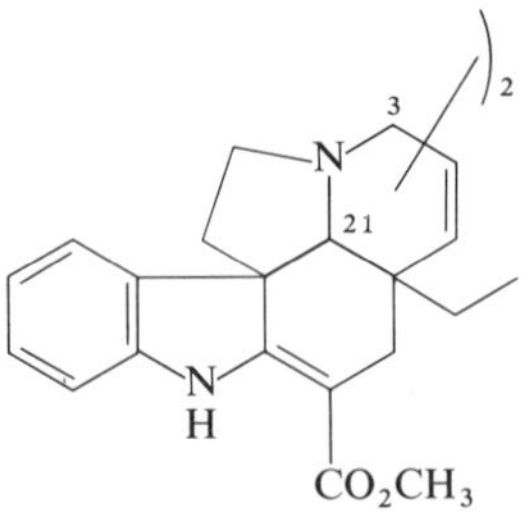

373 Bistabersonine

B. 10,10′-BIS(N-ACETYL)-11,12-DIHYDROXYASPIDOSPERMIDINE (327)

The Brazilian tree *Aspidosperma melanocalyx* Muell-Arg. has been studied by Miranda and Gilbert. From the bark, they obtained and subsequently characterized a novel dimeric alkaloid (*191*).

The crystalline alkaloid, mp 274°–275°, $[\alpha]_D$ +108°, gave a blue–green color with the ceric reagent following iodine treatment, and high-resolution mass spectrometry on the molecular ion at *m/e* 710 established the molecular formula $C_{42}H_{54}N_4O_6$.

The UV spectrum (λ_{max} 236, 297 nm) was similar to that of **374** and was unchanged on the addition of acid. However, on the addition of base a shift was observed to λ_{max} 243, 302, and 335 nm.

The NMR spectrum indicated that the molecule was a true dimer, all the signals being doubled. Thus a triplet was observed at 0.67 ppm integrating for six protons, and a singlet at 2.33 ppm also integrated for six protons, indicating the presence of two *N*-acetyl groups. A broad quartet at 4.20 ppm integrated for two protons and could be assigned to two C-2 protons. Two aromatic protons were observed as a singlet at 7.11 ppm.

Acetylation gave a tetraacetate derivative, M^+ 878, showing four successive losses of ketene, a base peak at *m/e* 124 (**375**), and other important fragment ions at *m/e* 138, 152, and 516. The last ion was present in the spectrum of both the initial alkaloid and the tetraacetate derivative and is regarded as having structure **376**.

The position of the linkage between the two units could be deduced from the NMR spectrum. In the aromatic region of monomer **374**, doublets were observed at 6.76 and 7.17 ppm for protons 10 and 9, respectively. Thus the two-proton singlet in the spectrum of the dimer should be due to two C-9

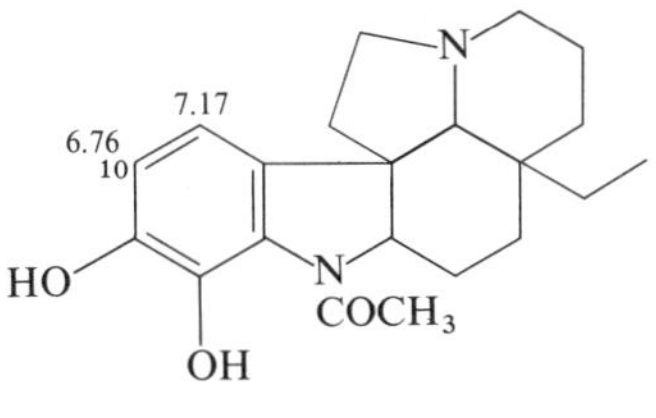

7.17
6.76
10
HO
N
COCH$_3$
OH

374

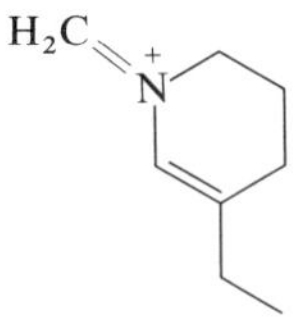

H_2C
+
N

375 *m/e* 124

H_2C
+
N
H
OH
OH HO
N
OH
COCH$_3$

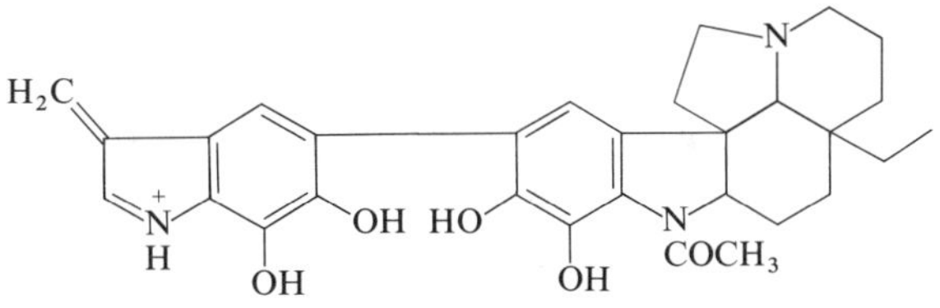

376 *m/e* 516

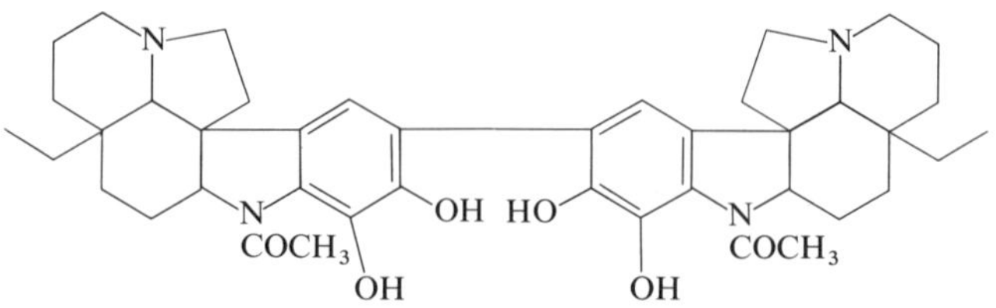

N COCH3 OH OH HO OH N COCH3

377

protons, and the linkage should be between two C-10 positions. In addition, comparison of the UV spectrum with that of **374** indicated a substantial increase in intensity and a bathochromic shift, a situation not possible except when the nuclei are 10,10′-linked. The alkaloid was therefore assigned the structure 10,10′-bis-*N*-acetyl-11,12-dihydroxyaspidospermidine (**377**) (*191*).

C. Vindolicine (382)

Although most of the bisindole alkaloids of *Catharanthus* are of the *Aspidosperma*–cleavamine type (see Section XXV), *C. roseus* (L.) G. Don (*192*, *193*) and *C. longifolius* (Pichon) Pich. (*194*) have also afforded an *Aspidosperma–Aspidosperma*-type dimer.

Vindolicine, mp 265°, was obtained from the strongly basic alkaloid fraction of *C. longifolius* stem and leaf material. The UV spectrum was typical of an indoline, and the IR spectrum indicated the presence of both hydroxyl and ester absorptions. The mass spectrum proved to be instructive in determining the structure.

A molecular ion was observed at *m/e* 924 and showed several low-molecular-weight fragments (*m/e* 107, 121, 122, 135, 149, 188, 278, 279, and 282) characteristic of the fragmentation of vindoline (**378**).

Losses of 30 (CH_2O), 59 (—CO_2CH_3), and 118 mu (—CO_2CH_3 and —$OCOCH_3$) from the molecular ion were also found, together with the larger losses of 159, 160, and 161 mu corresponding to the retro-Diels–Alder-derived loss of unit **379**.

The molecular ion would be in agreement with a compound comprised of two vindoline units (each of molecular weight 456) plus 12 mu, possibly as a methylene joining the two vindoline nuclei at some point.

Conclusive evidence for the structure of vindolicine could not be obtained. However, evaluation of two ions at *m/e* 308 and 469, formulated as **380** and **381**, respectively, suggested that vindolicine has structure **382** (*194*). Supporting evidence for this structure has recently been obtained by ^{1}H- and ^{13}C-NMR spectroscopic examination of vindolicine (*193*).

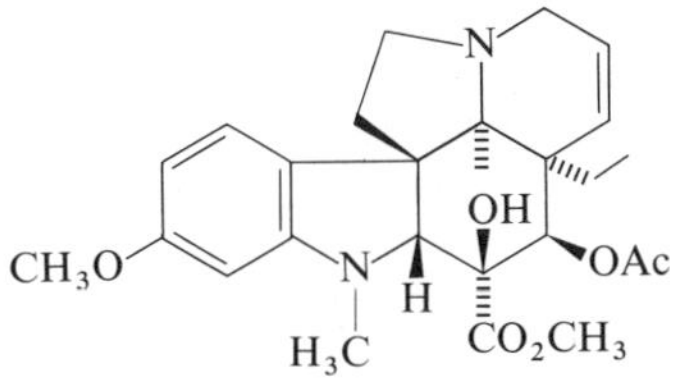

378 Vindoline
385 14,15-dihydro-

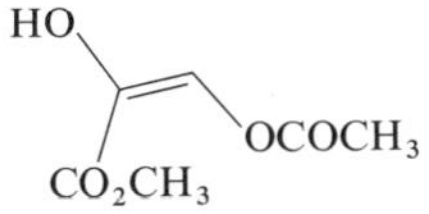

379 mass 160

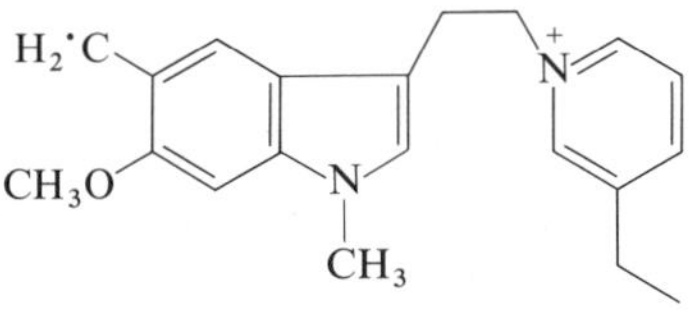

380 *m/e* 308

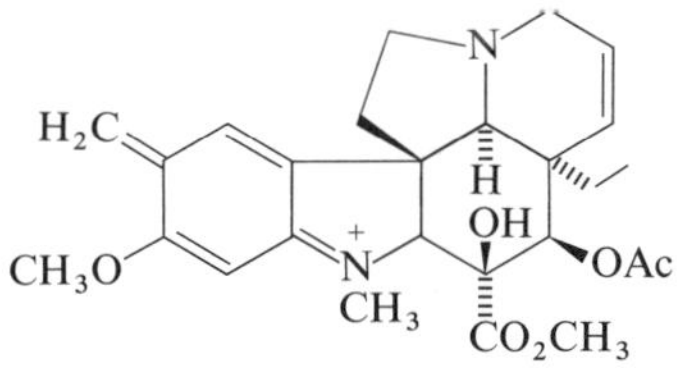

381 *m/e* 469

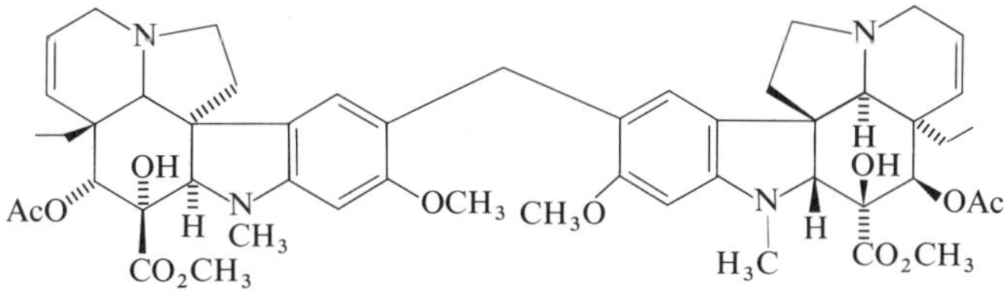

382 Vindolicine

D. A Vindoline Dimer Produced from *Streptomyces griseus*

There has been some recent interest in the microbial transformation of antitumor agents. The bisindole alkaloids of *Catharanthus roseus* (see Section XXV) are clearly prime objects for such manipulation. Initial work with vindoline (**378**) by the Lilly group (*195*) afforded several modified monomers, but subsequent work by Rosazza and co-workers (*196*) with the fungus *Streptomyces griseus* produced a new dimeric indole alkaloid.

The mass spectrum indicated a molecular ion at *m/e* 908 analyzing for $C_{50}H_{60}N_4O_2$ (=2 × vindoline − 4H). The ^{1}H-NMR spectrum indicated a close similarity to vindoline, with two *N*-methyl groups at 2.71 and 2.78 ppm and two acetyl singlets at 1.96 and 2.05 ppm. Doublets ($J = 8$ Hz) for two C-9 protons were observed at 6.93 and 6.96 ppm, indicating that the linkage did not involve the aromatic units. In agreement with this, the UV spectrum corresponded to that of vindoline.

Sodium borohydride reduction gave a dihydro derivative, and the absence of a band originally at 1653 cm^{-1} indicated this to be due to reduction of an enamine. At this point the ^{13}C-NMR spectrum (Table XI) was examined; in

TABLE XI
^{13}C-NMR DATA FOR THE DIMERIC MICROBIAL TRANSFORMATION PRODUCT OF VINDOLINE

Carbon	δ (ppm)	
	DHVE (**383**)	Microbial product
C-2	84.89	84.76, 83.69
C-3	46.62[a,b]	46.00
C-3 Olefin	—	132.25
C-5	50.91[b]	52.01, 51.07
C-6	46.30[a]	49.12, 46.00
C-7	52.34[a]	53.31, 51.46
C-8	130.28[b]	130.50, 130.31
C-9	120.85	121.73, 121.05
C-10	103.21	103.67, 103.21
C-11	160.91	161.01, 160.94
C-12	94.60	95.03, 94.67
C-13	150.52	151.72, 150.48
C-14 Olefin	—	111.95
C-14	22.25[b]	22.32
C-15	77.58	79.30, 76.77
C-16	88.11[b]	88.30, 85.80
C-17	74.82	74.79, 74.14
C-18	9.03	9.10, 8.51
C-19	24.76	32.55, 29.66
C-20	41.45[a]	48.50, 46.56
C-21	67.09	67.64, 66.18
Acetyl—C=O	169.78	170.17, 169.91
Ester—C=O	168.94	169.29, 168.94
Aryl—OCH_3	55.20	55.26, 55.20
Ester—OCH_3	54.38	52.40, 52.99
NCH_3	36.00	36.91, 36.00
Acetyl-CH_3	20.79	20.92, 20.86

[a] Chemical shift assignments are based on comparisons of ^{13}C spectral data for related compounds, and may be interchangeable.

[b] Assignments for these signals in DHVE are based on comparisons with data recorded for cathovaline (**384**).

particular, comparison was made with another of the metabolities, dihydrovindoline ether (DHVE) (**383**), as well as with cathovaline (**384**), vindoline (**378**), and dihydrovindoline (**385**). Signals for 49 of the 50 carbons were observed, and of these, 26 were found as 13 distinct pairs of signals. The data established the main features of the dimer as determined previously, but additionally confirmed the presence of an enamine double bond at carbons 3 and 14. By determining the number of attached protons it was established that the second DHVE unit should be attached at C-14. This could be confirmed by the ^{1}H-NMR spectrum, which showed singlets for the 3-H at 6.11 ppm and at 4.26 ppm for the 15-H.

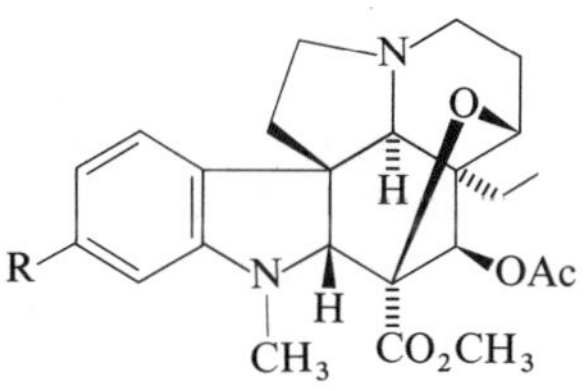

383 R = OCH_3
384 R = H Cathovaline

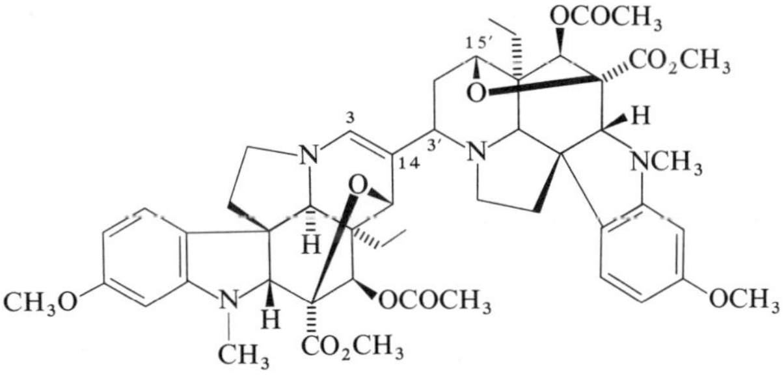

386

The point of attachment to the DHVE unit was suggested by the multiplicity of H-15′, which indicated coupling to two methylene protons. On this basis, structure **386** was proposed for the isolate (*196*). The mechanism of formation of the isolate from vindoline (**378**) was suggested to be along the lines shown in Scheme 15, in which the enamine **387** attacks its corresponding iminium ion.

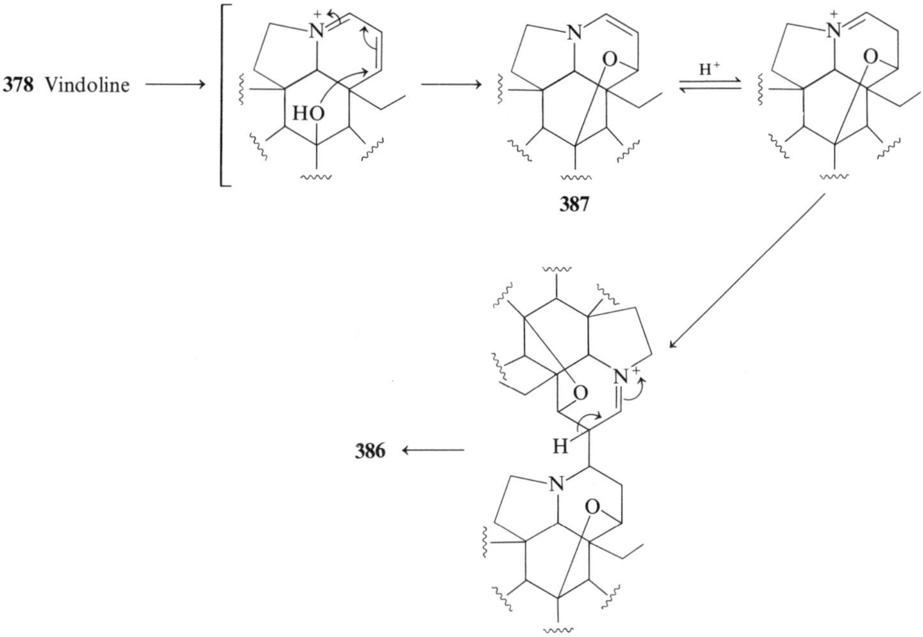

378 Vindoline

387

386

SCHEME 15.

E. Vobtusine (**390**)

The bisindole alkaloid vobtusine (*12*) is particularly widespread in the genus *Voacanga*, and in several instances considerable quantities have been obtained, for example, 7.5 g/kg from the root bark of *Voacanga chalotiana* (*197*).

Since the original postulation of a partial structure (*198*) and then the complete structure **388** (*199*) by French and Swiss groups respectively, it has been found that vobtusine is the parent compound of several closely related alkaloids (see Table XXI), and it is in this area where most of the new work has been reported. However, in the interim it was found necessary to revise the structure of vobtusine on the basis of X-ray crystallographic data (*200*).

Crystals of 10,10′-dibromovobtusine (**389**) were monoclinic in the space group $P2_1$, having cell dimensions $a = 13.80$, $b = 12.86$, and $c = 11.86$ Å, $\beta = 107.8°$, and two molecules per unit cell. The crystal structure was solved by the heavy atom method from 1900 observed reflections (*200*).

The structure deduced for vobtusine agreed almost completely with the structure proposed by Gorman *et al.* (*199*), with the very important exception that the tertiary hydroxy group previously placed at C-16 was shown to be located at C-2. The correct structure of vobtusine, including the C-14′, C-15, and C-15′ stereochemistries which were also determined from X-ray analysis, is as shown in **390**. Several other features of the molecule were also deduced from this analysis, including the presence of an internal hydrogen bond between the NH and the carbonyl of the β-anilinoacrylate unit and a very short distance (3.6 Å) between the C-14 methylene and the anilinoacrylate carbonyl group. The bond distances from C-14′ are also not equal. Thus, although the distances to C-3′, C-15′, and C-22 are within a normal range (1.50–1.57 Å), the C-14′–C-23 bond is somewhat longer, at 1.65 Å (*200*).

With the revision of the structure of vobtusine it is worthwhile to reconsider the chemical evidence in light of the new structure. Heating vobtusine in mineral acid gave anhydrovobtusine (**391**) by trans elimination of water. The UV spectrum of the product indicated the summation of a 2-methylene-7-methoxyindoline and a β-anilinoacrylate chromophore; consequently, the hydroxy group could be placed at either C-2 or C-16. The possibility of substitution at the latter position was thought to have been eliminated by a resistance of the carbinolamine to reduction with LAH. It is now realized that such a group may not be reduced quite as easily as first thought (*201*). In addition, the formation of ions at *m/e* 174 (**392**) and 188 (**393**) was thought to parallel the formation of *m/e* 160 (**394**) and 174 (**395**) in beninine (**396**). Now, these corresponding ions should be reformulated at *m/e* 190 (**397**) and 204 (**398**), but no such ions have been reported for vobtusine.

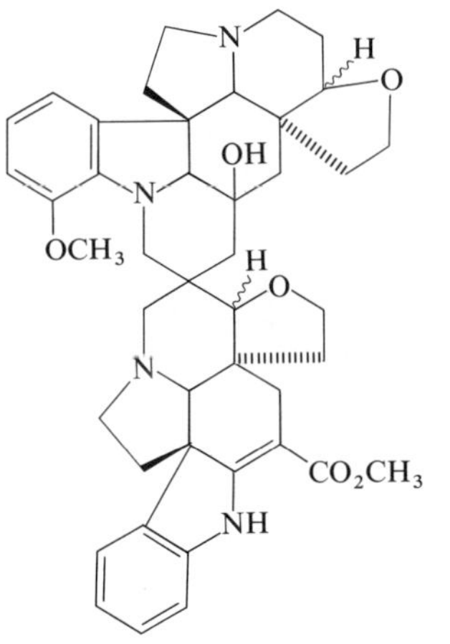

388 Vobtusine (original structure)

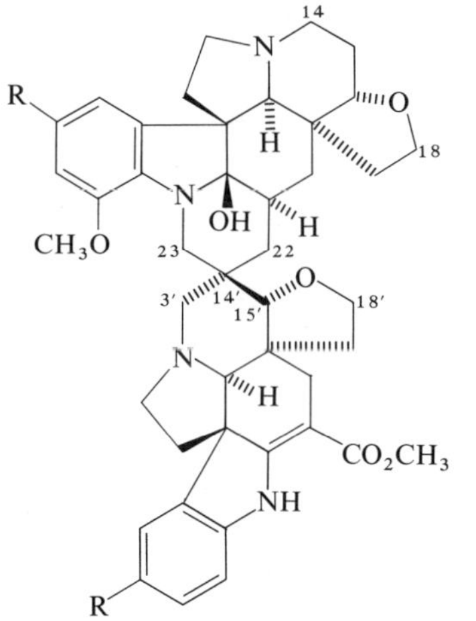

389 R = Br
390 Vobtusine (revised structure) R = H

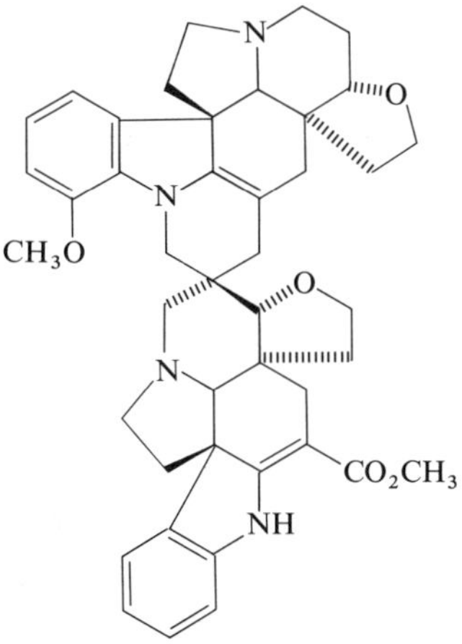

391 Anhydrovobtusine

One possibility for the ion at *m/e* 188 is that shown in Scheme 16, which involves an internal hydrogen migration from the hydroxy group in the ion $M^+ - 138$ in the vobtusine series. The ion at *m/e* 174 has not yet been rationalized in terms of the new structure of vobtusine.

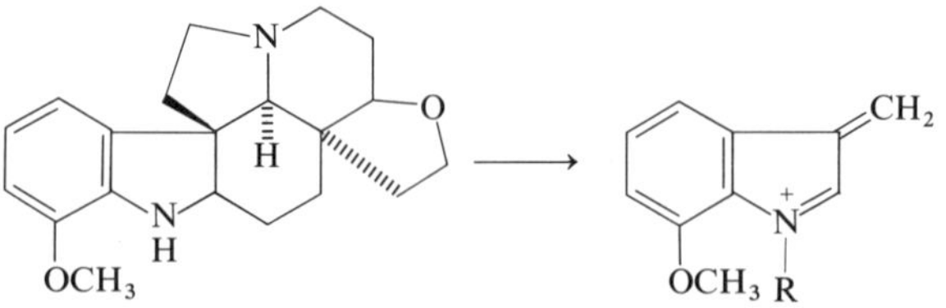

396 Beninine

392 *m/e* 174, R = CH_3
394 *m/e* 160, R = H

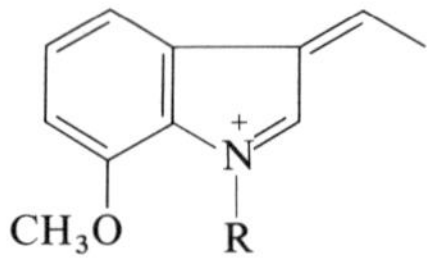

393 *m/e* 188, R = CH_3
395 *m/e* 174, R = H

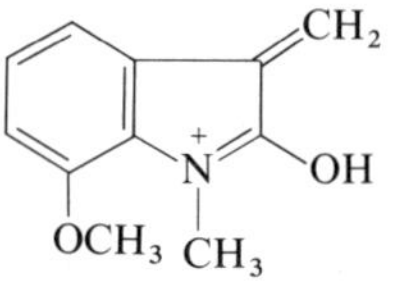

397 *m/e* 190

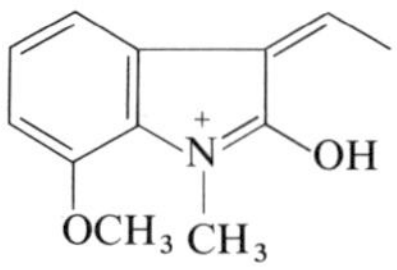

398 *m/e* 204

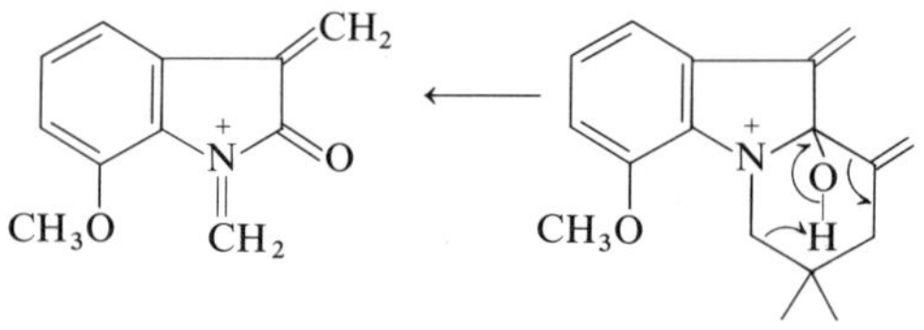

m/e 188

$M^+ - 138$

SCHEME 16.

The alkaloids in the vobtusine series show diagnostic similarities in their mass spectral fragmentation patterns. The derivation of these ions, which will be referred to frequently in discussing these alkaloids, is shown in Scheme 17, and their significance in the mass spectra of vobtusine and its derivatives is displayed in Table XII.

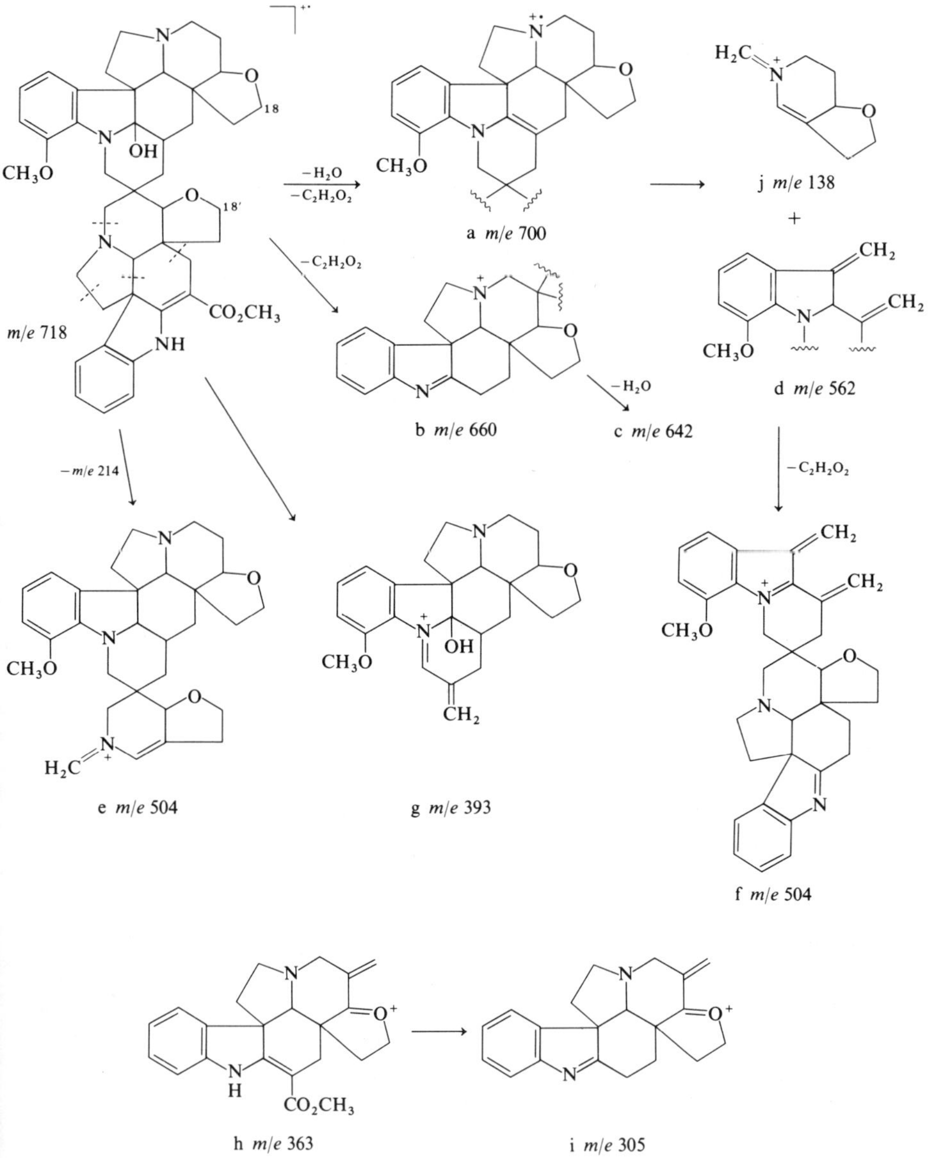

m/e 718

$-H_2O$, $-C_2H_2O_2$ → a m/e 700 → j m/e 138 + d m/e 562

$-C_2H_2O_2$ → b m/e 660; $-H_2O$ → c m/e 642

d m/e 562; $-C_2H_2O_2$ → f m/e 504

$-m/e$ 214 → e m/e 504

g m/e 393

h m/e 363 → i m/e 305

SCHEME 17. *Summary of the mass spectrum of vobtusine.*

TABLE XII

PRINCIPAL MASS SPECTRAL FRAGMENTS OF VOBTUSINE AND RELATED ALKALOIDS

Compound	Ion (mu)										
	M^+	a	b	c	d	e	f	g	h	i	j
Vobtusine (**390**)	718	700	660	642	562	504	504	393	363	305	138
12-Demethylvobtusine (**399**)	704	686	646	628	548	490	490	379	—	—	138
2-Deoxyvobtusine (**400**)	702	—	—	—	—	—	488	377	363	—	138
18-Oxovobtusine (**408**)	732	714	674	656	562	518	504	407	363	305	152
2-Deoxy-18-oxo-vobtusine (**409**)	716	—	658	—	—	502	—	391	363	305	152
3′-Oxovobtusine (**416**)	732	714	674	—	576	518	518	—	—	—	138
3′-Oxovobtusine *N*-oxide (**417**)	748	730	674	656	576	518	518	—	—	—	138
2-Deoxy-3′-oxo-vobtusine (**418**)	716	—	658	—	—	—	—	391	377	—	138
2,16-Anhydrovobtusine (**391**)	700	—	—	642	562	—	504	375	363	302	138
2,16-Anhydrodihydro-amataine (owerreine) (**420**)	700	—	—	642	562	—	504	375	363	305	138
Quimbeline (**421**)	716	—	—	—	—	502	—	391	363	305	138
Amataine (**426**)	716	—	—	—	578	502	—	391	363	305	138
18-Oxoamataine (**431**)	730	—	—	—	—	516	—	—	363	—	152

F. 12-DEMETHYLVOBTUSINE (**399**)

12-Demethylvobtusine, mp 295° (dec.), $[\alpha]_D$ −273°, was obtained initially from the roots of *Callichilia barteri* (*202*) and subsequently as alkaloid F from the leaves of *Voacanga thouarsii* (*203*). The structure was deduced independently by Swiss (*202*) and French (*204*) groups.

The alkaloid gave a molecular ion in the mass spectrum at *m/e* 704 which analyzed for $C_{42}H_{48}N_4O_6$ (vobtusine − CH_2). The UV and ORD spectra were similar to those of vobtusine (λ_{max} 263, 300, and 326 nm). However, addition of base to the UV spectrum caused a bathochromic shift in the absorption at 263 nm, indicating the presence of a hydroxyindoline unit. Treatment of the alkaloid with acetic anhydride–pyridine afforded a monoacetate having an aryl ester grouping (ν_{max} 1760 cm^{-1}) (*202*).

No ion was observed at *m/e* 188, but one was apparently shifted to *m/e* 174, a loss of 14 mu. The ions at M^+ −18, M^+ −CO_2CH_2, M^+ −138 − 18 and M^+ −214 were also shifted by 14 mu. (*204*)

In the NMR spectrum no aromatic methoxy group was observed, but a new exchangeable proton was found at 8.62 ppm, which also disappeared on acetylation.

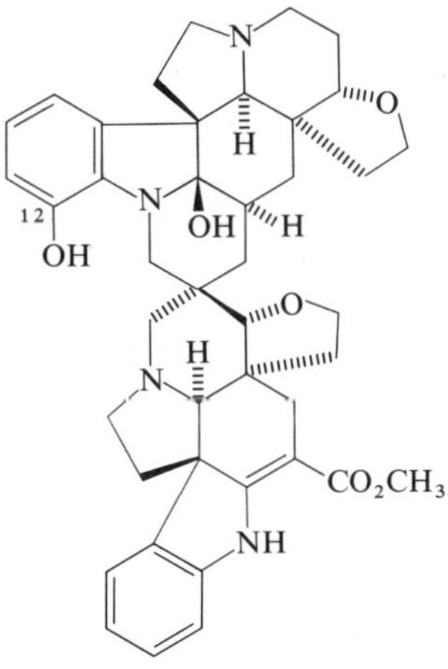

399 12-Demethylvobtusine

Methylation of the sodium salt of demethylvobtusine gave vobtusine (**390**), and the reverse reaction could be carried out with boron trifluoride etherate at $-75°$. Demethylvobtusine therefore has structure **399** (*202*, *204*).

G. 2-DEOXYVOBTUSINE (400)

2-Deoxyvobtusine, mp 305° (dec.), $[\alpha]_D$ $-355°$, was first obtained in low yield from the leaves of *Voacanga africana* (*205*), and subsequently also from *V. grandifolia* (*206*).

A molecular ion was observed at *m/e* 702 which analyzed for $C_{43}H_{50}\cdot N_4O_5$, a difference of one oxygen atom between the isolate and vobtusine. The UV spectrum (λ_{max} 224, 267, 303, and 327 nm) was very similar to that of vobtusine, and the IR spectrum displayed bands for NH (3370 cm^{-1}) and β-anilinoacrylate (1675 and 1605 cm^{-1}) functionalities.

The NMR spectrum also indicated a close similarity to vobtusine, showing an aromatic methoxy group at 3.78 ppm, a carbomethoxy group at 3.72 ppm, and the NH proton of the anilinoacrylate unit at 8.91 ppm.

Crucial in determination of the gross structure was the observation of particular fragment ions in the mass spectrum, because they revealed which of the oxygen atoms was missing compared to vobtusine. Thus comparison of ions e, g, and h and the absence of any $M^+ - 18$ ion indicated the compound to be a 2-deoxyvobtusine (**400**) (*205*).

At about this time, the Swiss group obtained an alkaloid, goziline, from the root bark (*207*) and leaves (*202*) of *Callichilia barteri*. The gross structure was determined originally on 2.8 mg of material. Its molecular formula, derived from the molecular ion at *m/e* 702, was $C_{43}H_{50}N_4O_5$. Principal fragment ions were observed at *m/e* 644, 488, 377, 363, 214, 188, and 174, with a base peak at *m/e* 138. The UV spectrum was superimposable on that of vobtusine, and the IR spectrum showed the characteristic bands for NH (3390 cm^{-1}) and β-anilinoacrylate (1669 and 1609 cm^{-1}) groups (*207*). These data indicated goziline to have the gross structure of 2-deoxyvobtusine (**400**).

Subsequently, a larger quantity of goziline was obtained (*202*), and an NMR spectrum made possible. A singlet was observed for the anilinoacrylate NH at 8.95 ppm and the carbomethoxy group at 3.73 ppm. The aromatic methoxy group was found at 3.79 ppm. In addition, there was a doublet ($J = 14$ Hz) at 5.36 ppm attributed to one of the C-8 protons. A similar signal in vobtusine occurs at 5.12 ppm (*202*).

Comparison was then made with *cis*-2,16-dihydroanhydrovobtusine (**401**), a compound obtained by catalytic reduction of anhydrovobtusine (**391**). Although this compound was quite similar to 2-deoxyvobtusine, a number of the properties were distinctly different. For example, goziline had mp 270° (dec.) while the vobtusine derivative **401** was amorphous. The $[\alpha]_D$ of goziline was $-337°$; that of the vobtusine derivative **401** was $-277°$ at similar concentrations. Also, the mass spectra of the two compounds, although qualitatively similar, were quantitatively different. Thus for goziline, *m/e* 138 was the base peak; for **401**, the base peak was the molecular ion at *m/e* 702. There was in addition a substantial difference in the intensities of the ions corresponding to ion g in the two spectra, 44% of the base peak in goziline and 78% in the case of **401**. These compounds are therefore stereoisomers at either C-16 or C-14′, or at both positions.

An effort to resolve this problem was made by Majumder *et al.* (*206*), who obtained deoxyvobtusine from the leaves of *V. grandifolia*. The isolate, mp 290° (dec.), $[\alpha]_D$ $-303°$, showed M^+ *m/e* 702 which was also the base peak, and the ion g-16 (*m/e* 377) was 65% of the base peak.

From the blue color reaction of goziline and the deoxyvobtusine to ceric (IV), it appears that these compounds could be identical, although no direct comparison had been made between the isolates.

Acid-catalyzed decarboxylation of deoxyvobtusine (*206*) followed by zinc–acid reduction gave a dihydrodecarbomethoxydeoxyvobtusine (**402**). In addition, anhydrovobtusine (**391**) on similar treatment gave tetrahydrodecarbomethoxyanhydrovobtusine (**403**), isomeric with **402** and differing only in the stereochemistry of C-16.

Reduction of deoxyvobtusine with zinc and 10% methanolic sulfuric acid gave a mixture of 2′,16′-dihydroisomers **404** and **405**; similarly, reduction of anhydrovobtusine (**391**) gave a mixture of **406** and **407**. These four isomers were chromatographically distinct (*206*). The key to the relationship

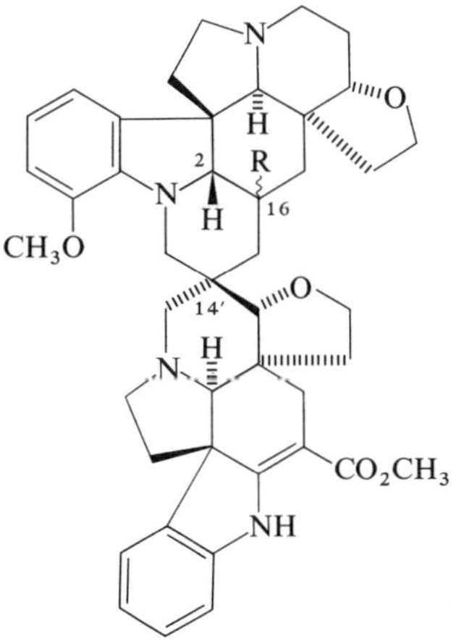

400 R = α-H 2-Deoxyvobtusine
401 R = β-H

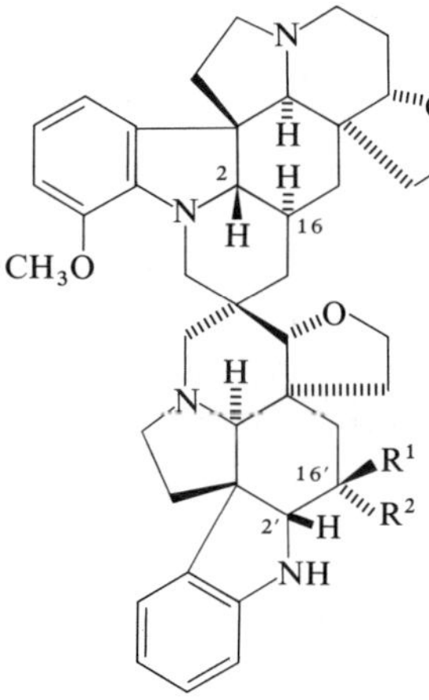

402 $R^1 = R^2 = H$
404 $R^1 = H, R^2 = CO_2CH_3$ (major)
405 $R^1 = CO_2CH_3, R^2 = H$ (minor)

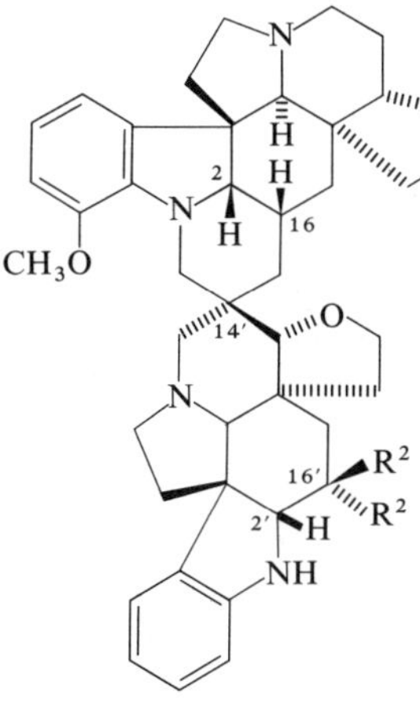

403 $R^1 = R^2 = H$
406 $R^1 = H, R^2 = CO_2CH_3$ (major)
407 $R^1 = CO_2CH_3, R^2 = H$ (major)

of vobtusine and deoxyvobtusine was that their ORD curves were very similar. In addition, the ORD curves of goziline and *cis*-2,16-dihydro-anhydrovobtusine (**401**) were similar. This indicates that in all these compounds the stereochemistry at C-14′ is the same as that in vobtusine and that the stereoisomerism must be due to a changing stereochemistry at C-16.

Thus, goziline (*202*, *207*), the deoxyvobtusine from *V. africana* (*205*), and the alkaloid from *V. grandifolia* (*206*) should all probably be regarded as

2-deoxyvobtusine, having structure **400**. *cis*-2,16-Dihydroanhydrovobtusine (**401**) is better named 2-deoxy-16-epivobtusine.

H. 18-Oxovobtusine (408) and 2-Deoxy-18-oxovobtusine (409)

18-Oxovobtusine (vobtusine lactone) and a 2-deoxy-18-oxovobtusine (2-deoxyvobtusine lactone) have been obtained thus far only from *Voacanga* species (see Table XXI). The alkaloids were originally isolated from the leaves of *Voacanga africana* (*208*); their structures were deduced subsequently (*209*). 18-Oxovobtusine, mp 310° (dec.), $[\alpha]_D$ −320°, showed a molecular ion at *m/e* 732 analyzing for $C_{43}H_{48}N_4O_7$, indicating a loss of two hydrogens and the addition of one oxygen in comparison with vobtusine. The UV spectrum was superimposable on that of vobtusine, and the NMR spectrum indicated the presence of aromatic methoxy (3.78 ppm), carbomethoxy (3.73 ppm), and NH (8.98 ppm) groups. The clue to the nature of the oxygen function came from the IR spectrum which, in addition to the anilinoacrylate ester absorption at 1685 cm^{-1}, showed a new carbonyl band at 1795 cm^{-1} for a γ-lactone moiety.

There are clearly two positions where such a lactone functionality may be placed in the vobtusine molecule, namely, C-18 or C-18′. A distinction between these two positions is most easily made from the mass spectrum. Thus ions a, b, c (b − H_2O), e, g, and j were all shifted by 14 mu, but those ions not containing the piperidine moiety of unit A, such as d, f, h, and i, were not shifted. These data clearly indicate the lactone carbonyl group to be located at C-18 and 18-oxovobtusine to have the structure **408** (*209*).

2-Deoxy-18-oxovobtusine, mp 305° (dec.), $[\alpha]_D$ −348°, displayed a molecular ion at *m/e* 716 which analyzed for $C_{43}H_{48}N_4O_6$, a loss of one oxygen atom by comparison with 18-oxovobtusine (**408**). The UV spectrum was identical to that of vobtusine and the NMR spectrum showed the characteristic signals for aromatic methoxy (3.78 ppm), carbomethoxy (3.73 ppm), and NH (8.98 ppm) groups. Two carbonyl bands were again observed at 1685 and 1795 cm^{-1}, the latter indicating the presence of a lactone moiety.

A comparison of the mass spectra of vobtusine (**390**), 18-oxovobtusine (**408**), and the new alkaloid (see Table XII) indicated that compared with **408**, the ions M^+, b, e, and g occurred at 16 mu *less*, whereas h, i, and j all appeared at the same mass. Of particular significance was the absence of ions a, c, and d in the spectrum of the new alkaloid; all these ions are associated with the loss of water from C-2 and C-16. The new compound is therefore a 2-deoxy-18-oxovobtusine (**409**) (*209*).

The structures of these alkaloids were at least partially confirmed by correlation with vobtusine (**390**). Acid-catalyzed dehydration of 18-

oxovobtusine followed by decarbomethoxylation and zinc–acid reduction gave the bisindoline lactone **410**, having 2β,16β-stereochemistry. Reduction with LAH gave the corresponding diol **411**, which could be recyclized with hydrogen bromide to afford a bisindoline bisether (**412**) identical with the compound obtained from vobtusine (**390**).

Treatment of 2-deoxy-18-oxovobtusine (**409**) in the same way gave an intermediate diol (**413**) which, although having a mass spectrum identical to that of the diol (**411**) from vobtusine lactone, was chromatographically

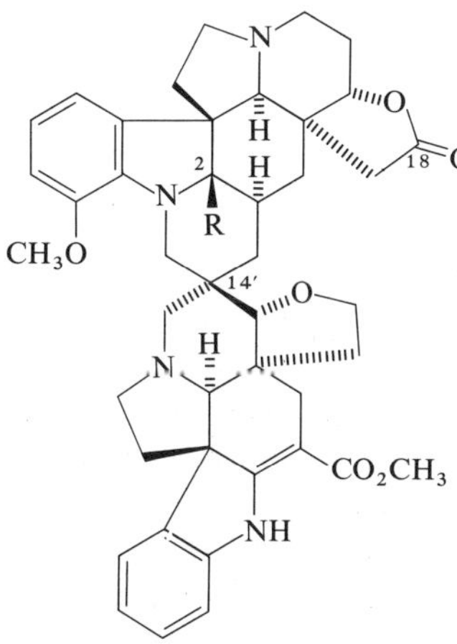

408 R = OH 18-Oxovobtusine
415 R = H

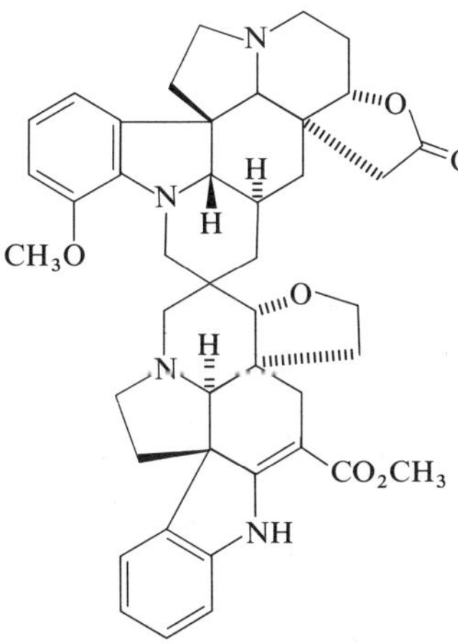

409 2-Deoxy-18-oxovobtusine

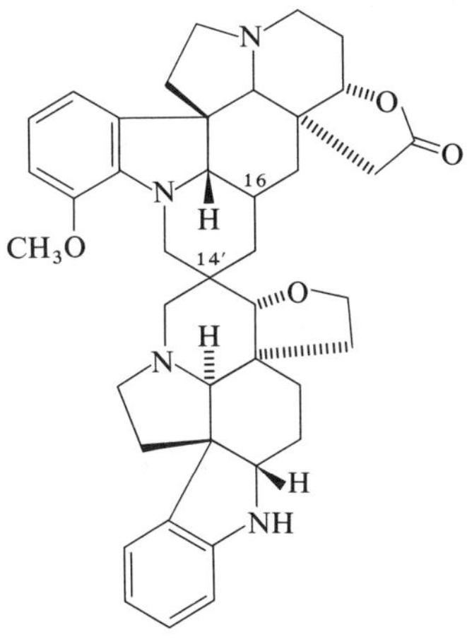

410 14′*S*, 16β-H

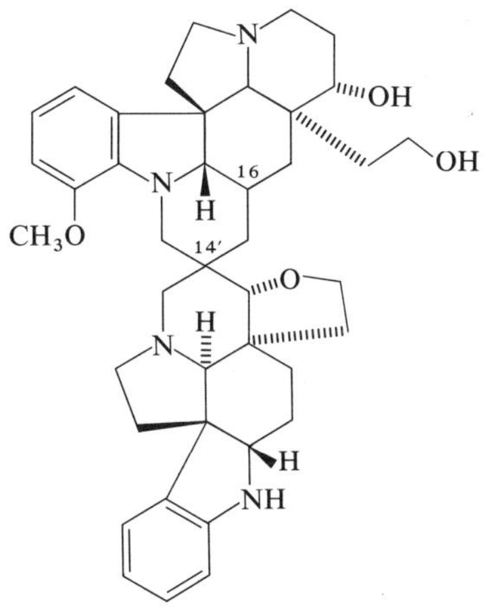

411 14′*S*, 16β-H
413

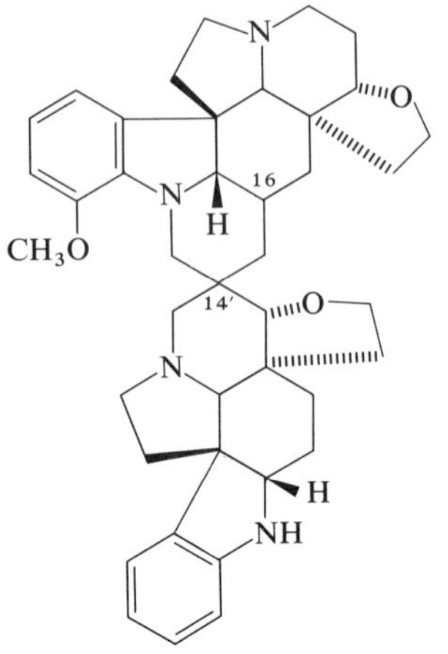

412 14′S, 16β-H
414

different and displayed different physical properties. Similarly, when the diol was cyclized, ether **414**, although similar to the ethers from both vobtusine (**390**) and 18-oxovobtusine (**408**), was distinctly different. Some of these physical properties are shown in Table XIII (*209*).

It was not clear where the difference in stereoisomerism lies between 18-oxovobtusine and 2-deoxy-18-oxovobtusine. Two possibilities were C-16 and C-14′, but a distinction between these two was not made at this time. Thus although the structure of 18-oxovobtusine may be represented by **408**, that of the corresponding 2-deoxy derivative is more correctly assigned as **409**, in which the C-16 and C-14′ stereochemistries remain to be determined (*209*).

In a subsequent paper (*210*) dealing with the ^{13}C-NMR spectra of vobtusine derivatives, the structure of 2-deoxy-18-oxovobtusine was given as **415**,

TABLE XIII

PHYSICAL PROPERTIES OF THE DIOLS AND ETHERS FROM 18-OXOVOBTUSINE (**408**) AND 2-DEOXY-18-OXOVOBTUSINE (**409**)

Compound	mp (dec.)	$[\alpha]_D$	UV Spectrum (λ_{max}, nm)
Diol from 18-oxovobtusine (**411**)	185°–188°	−3°	215, 260, 302
Diol from 2-deoxy-18-oxovobtusine (**413**)	275°–280°	−108°	220, 250(sh), 266, 303
Ether from 18-oxovobtusine (**412**)	292°–296°	−36°	215, 258, 302
Ether from 2-deoxy-18-oxovobtusine (**414**)	300°–303°	−125°	222, 250(sh), 266, 303

although no comparison was made with the data of 2-deoxyvobtusine. This structure would appear to negate the differences in properties of the derivatives (Table XIII). This point is discussed further in Section XXI,R.

I. 3′-Oxovobtusine (416), 3′-Oxovobtusine *N*-Oxide (417), and 2-Deoxy-3′-oxovobtusine (418)

Of the 12 bisindole alkaloids isolated from the leaves of *Voacanga thouarsii* Roem. et Schult. (syn. *V. obtusa* K. Schum.), nine were new (*203*). Three of these new alkaloids were found to be the closely related compounds 3′-oxovobtusine (vobtusine lactam) (**416**), 3′-oxovobtusine *N*-oxide (vobtusine lactam *N*-oxide) (**417**), and 2-deoxy-3′-oxovobtusine (2-deoxyvobtusine lactam) (**418**) (*204*).

3′-Oxovobtusine crystallized from methanol, mp 282°–285° (dec.), $[\alpha]_D$ −130°, and displayed λ_{max} 263, 292, and 328 nm characteristic of the summation of 7-methoxyindoline and β-anilinoacrylate chromophores present in vobtusine. The NMR spectrum confirmed the presence of an aromatic methoxy group (singlet, 3.78 ppm) and also of a β-anilinoacrylate unit (NH at 8.65 ppm, carbomethoxy at 3.75 ppm). The latter group was verified by the IR spectrum (ν_{max} 3400, 1680, and 1615 cm^{-1}), and this technique also indicated an additional carbonyl absorption at 1650 cm^{-1}, typical of a δ-lactam. This group may be placed at C-3, C-3′, or C-23. Location of the carbonyl at C-23 could be eliminated by the UV spectrum (not an *N*-acylindoline); also, the NMR spectrum, which showed a doublet ($J = 14$ Hz) at 4.55 ppm attributed to one of the C-3 protons (but see later), appeared to limit substitution to C-3′. Confirmation of this assignment came from both spectral analysis and chemical correlation.

In the mass spectrum of 3′-oxovobtusine (see Table XII), a molecular ion was observed at *m/e* 732 analyzing for $C_{43}H_{48}N_4O_7$ and indicating the alkaloid to contain one additional oxygen and two less hydrogens than vobtusine. The presence of a substantial $M^+ - 18$ ion confirmed a 2-hydroxy group in the isolate. Comparison of the ions d (*m/e* 576) and j (*m/e* 138) with the corresponding fragments in vobtusine indicates that the lactam moiety is located in the piperidine ring of the anilinoacrylate unit, i.e., at C-3′.

Reduction with sodium cyanoborohydride in acetic acid gave a 2′,16′-dihydro derivative which on LAH reduction gave the amorphous alcohol **419** (M^+, *m/e* 692). Successive treatment of vobtusine (**390**) in the same way gave an identical product. Therefore, the structure of the alkaloid was confirmed to be 3′-oxovobtusine (**416**) (*204*).

3′-Oxovobtusine *N*-oxide crystallized from acetone, mp 270°–290° (dec.), $[\alpha]_D$ −142°, and showed IR (ν_{max} 3400, 1685, 1650 and 1620 cm^{-1}) and UV (λ_{max} 263, 297, and 327 nm) spectral data very similar to those of **416**.

However, the molecular ion at *m/e* 748 suggested a formula $C_{43}H_{48}N_4O_8$, an addition of one oxygen compared with **416**. Losses of 16 and 17 mu in the mass spectrum suggested the presence of an *N*-oxide moiety, and this was substantiated when reduction with iron(II) in acetic acid at low temperature afforded 3′-oxovobtusine (**416**) identical to the natural product.

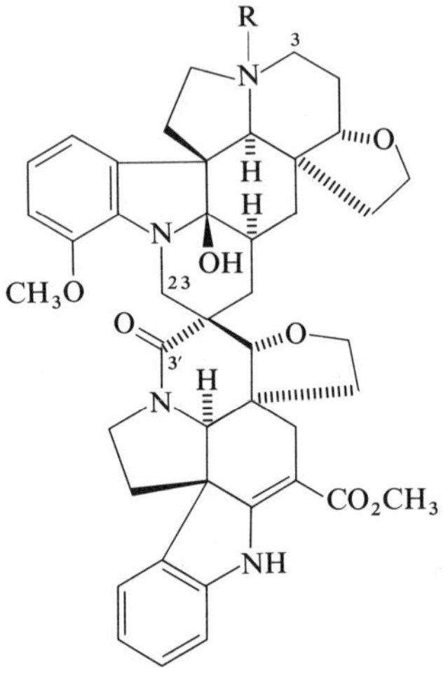

(i) $NaBH_3CN$
(ii) $LiAlH_4$

416 R = lone pair 3′-Oxovobtusine
417 R = O 3′-Oxovobtusine *N*-oxide

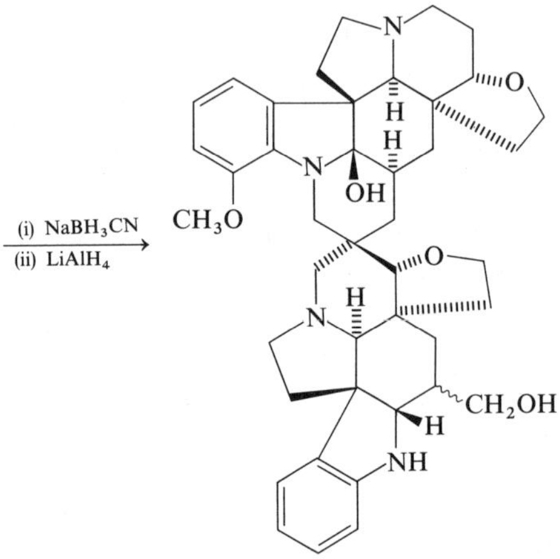

419

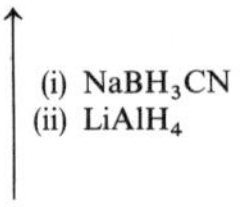

(i) $NaBH_3CN$
(ii) $LiAlH_4$

390 Vobtusine

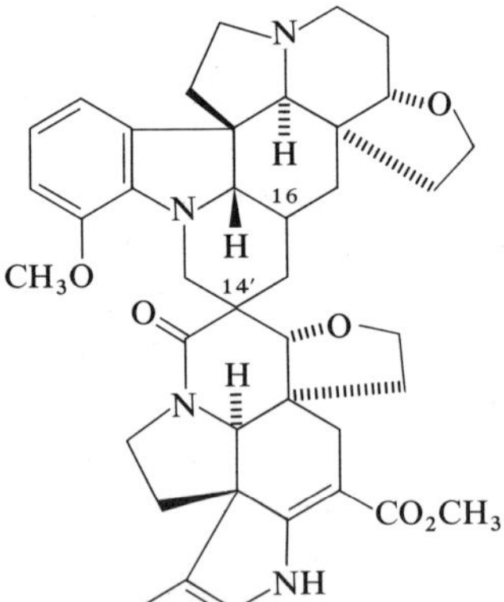

418 2-Deoxy-3′-oxovobtusine

The NMR spectrum of the parent compound showed singlets for the aromatic methoxy and carbomethoxy groups at 3.82 and 3.84 ppm and for the NH proton at 9.0 ppm and a doublet ($J = 14$ Hz) at 4.75 ppm for one of the C-3 protons. On this basis, the alkaloid was assigned the structure 3′-oxovobtusine *N*-oxide (**417**) (*204*).

3′-Oxo-2-deoxyvobtusine was not obtained in crystalline form but displayed typical UV (λ_{max} 265, 297.5, and 328 nm) and IR (ν_{max} 3400, 1680, and 1615 cm^{-1}) spectral properties for a compound in the vobtusine series. In addition, an absorption at 1655 cm^{-1} could be attributed to a δ-lactam group. The NMR spectrum confirmed the presence of aromatic methoxy and carbomethoxy groups (singlets at 3.77 and 3.80 ppm), an NH group (singlet at 8.85 ppm), and a doublet ($J = 13$ Hz) at 4.80 ppm.

The mass spectrum indicated a molecular ion at *m/e* 716 which by high-resolution mass spectrometry analyzed for $C_{43}H_{48}N_4O_6$, two hydrogens less than vobtusine. Because a lactam group was already indicated, it remained to determine the site of the lost oxygen atom.

No $M^+ - 18$ ion was observed in the mass spectrum of the isolate, and ions b, g, and h were all shifted by +14 mu (+O − 2H) in comparison with 2-deoxyvobtusine, whereas ion j remained at *m/e* 138. The compound could therefore be assigned the gross structure of a 2-deoxy-3′-oxovobtusine (**418**), although the stereochemistry of C-16 and C-14 remains unknown (*204*).

J. Owerreine (**420**)

Owerreine is a trace alkaloid obtained by the Swiss group from the root bark of *Callichilia barteri* (*207*). The alkaloid crystallized from methanol, mp 265° (dec.), $[\alpha]_D$ −496°, and showed λ_{max} (272(sh), 298 and 327, nm) and IR spectral absorptions for NH (3378 cm^{-1}) and β-anilinoacrylate units (1686 and 1610 cm^{-1}). An additional band at 1721 cm^{-1} was attributed to an enamine group.

The mass spectrum showed a molecular ion at *m/e* 700 analyzing for $C_{43}H_{48}N_4O_5$, a loss of H_2O compared with vobtusine. The fragment ions c, d, f, h, and i and the piperidinium ion *m/e* 138 all appeared at the same mass as in vobtusine, but the ion g was observed at 18 mu less, indicating the loss of water to be from the positions 2 and 16 of vobtusine.

These data would suggest that the compound was 2,16-anhydrovobutusine (**391**); however, comparison of physical data indicated that owerreine had a number of physical properties which distinguished it from the corresponding compound derived from vobtusine. For example, 2,16-anhydrovobutusine (**391**) shows $[\alpha]_D$ −453°, λ_{max} 294 and 324 nm, and bands in the carbonyl region at 1713, 1672, and 1608 cm^{-1}. The reaction of each compound to ceric(IV) sulfate is also different. Owerreine shows a bright green color but anhydrovobtusine a light blue fading to greenish yellow. The mass spectra, although qualitatively similar, are quantitively different. Thus, *m/e* 138 is the base peak in the spectrum of owerreine but *m/e* 562 ($M^+ - 138$) is the base peak in the spectrum of 2,16-anhydrovobtusine (**391**). The ions *m/e* 363 and 305 are also more intense in the spectrum of **391**. When the anilinoacrylate chromophore is subtracted from each spectrum the result is not the same. The anhydro derivative shows λ_{max} 275 and 295 nm, whereas owerreine shows 270 and 302 nm. The interactions between the chromophores in each compound are, therefore, somewhat different.

The compounds owerreine and 2,16-anhydrovobtusine (**391**) are clearly not identical, and the most probable explanation is that owerreine is the C-14′ diastereoisomer (**420**) of 2,16-anhydrovobtusine (**391**) (*207*).

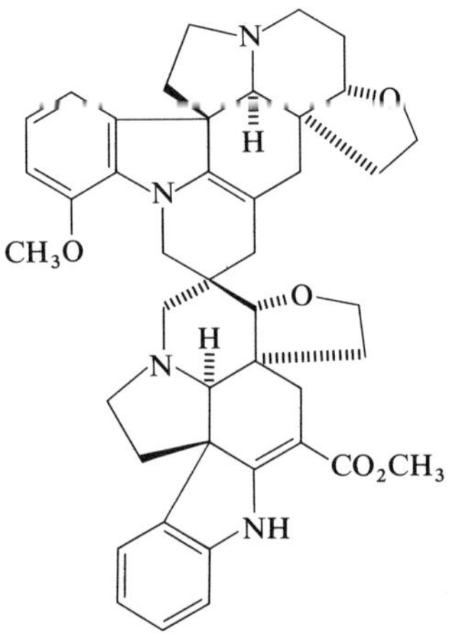

391 Anhydrovobtusine

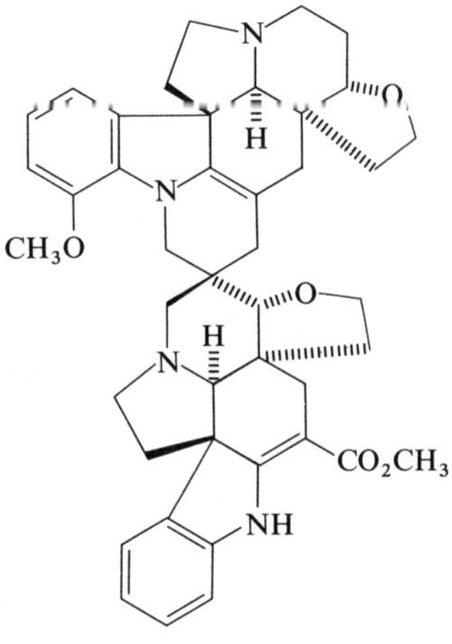

420 Owerreine

K. Quimbeline (**421**)

Quimbeline is an alkaloid obtained by Gabetta *et al.* from the root bark of *Voacanga chalotiana* Pierre ex Stapf (*197*). Its structure was deduced subsequently (*211*).

The crystalline alkaloid, mp 270°, $[\alpha]_D$ −195°, showed a molecular ion at *m/e* 716 analyzing for $C_{43}H_{48}N_4O_6$ and displayed λ_{max} at 263, 301, and 327 nm typical of the vobtusine system. The IR spectrum confirmed the presence of an anilinoacrylate system (ν_{max} 1675 and 1610 cm^{-1}), and this was substantiated by the NMR spectrum which showed the NH at 8.95 ppm and the carbomethoxy group at 3.74 ppm. The aromatic methoxy group appeared as a singlet at 3.80 ppm and the "vobtusine doublet" at 4.95 ppm ($J = 14$ Hz).

The mass spectrum, in which ions e and g are shifted by 2 mu compared to vobtusine and ions h, i and j which are not, indicates that the two hydrogens are missing from the ring linking the two units. Quimbeline does not lose water in the mass spectrum and shows no olefinic protons in the NMR and no additional carbonyl groups in the IR spectrum. The loss of two hydrogens should therefore be by the formation of an ether function.

Hydrogenolysis of quimbeline with palladium–charcoal for 5 days afforded vobtusine (**390**), indicating that one end of the ether bridge was located at C-2. The other linkage point could be C-6, C-16, C-17, C-22, or C-23. The latter position could be excluded because H-23 should then appear as a singlet downfield if it were the correct structure.

Examination of the ^{13}C-NMR spectrum of quimbeline indicated the second point of attachment to be C-22. On comparison with the spectrum of vobtusine an additional oxymethine carbon was observed at 82.0 ppm, and because both C-14′ (43.3 ppm) and the doublet for C-16 (58.3 ppm) had moved downfield, it was suggested that quimbeline had the novel oxetane structure **421** (*211*). The interpretation of some of these data has been challenged by Poisson and Wenkert (*210*).

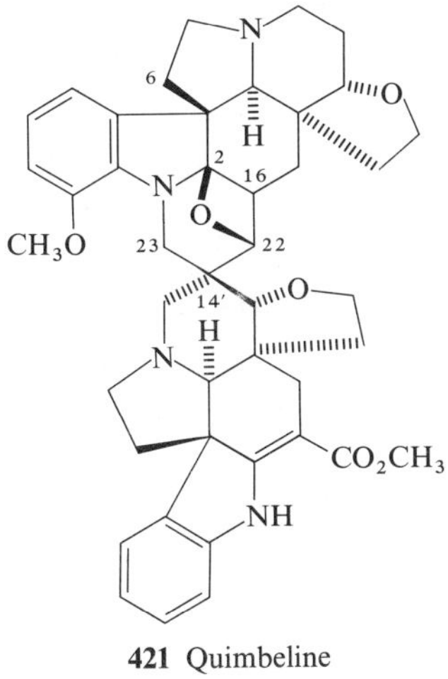

421 Quimbeline

L. Amataine (**426**)

The alkaloid amataine has been obtained as grandifoline from *Voacanga grandifolia* (Miq.) Rolfe (*212*), as subsessiline from *Callichilia subsessilis* Stapf (*205*), and as amataine from *Hedranthera barteri* (Hook F.) Pichon (*207*) and *Voacanga chalotiana* (*197*). It crystallized from hexane, mp 216°–221°, $[\alpha]_D$ $-262°$, and showed λ_{max} 262, 298, and 326–328 nm typical of the vobtusine-type alkaloids. The IR spectrum substantiated the presence of the β-anilinoacrylate unit (ν_{max} 3400, 1684, and 1613 cm^{-1}) and indicated that no other carbonyl groups were present. The NMR spectrum showed two three-proton singlets for the carbomethoxy and aromatic methoxy groups at 3.77 and 3.76 ppm, an NH proton at 8.98 ppm, and two one-proton singlets at 4.75 and 4.18 ppm. No doublet ($J = 14$ Hz) was observed in the region of 4.8–5.2 ppm (*207*).

A molecular ion, which was also the base peak, appeared at *m/e* 716 and analyzed for $C_{43}H_{48}N_4O_6$, two hydrogens less than vobtusine. The major fragment ions e and g were shifted by 2 mu to lower mass, but ions h, i, and j were unchanged compared to vobtusine, suggesting that the ring joining the two groups contained the oxygen function as an ether linkage (*207*).

Potassium borohydride reduction afforded a dihydro derivative which was isomeric with vobtusine and was named isovobtusine. Such a reduction is indicative of the presence of a carbinolamine ether moiety. When sodium borodeuteride was used as the reducing agent, ions h and i were increased by 1 mu, demonstrating that the reduced group was located in this part of the molecule. On this basis, the alkaloid was assigned structure **422** (*205*), wherein C-14′ had a configuration opposite to that in vobtusine.

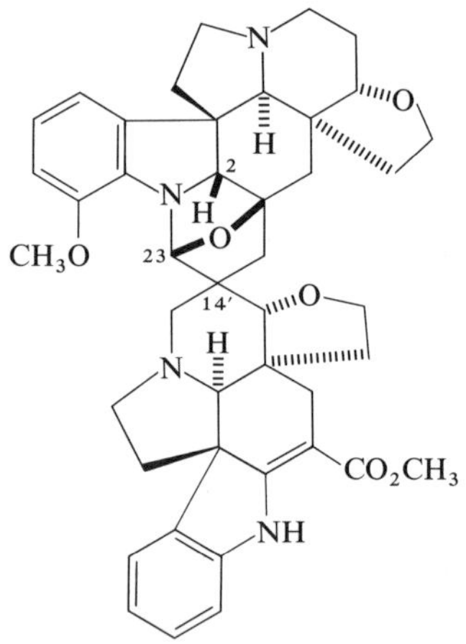

422 Amataine (original)

After the structure of vobtusine was revised to **390**, it was suggested that the ether linkage was either between C-2 and C-23 or between C-2 and C-3′ (*197*). Further work (*210*, *213*, *214*) has proved the latter to be the correct structure and has also provided evidence for the C-14′ stereochemistry.

Subtraction of the β-anilinoacrylate chromophore from the spectrum of amataine indicated it to be slightly different from the subtracted UV spectrum of vobtusine (**390**). Thus, whereas the latter showed λ_{max} 265 and 302 nm, the former showed λ_{max} 260 and 299 nm. When a solution of amataine was allowed to stand for 2 hr in concentrated HCl, the UV spectrum resembled that of anhydrovobtusine (**391**) in acid (λ_{max} 296 and 332 nm) (*213*).

Heating amataine with 70% phosphoric acid afforded formaldehyde and a 40% yield of 1,2-dehydrobeninine (**423**) (*213*). This is the only reported cleavage of an alkaloid in the vobtusine series, and is of interest because vobtusine does not undergo this reaction. The mechanism shown in Scheme 18 accounts for the formation of the observed products.

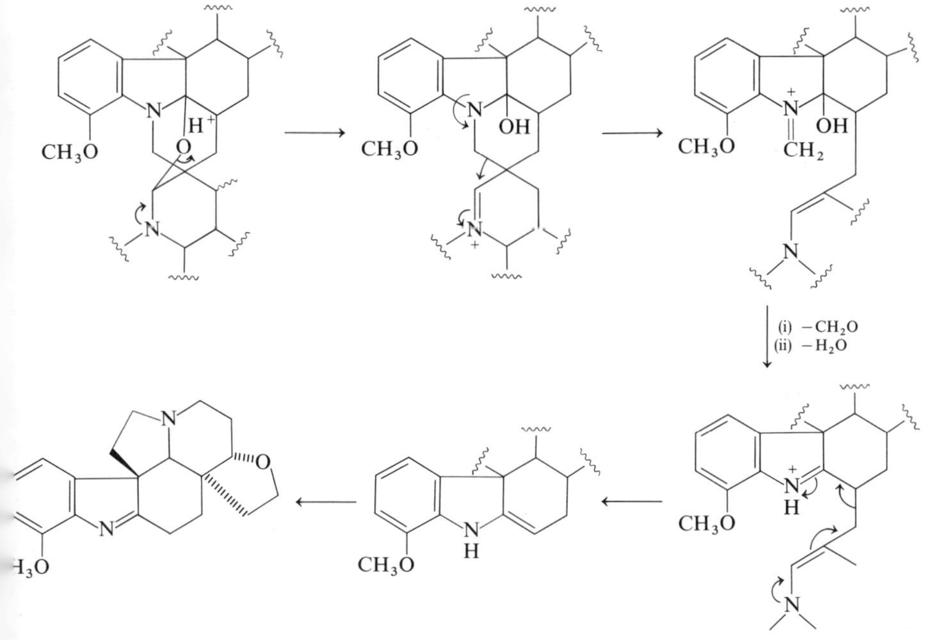

(i) $-CH_2O$
(ii) $-H_2O$

423 1,2-Dehydrobeninine

SCHEME 18.

Catalytic reduction of amataine with platinum oxide in 10% methanolic acetic acid (*213*) or with sodium borohydride (*213*, *214*) afforded dihydroamataine, $C_{43}H_{50}N_4O_6$, which gave a reddish–blue color reaction with ceric sulfate (vobtusine gives blue). The product showed IR and UV spectra very similar to those of vobtusine (**390**), and in the NMR spectrum the "vobtusine doublet" had reappeared at 4.70 ppm ($J = 14$ Hz). This signal appears at 5.13 ppm ($J = 14$ Hz) in vobtusine (**390**). The mass spectrum showed the typical ions a–c for vobtusine. d_1-Dihydroamataine produced from amataine by reduction with sodium borodeuteride gave ions m/e 718,

700, 660, 642, 504, and 393 at 1 mu greater than in dihydroamataine, and the peak normally at *m/e* 305 was observed as a mixture of *m/e* 305 and 306 in the ratio 0.8:1.

This ion (*m/e* 305) is not present in the spectrum of compounds such as *cis*-2,16-dihydroanhydrovobtusine (**401**) and is therefore derived from ion a ($M^+ - H_2O$) by way of ion d. To explain the observed formation of *m/e* 305 and 306, it was postulated (*213*) that two possible fragmentation pathways are available for ion f (*m/e* 505), as shown in Scheme 19, depending on the site of hydrogen radical abstraction. In either case, however, C-23 is present in this species—and this in itself is extremely valuable information. Treat-

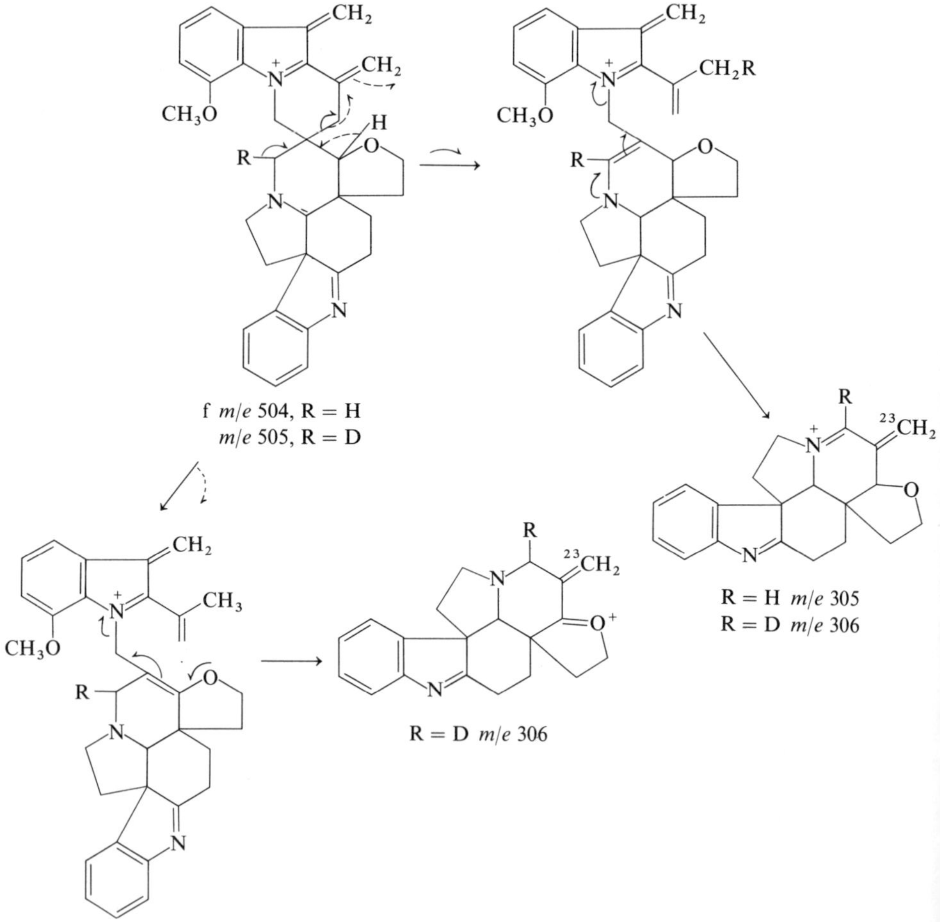

f *m/e* 504, R = H
m/e 505, R = D

R = H *m/e* 305
R = D *m/e* 306

R = D *m/e* 306

SCHEME 19. *Formation of ion i in the mass spectrum of vobtusine derivatives.*

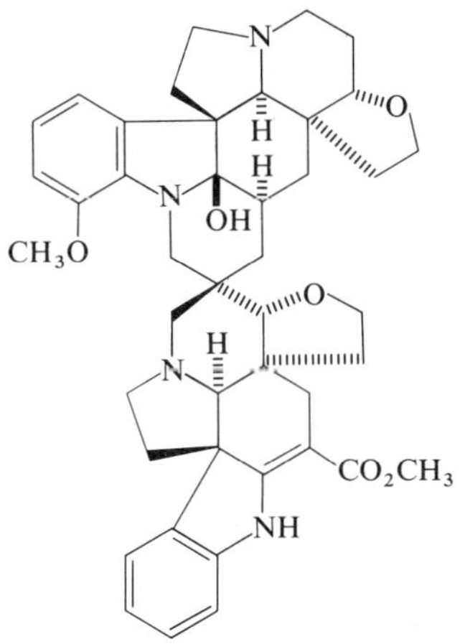

424 Dihydroamataine (isovobtusine)

ment of dihydroamataine (isovobtusine) (**424**) with 6 *N* HCl at 25° for 15 hr afforded anhydrodihydroamataine (owerreine) (**420**), which gave a blue–violet color with the ceric(IV) reagent (*213*). Both the IR (ν_{max} 3356, 1678, and 1608 cm^{-1}) and UV (λ_{max} 296 and 325 nm) spectra were similar to those of anhydrovobtusine (**391**), and the mass spectrum showed M^+ *m/e* 700 and principal fragment ions at *m/e* 642, 562, 504, 486, 305, and 138. The NMR spectrum displayed a one-proton singlet for the NH proton at 8.88 ppm and singlets for the aromatic methoxy and carbomethoxy groups at 3.82 and 3.76 ppm, respectively. The vobtusine doublet was observed at 4.55 ppm ($J = 12$ Hz) coupled to a proton at 2.63 ppm. A second doublet at 3.24 ppm ($J = 11$ Hz) was assigned to H-3′, and a singlet at 3.27 ppm to either H-5′, H-21, or H-21′. These data are in agreement with structure **420** for anhydrodihydroamataine, and it was suggested (*213*) that owerreine, an alkaloid of *Hedrantha barteri* (*207*), has the same structure.

Reaction of amataine with 0.01 *N* HCl in 50% aqueous dioxan at 80° for 2 hr gave hydratoamataine (3′-hydroxyvobtusine) (**425**) in 60% yield. This compound gave a blue color reaction with ceric(IV) sulfate. The NMR spectrum indicated the presence of aromatic methoxy (3.79 ppm) and carbomethoxy (3.73 ppm) groups, an NH group (8.98 ppm), a doublet ($J = 14$ Hz) at 5.07 ppm, and a doublet ($J = 10$ Hz) at 4.53 ppm. Irradiation of the hydroxy group at 0.96 ppm collapsed the doublet at 4.53 ppm to a singlet, indicating that it was the carbinolamine methine proton at C-3′ (*213*).

Hydratoamataine (**425**) could be reduced with sodium borohydride in methanol to yield vobtusine (**390**) in good yield, identified by its chromatographic and spectral properties.

Alternatively, pyrolysis of **425** at 340° for about 2 sec gave an almost quantitative yield of amataine (**426**) (*213*). These two transformations, together with the formation of hydratoamataine (**425**) from amataine (**426**), were suggested to take place by way of the mechanism shown in Scheme 20.

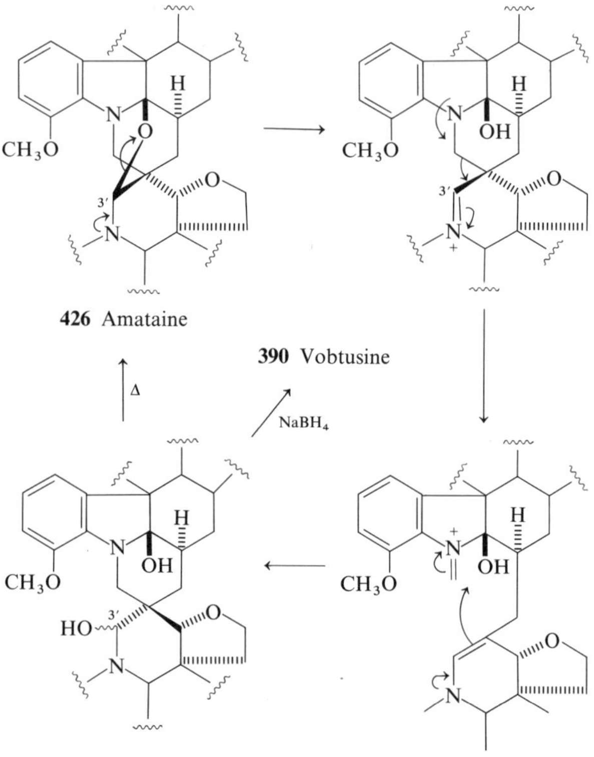

426 Amataine

390 Vobtusine

$NaBH_4$

Δ

425 Hydratoamataine

SCHEME 20.

When the extraction of *H. barteri* is carried out with acetone, the yield of amataine is reduced at the expense of the formation of a condensation product (*213*). The structure of the condensation product was shown (*213*) by its spectral properties to be **427**. Heating **427** at 300° for 10 sec in high vacuum gave a product showing M^+ at *m/e* 698, and having a UV spectrum [λ_{max} 269 and 310(sh) nm] similar to that of anhydrodecarbomethoxyvobtusine (**428**). The IR spectrum showed absorption for a ketone (1706 cm^{-1}), an indoline (1600 cm^{-1}), and an imine (1579 cm^{-1}). In the NMR spectrum no carbomethoxy group was observed, just a singlet for the aromatic methoxy group at 3.85 ppm. A singlet for a methyl ketone was observed at 2.28 ppm and a doublet ($J = 11$ Hz) at 5.05 ppm. These data are in agreement with structure **429** for this pyrolysis product (*213*).

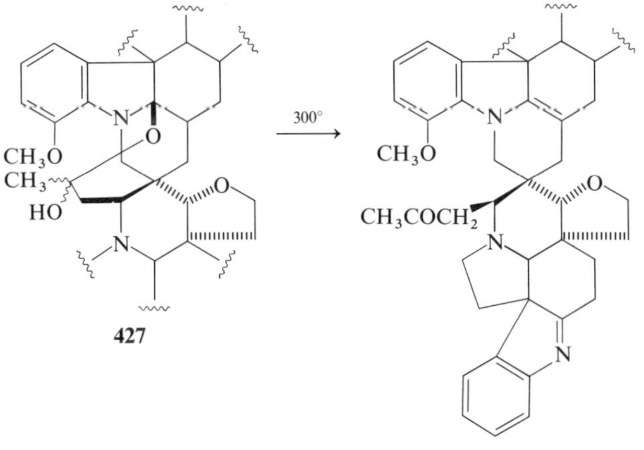

427 $\xrightarrow{300^\circ}$ **429**

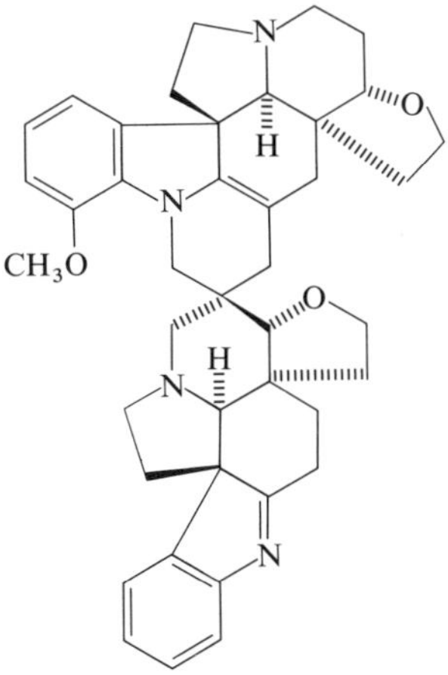

428

The above reactions and derivatives of amataine—in particular, the reduction to dihydroamataine, the formation of hydratoamataine, and the rearrangement of the latter compound to vobtusine (**390**)—indicate that amataine (grandiflorine, subsessiline) has the complete structure and stereochemistry shown in **426** (*213*).

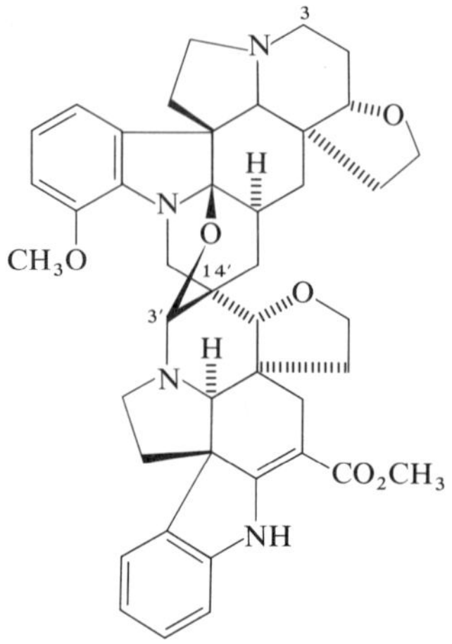

426 Amataine

M. 18-OXOAMATAINE (**431**)

18-Oxoamataine (subsessiline lactone) was first isolated from the leaves of *Voacanga thouarsii* as alkaloid H (*203*). The alkaloid crystallized from acetone, mp 237°–239°, $[\alpha]_D$ −292°, and exhibited UV (λ_{max} 220, 263, 295, and 328 nm) and IR (ν_{max} 3400, 1680, and 1610 cm^{-1}) spectra typical of alkaloids in the vobtusine series. In addition, the IR spectrum showed a band at 1780 cm^{-1} for a δ-lactone. The β-anilinoacrylate moiety was confirmed by the NMR spectrum, which showed singlets at 8.96 ppm for the NH group and at 3.80 and 3.82 ppm for the carbomethoxy and aromatic methoxy groups, respectively (*203*).

The molecular ion and base peak at *m/e* 730 in the mass spectrum were in agreement with a molecular formula $C_{43}H_{46}N_4O_7$, supporting the addition of one oxygen atom and the loss of two hydrogens required for a lactone moiety. Such a group could be reasonably located at either C-18 or C-18′. Major fragments were observed at *m/e* 516 (ion e), *m/e* 363 (ion h), and *m/e* 152 (ion j). Comparison with the mass spectra of both vobtusine (**390**) and amataine (**426**) suggested that the lactone moiety was located at C-18, as it is in 18-oxovobtusine (**408**) (*214*).

Reduction of 18-oxoamataine with sodium borohydride gave a product having spectral properties very similar to those of 18-oxovobtusine (**408**) but having different chromatographic properties. The mass spectrum indicated an identical fragmentation pattern, but several intensities were somewhat different. For example, in the spectrum of 18-oxovobtusine (**408**) the ion at *m/e* 518 was the base peak and the ions at *m/e* 714, 562, and 407 were considerably less intense than those in the reduction product from 18-oxoamataine (*214*).

By analogy with the reduction product of amataine, which was isovobtusine, the reduction product of 18-oxoamataine could be assigned the structure 18-oxoisovobtusine (isovobtusine lactone) (**430**), and the parent

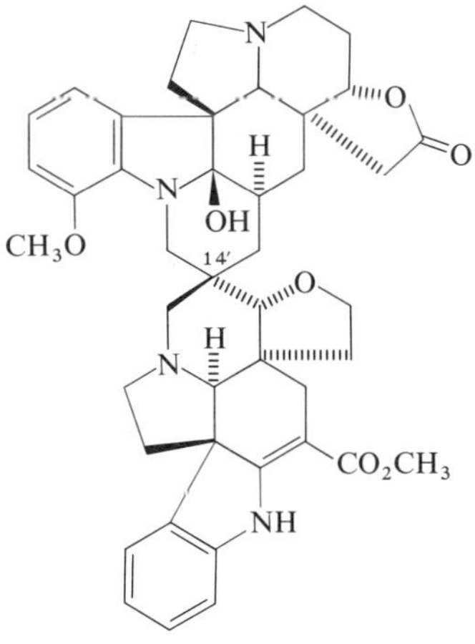

430 18-Oxoisovobtusine

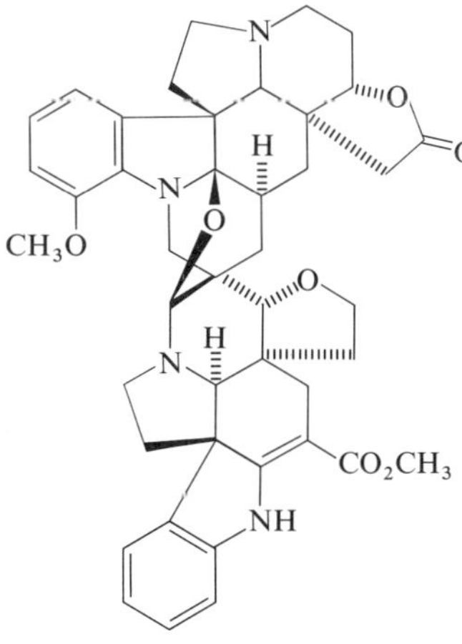

431 18-Oxoamataine

compound, 18-oxoamataine (subsessiline lactone), could be assigned structure **431** having a 14*R* configuration (*214*). Supporting evidence came from an independent examination of the ^{13}C-NMR spectrum of 18-oxoamataine (*210*), although an opposite stereochemistry for the center C-3′ was deduced for each compound.

N. Callichiline

Callichiline, $C_{42}H_{48}N_4O_5$, was briefly mentioned previously in this series (Volume XI, p. 302), and a proposed structure was indicated. Callichiline, mp 208°–210°, $[\alpha]_D$ −437°, co-occurs with vobtusine in *Callichilia subsessilis* (*215*) and *C. barteri* (*207, 216–218*). Its molecular formula was established via mass spectrometry (*217, 218*). It differs from vobtusine by a CH_2O unit, but like vobtusine it resists cleavage to its component parts and cannot be formylated, acetylated, or catalytically hydrogenated.

The UV spectrum (λ_{max} 252, 298, and 326 nm) establishes the presence of β-anilinoacrylate and methoxyindoline units, although because of intramolecular interactions the nature of the methoxyindoline cannot be obtained by subtraction.

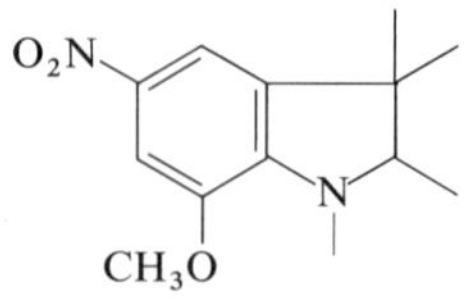

O_2N CH_3O N

432

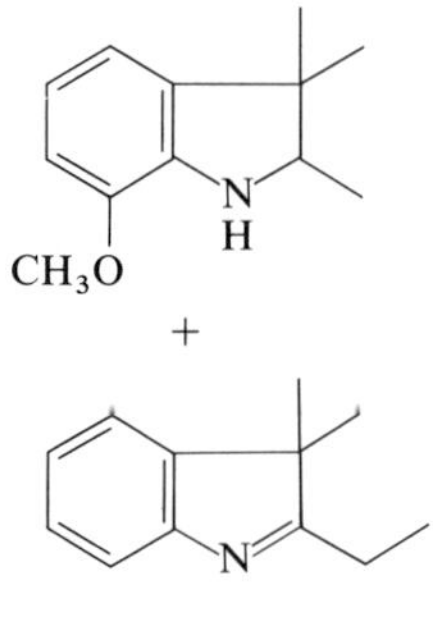

CH_3O N H + N **433**

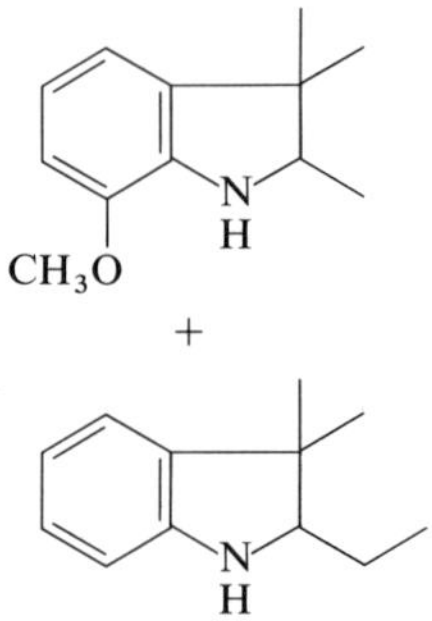

CH_3O N H + N H **434**

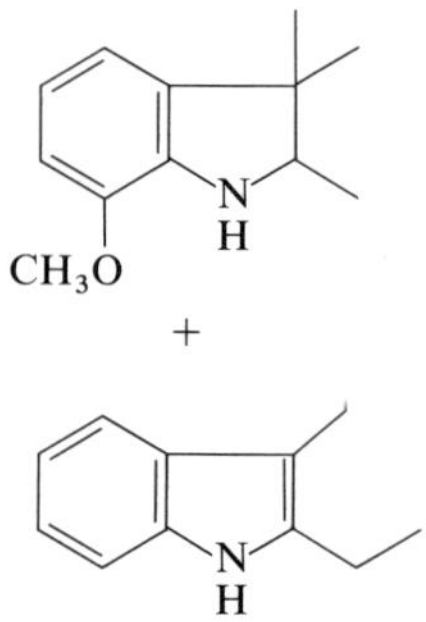

CH_3O N H + N H **435**

H_2C $\overset{+}{N}$ O

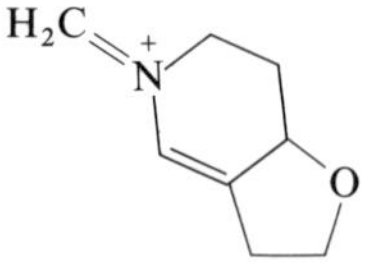

436 *m/e* 138

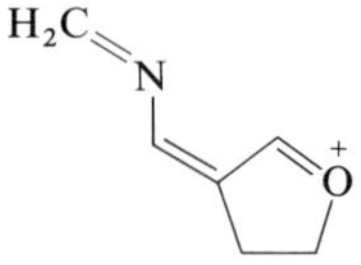

H_2C N $\overset{+}{O}$

437 *m/e* 110

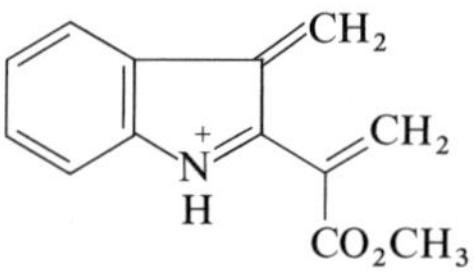

CH_2 CH_2 $\overset{+}{N}$ H CO_2CH_3

438 *m/e* 214

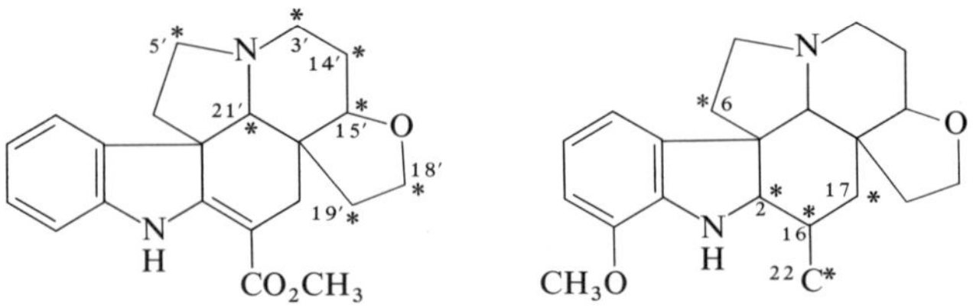

5′ N 3′ 14′ 21′ 15′ O 18′ 19′ N H CO_2CH_3

Unit A

N 6 O N H 2 17 16 CH_3O 22 C*

Unit B

439

SCHEME 21.

However, nitration under mild conditions gives a mononitro derivative whose UV and NMR spectra indicate close similarity with 10′-nitro-vobtusine. Nitrocallichiline is, therefore, established to be a 5-nitro-7-methoxyindoline (**432**). The NH resonance of the β-anilinoacrylate unit was observed at 8.98 and 8.99 ppm in callichiline and 10-nitrocallichiline, respectively. In the spectrum of the latter compound an additional, exchangeable NH function was observed at 4.60 ppm.

Some of the simple chemical transformations of callichiline are shown in Scheme 21. Treatment with acid gave a decarbomethoxy derivative (**433**), indicating that the missing oxygen atom is the one at C-2 in vobtusine. As expected, sodium borohydride reduction of **433** gave two products, a bis-indoline (**434**) and an indole–indoline (**435**), the latter being produced by a retro-Mannich reaction resulting in Smith cleavage of the 7,21-bond. The remaining two oxygen atoms were deduced to be ether linkages from spectral evidence and resistance to chemical (LAH and zinc–acetic acid) reduction.

Like vobtusine, callichiline shows fragment ions at *m/e* 138 (**436**), *m/e* 110 (**437**), and *m/e* 214 (**438**), indicating the presence of a "beninine (**396**) half" and a "vincadifformine-like half." The difference of one carbon atom between vobtusine and callichiline is accounted for by C-23, for callichiline shows two NH protons. The aromatic region of the NMR spectrum is similar to that of vobtusine, and this excludes any linkages to the aromatic nucleus. Thus, on biogenetic grounds, callichiline was suggested to be **439**, in which the two units are joined with the formation of an additional ring through any of the starred carbons, as indicated. The further arguments for the structure of callichiline are based mainly on a detailed evaluation of mass spectral data (*217*).

The presence of an ion at *m/e* 144, regarded as **440**, eliminates C-5′ as a linking point. In addition, there is no $M^+ - 28$ ion (loss of C-16 and C-17 from beninine unit) or fragments which can be formulated as **394** and **395**. Thus either C-2 or C-6 is linked with C-16, C-17, or C-22 to unit A.

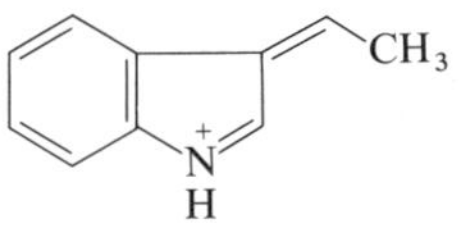

440 *m/e* 144 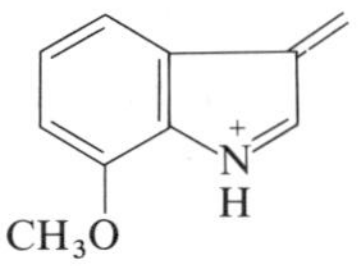 **394** *m/e* 160 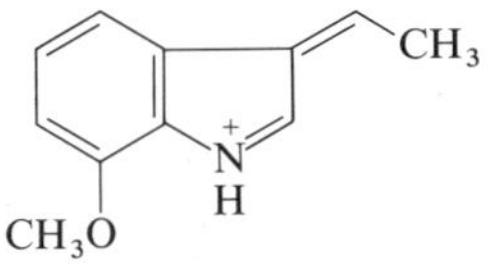 **395** *m/e* 174

In the spectrum of callichiline and derivatives there is an ion at *m/e* 324 which is attributed structure **441**. The easy loss of C-22 indicates that it is not bound to any carbon other than C-16 in unit B, and the resistance of the indoline nitrogen to formylation suggests that C-2 is fully substituted.

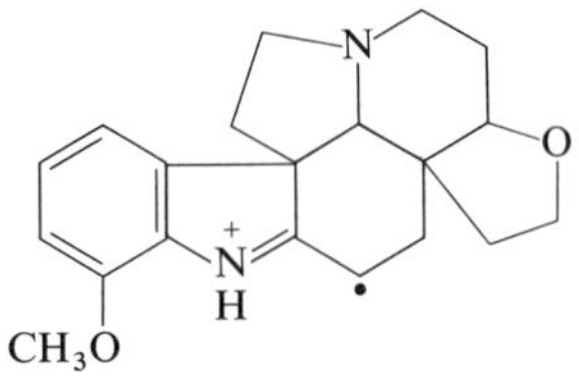

441 *m*/*e* 324

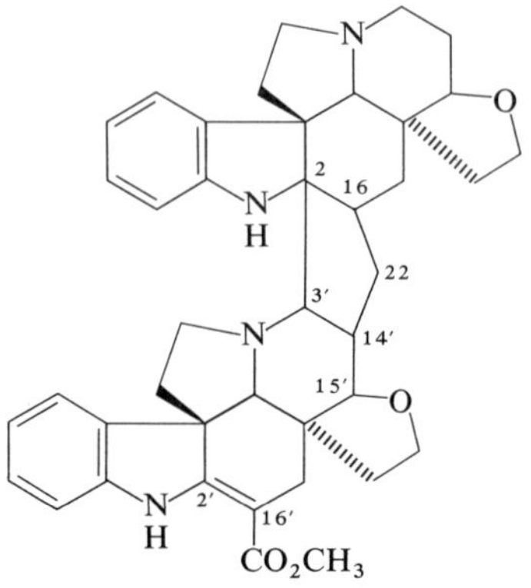

442 Callichiline

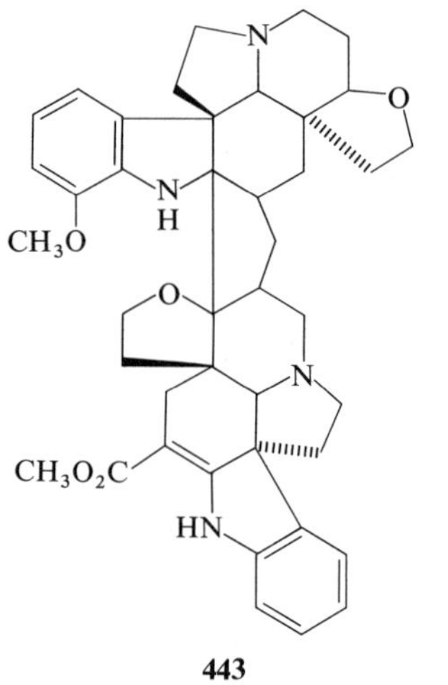

443

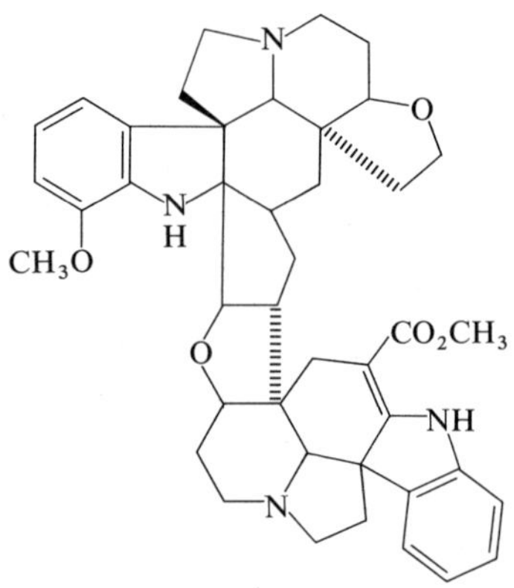

444

Thus, C-2 and C-22 are considered the points of linkage between unit A and unit B, and three structures (**442**–**444**) were proposed for callichiline, of which again structure **442** was preferred.

Although callichiline has no ethyl side chain, losses of ethyl radical with supporting metastable ions are found in many of the derivatives of callichiline. Neither beninine (**396**) nor vobtusine (**390**) shows such a loss, indicating that it probably arises because of the nature of the linkage between the two units. Using structure **442**, a mechanism has been proposed for this process (Scheme 22) and has been cited as evidence that C-2 is fully substituted.

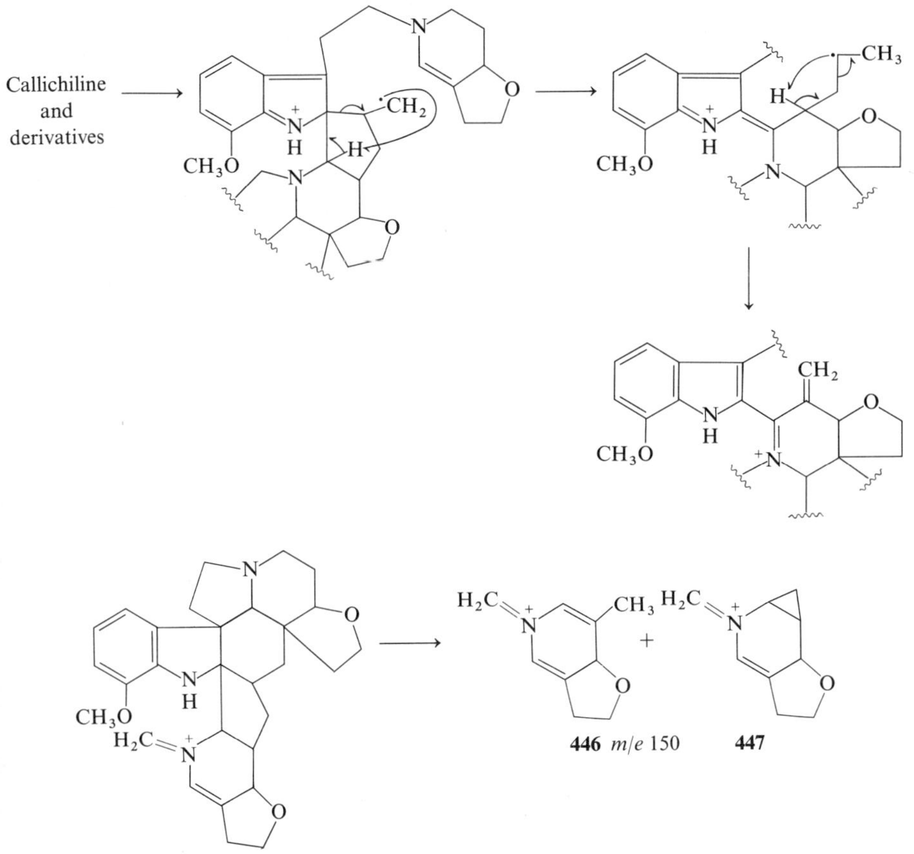

Callichiline and derivatives

445 *m/e* 474

446 *m/e* 150 447

SCHEME 22.

Evidence for the presence of a five-membered ring ether in unit A is also derived from the mass spectrum. Loss of *m/e* 214 (**438**) leads to an ion at *m/e* 474 (**445**) which fragments directly to an ion at *m/e* 150 (metastable at *m/e* 47.4). This ion may be formulated as either **446** or **447**, depending on the fate of the C-14′ proton. In addition, there is an intense fragment at *m/e* 349 in the spectrum of callichiline which shifts to *m/e* 265 in the spectrum of 2′,16′-dihydrocallichiline. This shift may be rationalized in terms of structure **448** for *m/e* 349 and structure **449** for *m/e* 265 (Scheme 23), and supports a five-membered ether rather than a bridged ether to some other position.

Thus, although structure **442** is favored for callichiline, no definitive proof is available (*217*, *218*).

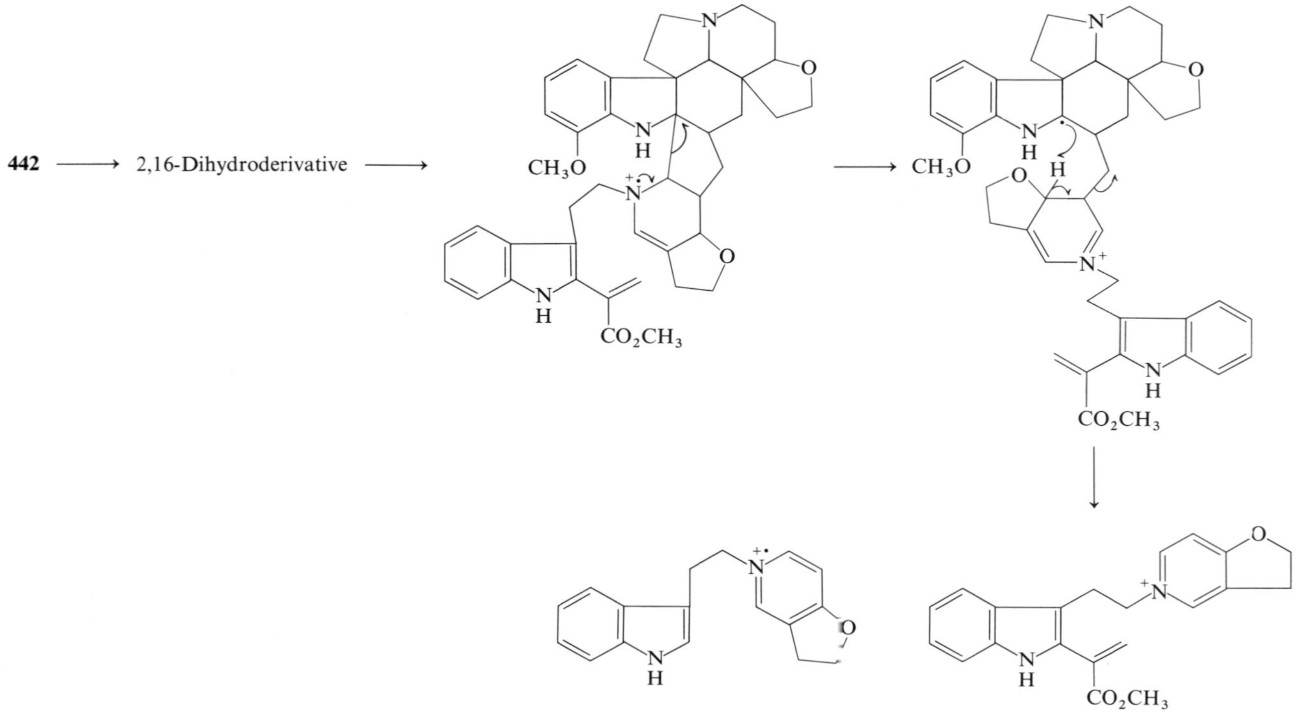

442 ⟶ 2,16-Dihydroderivative ⟶

CH_3O N H $\overset{+\cdot}{N}$ N H CO_2CH_3 O O N

⟶ CH_3O N H H O N^+ N H CO_2CH_3 O N

449 m/e 265

448 m/e 349

SCHEME 23. *Summary of the mass spectrum of calechiline.*

O. VOAFOLIDINE (**454**)

The leaves of *Voacanga africana*, in addition to yielding such bisindole alkaloids in the vobtusine series as 2-deoxyvobtusine, 2-deoxy-18-oxovobtusine, 18-oxovobtusine, and vobtusine, also afford a series of structurally slightly different bisindole alkaloids (*208*).

Voafolidine (alkaloid IX), obtained as a minor constituent, crystallized from methanol, $[\alpha]_D$ $-300°$, but as a result of decomposition at about 200°, it showed no melting point. The UV spectrum was similar to those of members of the vobtusine series (λ_{max} 260, 300, and 325 nm), and the IR spectrum indicated the presence of a β-anilinoacrylate moiety (ν_{max} 3410, 1690, and 1615 cm^{-1}) as well as a hydroxy group (3450 cm^{-1}) (*208*).

In the NMR spectrum, a three-proton singlet was observed for the carbomethoxy group at 3.80 ppm, but no aromatic methoxy group was observed. Instead, eight aromatic protons appeared in the region 6.3–7.3 ppm, demonstrating that the indoline and anilinoacrylate nuclei were not further substituted. The NH proton was observed as a singlet at 8.8 ppm and the "vobtusine doublet" at 4.4 ppm ($J = 13$ Hz), itself an excellent indication of the general structure class. The main difference in the aliphatic region on comparison with vobtusine was the observation of a triplet ($J = 7$ Hz) at 0.88 ppm indicating the presence of an ethyl side chain. Biogenetically, this unit should be attached at C-20.

The mass spectrum supplied the key to establishing the relationship of voafolidine to alkaloids in the vobtusine series, for analogous fragmentation ions were observed.

The molecular ion at *m/e* 688 ($C_{42}H_{48}N_4O_5$) is 30 mu less than that of vobtusine (loss of aromatic methoxy group) and suggests that the remaining unit is isomeric with vobtusine. Ions b′, c′, e′, and g′ were also shifted 30 mu, and the $M^+ - 18$ ion (ion a) gave ions d′ and f′ also shifted by 30 mu. Ions h′, i′, and j′ (*m/e* 363, 305, and 138, respectively) were not shifted, demonstrating that the β-anilinoacrylate unit is probably similar to the one in vobtusine and that ion j′ must be isomeric with the corresponding ion from vobtusine (*205*) (see Scheme 24 and Table XIV).

Considering the presence of an ethyl group in the molecule from the NMR spectrum and the co-occurrence (*208*) with the monomeric alkaloid voaphylline (**450**), ion j′ was envisaged as having structure **451**.

Treatment of voafolidine with concentrated HCl gave a product whose UV spectrum (λ_{max} 228 and 272 nm) indicated the presence of an indolenine and whose mass spectrum indicated that hydrolysis of the epoxide had occurred with introduction of chlorine. On this basis and on comparison with the former structure of vobtusine, voafolidine was assigned structure **452**, and the indolenines produced by hydrolysis were assigned structures

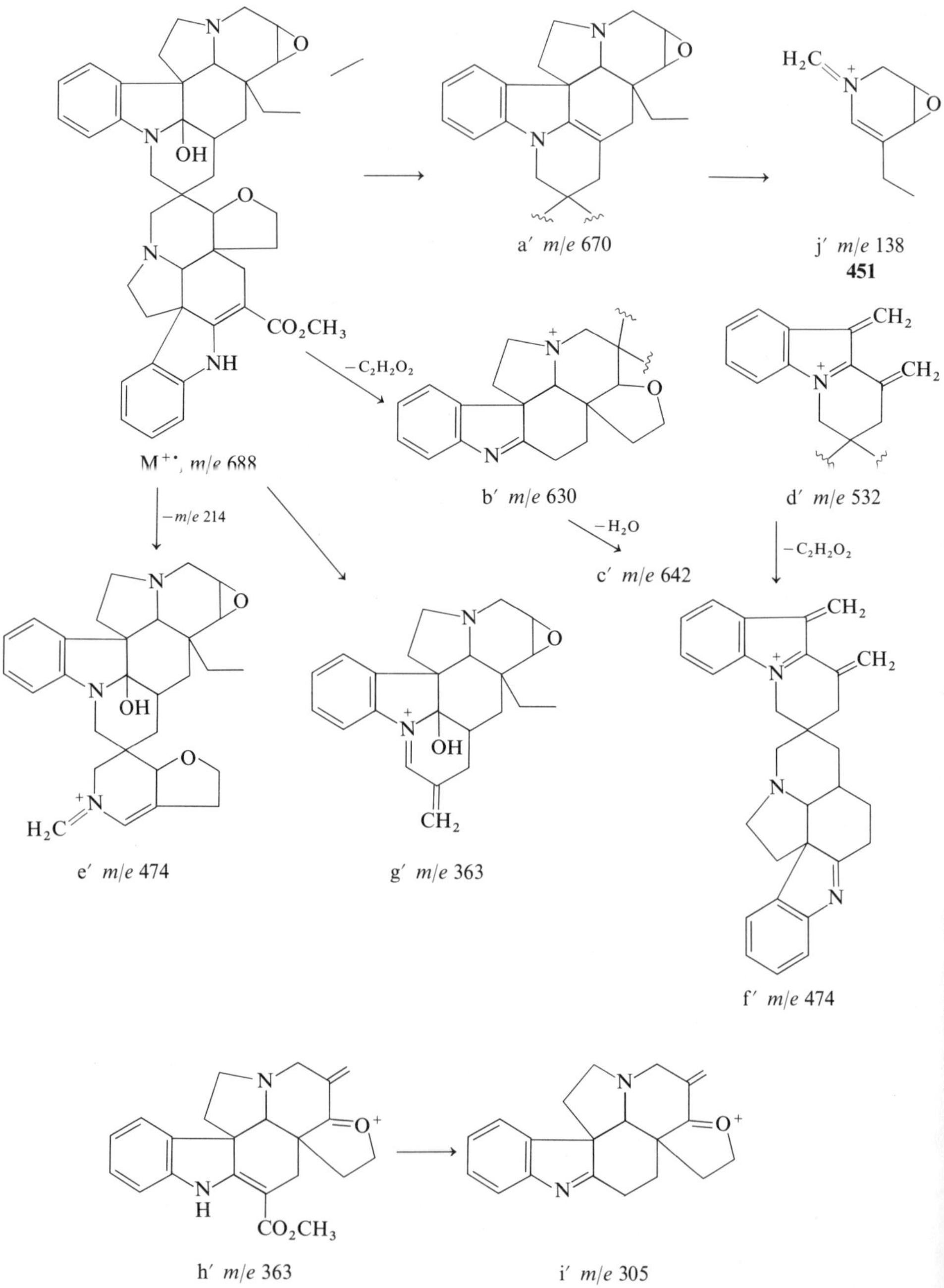

N O N OH O N CO_2CH_3 NH

M$^{+\cdot}$, m/e 688

N O N

a′ m/e 670

H_2C $\overset{+}{N}$ O

j′ m/e 138

451

$-C_2H_2O_2$

$\overset{+}{N}$ O N

b′ m/e 630

CH_2 $\overset{+}{N}$ CH_2

d′ m/e 532

$-m/e$ 214

$-H_2O$

c′ m/e 642

$-C_2H_2O_2$

N O N OH O $\overset{+}{N}$ H_2C

e′ m/e 474

N O $\overset{+}{N}$ OH CH_2

g′ m/e 363

CH_2 $\overset{+}{N}$ CH_2 N N

f′ m/e 474

N O^+ N H CO_2CH_3

h′ m/e 363

N O^+ N

i′ m/e 305

SCHEME 24.

TABLE XIV

PRINCIPAL MASS SPECTRAL FRAGMENTS OF VOAFOLIDINE AND RELATED ALKALOIDS

Compound	M^+	a′	b′	c′	d′	e′	f′	g′	h′	i′	j′
Voafolidine (**455**)	688	670	630	612	532	474	474	363	363	305	138
Voafoline (**464**)	672	—	—	—	—	458	—	347	363	—	138
Isovoafoline (**465**)	672	—	614	—	—	458	—	347	363	305	138
Folicangine (**468**)	686	—	628	—	—	472	—	—	363	305	138

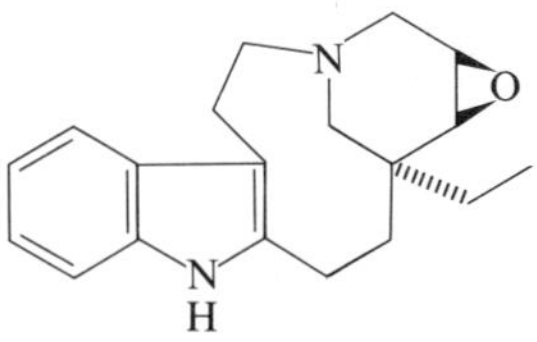

450 Voaphylline

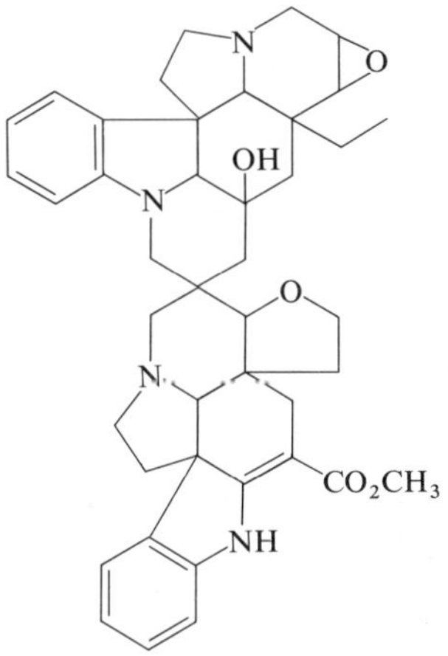

452 Voafolidine (original structure)

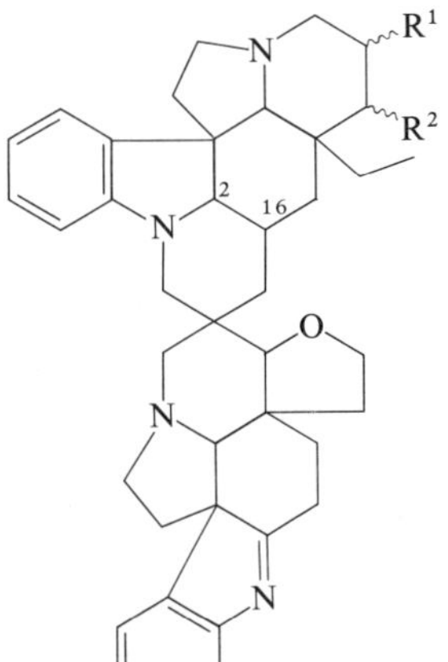

453/454 $R^1 = OH, R^2 = Cl$ or $R^1 = Cl, R^2 = OH$
459/460 $R^1 = OH, R^2 = Cl$ or $R^1 = Cl, R^2 = OH$

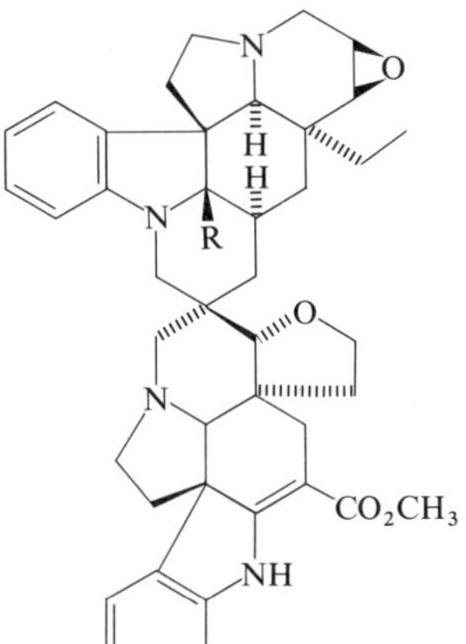

455 R = OH Voafolidine (revised)
464 R = H Voafoline (revised)

453 and **454**. No stereochemical assignments were made (*205*). The structure of voafolidine was subsequently revised, and its stereochemistry determined to be that shown in **455** (*210*).

P. VOAFOLINE (464) AND ISOVOAFOLINE (465)

Voafoline and isovoafoline were obtained from the leaf bases of *Voacanga africana*. The latter alkaloid is one of the major bases obtained (*205*).

Isovoafoline crystallized from methanol, mp 232° (dec.), $[\alpha]_D$ −567°, and, like voafolidine, showed a UV spectrum (λ_{max} 262, 303, and 328 nm) quite similar to that of vobtusine. From the difference UV spectrum with tabersonine, the presence of an *N*-alkylindoline was deduced. The β-anilinoacrylate unit suggested by the UV spectrum was confirmed by the IR (ν_{max} 3390, 1675, and 1615 cm^{-1}) and NMR [singlets at 8.9 (1H, NH) and 3.7 (3H) ppm] spectra. Eight aromatic protons were observed in the region 6.6–7.3 ppm, indicating the presence of unsubstituted indoline and β-anilinoacrylate units. An ethyl side chain was deduced from the observation of a triplet ($J = 7$ Hz) at 0.91 ppm. The "vobtusine doublet" ($J = 14$ Hz) was observed at 4.15 ppm.

In the mass spectrum of isovoafoline a molecular ion was observed at *m/e* 672, analyzing for $C_{42}H_{48}N_4O_4$ and indicating a loss of one oxygen atom compared with voafolidine. No $M^+ - 18$ ion was observed, and ions e′ and g′ were apparent at 16 mu less than in **455**. On the other hand, ions h′, i′, and j′ showed no such loss; consequently, the oxygen atom should be missing from C-22. The presence of an ethyl side chain in the compound again suggested that a C-14,C-15-epoxide was present in the molecule's upper half, and consequently structure **456** was proposed for isovoafoline (*205*).

Treatment of isovoafoline (**456**) with acetic acid under reflux gave a mixture of two isomers having the molecular formula $C_{44}H_{52}N_4O_6$ and regarded as the hydroxyacetoxy derivatives **457** and **458** obtained by nucleophilic opening of the epoxide group (*205*). Isomer **457** showed mp 205°–210° and $[\alpha]_D$ −450°, whereas the other isomer, **458**, displayed mp 195°–200° and $[\alpha]_D$ −455°. Each compound indicated the presence of hydroxy (ν_{max} 3500 cm^{-1}) and acetoxy (1740 and 1245 cm^{-1}) groups and showed the presence of a —$CHOCOCH_3$ grouping in the NMR spectrum (one-proton multiplets at 5.0 and 4.8 ppm, singlets at 2.10 and 2.13 ppm). In their mass spectra, the derivatives showed addition of acetic acid (60 mu) in ions b, e, g, and c compared with isovoafoline (*205*).

In 11 *N* HCl on a water bath, isovoafoline gave a mixture of chlorohydrin isomers (**459** and **460**) in which both decarbomethoxylation and cleavage of the epoxide had occurred.

Voafoline crystallized from methanol, mp 325° (dec.), $[\alpha]_D$ $-310°$, was isomeric with isovoafoline, and exhibited a molecular ion (and base peak) at *m/e* 672 which analyzed for $C_{42}H_{48}N_4N_4$ (*208*). The UV and IR spectra established the presence of a nucleus resembling that in isovoafoline, showing λ_{max} 263, 300, and 328 nm and ν_{max} 3395, 1690, and 1615 cm^{-1}, respectively. The NMR spectrum again showed eight aromatic protons in the region 7.3–6.3 ppm and the presence of an ethyl side chain from a three-proton triplet ($J = 7$ Hz) at 0.85 ppm. The only distinguishing features were the characteristic doublet ($J = 14$ Hz) at 4.60 ppm and the singlets for the β-anilinoacrylate at 8.95 (1H) and 3.80 ppm (3H) (*205*, *208*).

The mass spectrum showed the same overall fragmentation pattern as that of isovoafoline, although the ion at *m/e* 458 (ion e′) was considerably more intense in the spectrum of isovoafoline than in that of voafoline (*205*).

It is worth mentioning at this point that one of the ways in which the voafolidine series differs from the vobtusine series is in the region below *m/e* 138. Thus, in the vobtusine series a prominent (20% or greater of base peak) ion is observed at *m/e* 110 and formulated as **461**. In the voafolidine series no such ion is observed; instead, the ion at *m/e* 108 is typically quite substantial. This ion has been suggested to have the structure **462**.

Treatment of voafoline with glacial acetic acid at reflux for 2 hr gave a single crystalline product, mp 250° (dec.), $[\alpha]_D$ $-221°$, having the same UV spectrum as voafoline but showing in the IR spectrum an additional ester carbonyl at 1740 cm^{-1}. The NMR spectrum was very similar to that of voafoline, except that a three-proton singlet was apparent at 2.10 ppm and a one-proton multiplet at 5.0 ppm. These data indicate that the epoxide

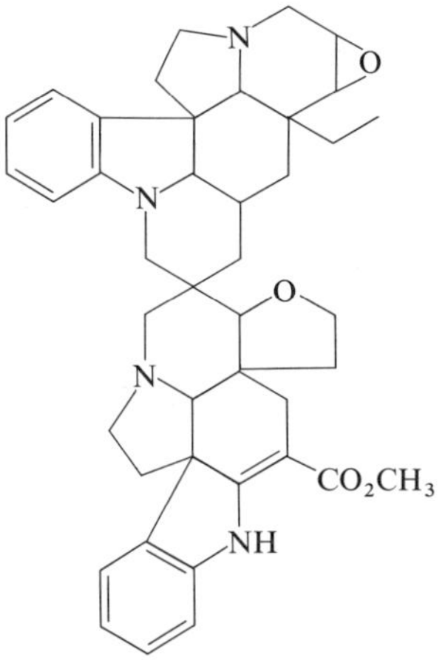

463 Voafoline (original structure)
456 Isovoafoline (original structure)

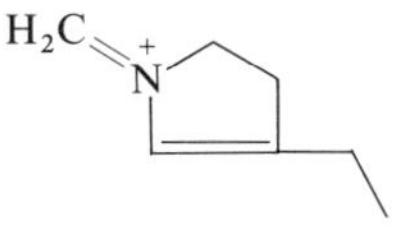

461 *m/e* 110

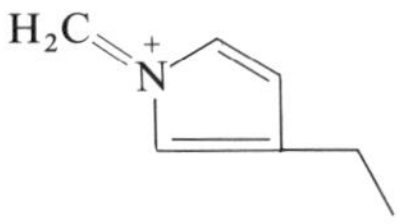

462 *m/e* 108

457 $R^1 = OCOCH_3$, $R^2 = OH$
458 $R^1 = OH$, $R^2 = OCOCH_3$

465 R = H Isovoafoline (revised)
475 R = OH Isovoafolidine (revised)

group has been cleaved with the addition of acetic acid and the molecular ion as well as ions e′ (base peak), g′, and l′ were shifted by 60 mass units, whereas ions h′ and i′ were unaffected (*205*). The compound did not correspond to either of the hydroxy acetates (**457** and **458**) from isovoafoline. These data indicated that voafoline (**463**) and isovoafoline (**456**) were stereoisomers, although the nature of the stereoisomerism could not be determined at the time. Subsequent examination of the ^{13}C-NMR spectra of these compounds (*210*) confirmed the structure and established the stereochemistry of voafoline to be **464** and that of isovoafoline to be **465**; i.e., they are stereoisomers at C-14′.

Q. Folicangine (468)

Folicangine, $[\alpha]_D$ $-271°$, from the leaves of *Voacanga africana* (*208*) crystallized from methanol but gave no melting point below the decomposition point of 200°. The IR spectrum, with absorptions at 3395 (NH) and 1685 and 1610 cm^{-1} ($C{=}C{-}CO_2CH_3$), indicated the presence of a β-anilinoacrylate moiety, and this was confirmed by the NMR spectrum which showed a singlet at 8.72 ppm for the NH and at 3.65 ppm for the carbomethoxy group (*205*, *208*).

The UV spectrum differed slightly from that of the other dimeric *V. africana* alkaloids in that the band at 260 nm had undergone a hypsochromic shift to 255 nm. The other longer wavelength absorptions were at their characteristic positions of 298 and 328 nm (*205*, *208*). Eight aromatic protons were observed in the region 6.3–7.3 ppm and a three-proton triplet ($J =$ 7 Hz) for the hydrogens on C-18′ at 0.82 ppm. Two one-proton singlets were observed at 3.96 and 4.60 ppm. The "vobtusine doublet" was not observed (*205*, *208*).

The molecular ion at *m/e* 686 analyzed for $C_{42}H_{46}N_4O_5$, two hydrogens less than voafolidine. Significantly, no water was lost from the molecular ion, and the ions b′ and e′ appeared at 2 mu less than the corresponding ions in voafolidine. Because ions h′, i′, and j′ remained at the same mass and no carbonyl or double bonds were observed, it was suggested that the oxygen and additional degree of unsaturation were located in the ring joining the two units. Consideration of the above evidence, and the knowledge at the time that voafolidine had structure **452**, led the French group to propose structure **466** for folicangine (*205*). The close relationship to the original structure proposal for amataine should be evident.

Such a structure should be readily attacked at the carbinolamine moiety by a hydride reducing reagent; indeed treatment with potassium borohydride gave a crystalline product, mp 232° (dec.), $[\alpha]_D$ −642°, in high yield.

The UV spectrum of the product now showed λ_{max} 258, 300, and 328 nm, quite close to that of the other dimeric *Voacanga* alkaloids. The mass spectrum indicated a molecular ion at *m/e* 688, isomeric with voafolidine (**452**); indeed, the spectra were extremely similar in their overall pattern. However, two ions—*m/e* 474 and 363 (e′/f′ and g′/h′)—were of quite different intensities. Thus in voafolidine, their percentages were 100% and 35%, respectively, but in the reduction product they were 9% and 100%, respectively. The NMR spectrum showed signals at 8.88 and 3.70 ppm for the β-anilinoacrylate and at 0.96 ppm for the C-18 proton. The only other relevant feature of the spectrum was a return of the characteristic doublet ($J = 14$ Hz) at 3.96 ppm, shifted upfield (by 0.4 ppm) from its position in voafolidine (*205*).

The compound was suggested to be an isomer of voafolidine (**452**) and was given the name isovoafolidine (**467**). The 300° difference in optical rotation between the two compounds was noted (*205*).

When the reduction of folicangine was carried out with sodium borodeuteride, a monodeutero product M^+ *m/e* 689 was obtained and ions a′–f′ were all shifted by 1 mu. However, although h′ was shifted, g′ was not. Because an $M^+ - H_2O$ ion was still observed in the spectrum of the monodeutero derivative, this finding specifically excludes a 2,16-oxygen bridge (*205*).

After the revision of the structure of vobtusine, attention was turned to other related alkaloids, including folicangine (*214*), for it became apparent that like vobtusine, voafolidine should have a hydroxy group at C-2′. Thus, attachment of the ether linkage at C-23 was thought unlikely, and positions 3′, 5′, and 21′ were considered possible sites for the carbinolamine ether. Location of oxygen at any one of these positions would be in agreement with the mass spectral data. However, molecular models indicated that of these only C-3′ was close enough to C-2 to form an ether bridge. In addition, *if* the 2-hydroxy group in voafolidine is β, such a bridge could only be formed if the spiro center at C-14′ has the *R*-configuration. On the basis of this reasoning, structure **468** was suggested for folicangine (*214*), without stereochemistry for the epoxide linkage.

Such a structure is in agreement with ^{1}H-NMR data, where a singlet at 4.6 ppm can now be attributed to the C-3′ α protons. This compares with chemical shifts of 4.75 and 4.78 ppm for this signal in amataine (**426**) and 18-oxoamataine (**431**), respectively (*214*).

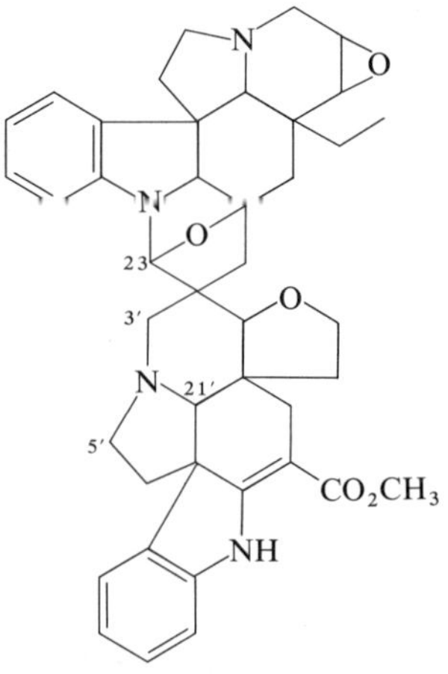

466 Folicangine (original)

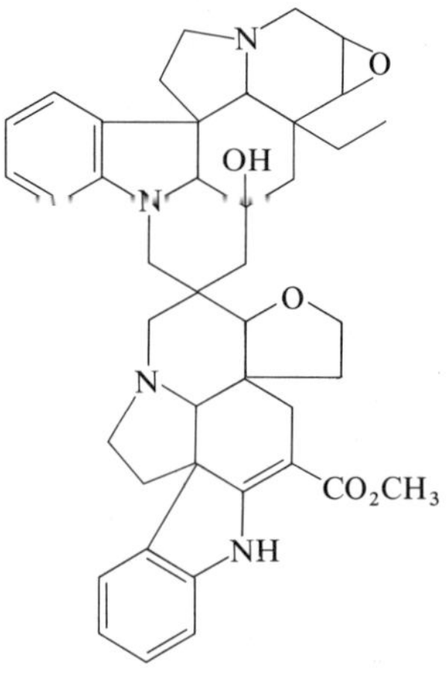

452 Voafolidine (original)
467 Isovoafolidine (original)

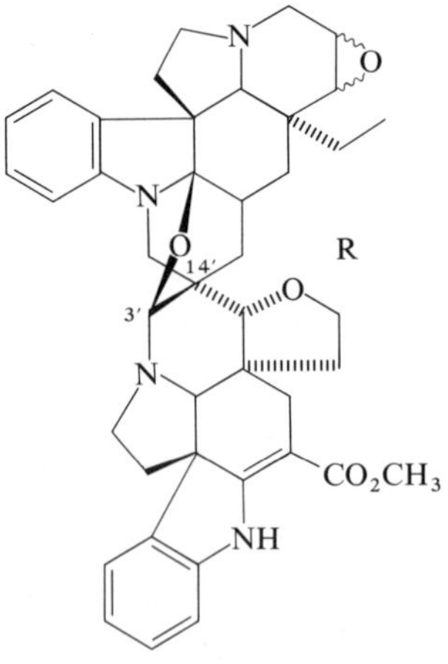

468 Folicangine (revised)

In a study of the structure of folicangine using ^{13}C-NMR, the 14,15-epoxide group was determined to be β, as was the C-3′ proton (*210*) (see Section XXI,R).

R. ^{13}C-NMR Spectra of Spiro *Aspidosperma–Aspidosperma* Alkaloids

Wenkert and co-workers, in collaboration with Poisson and co-workers, have reported on the ^{13}C-NMR spectra of the major compounds in this series (Table XV) (*210*).

In beginning to make assignments, comparison was made with vandrikine (**469**), tabersonine (**470**) and N_a-methyl-2β,16β-dihydrotabersonine (**471**), and by applying methoxy substitution parameters the ring A carbon shifts could be assigned.

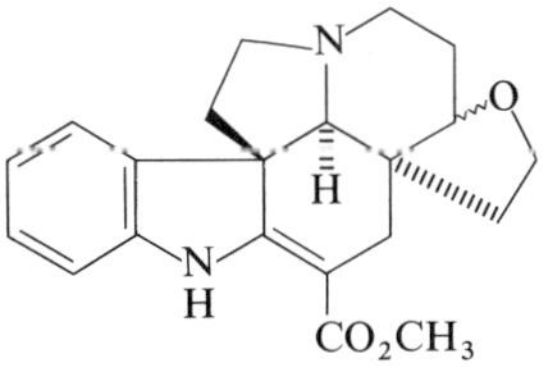

469 Vandrikine

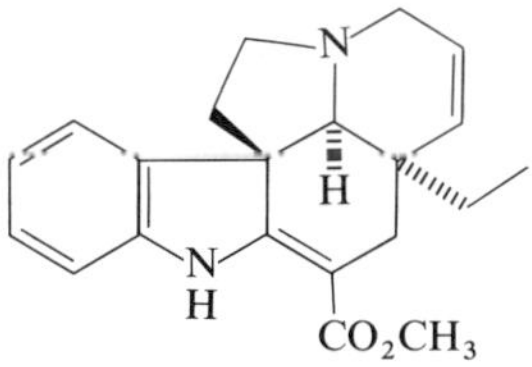

470 Tabersonine

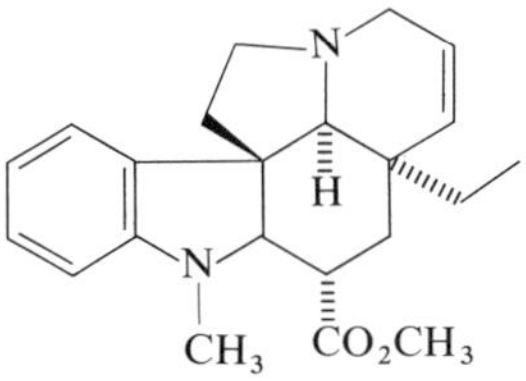

471

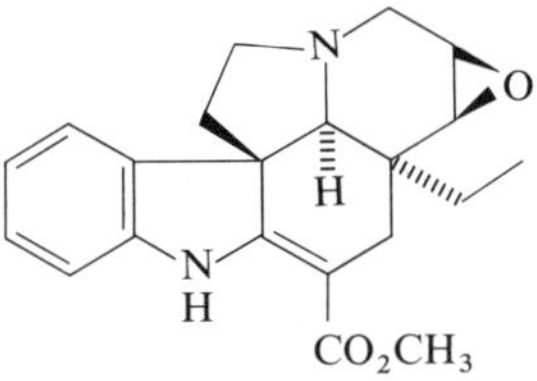

472 Pachysiphine

As in most of the terpenoid indole alkaloids, the major problem is making a distinction between the various methylene signals, both for the C—CH_2— and the N—CH_2— groups, e.g., C-6, C-17, C-19, C-22, and C-19′, and the series C-3, C-5, C-23, C-3′, and C-5′.

Comparison of the spectrum of vobtusine (**390**) with those of 18-oxovobtusine (**408**) and 2-deoxyvobtusine (**400**) permitted a distinction to be made for the most difficult assignments. Thus, whereas C-3 and C-23 have similar chemical shifts in vobtusine (**390**), only C-3 is shifted in 18-oxovobtusine (**408**). Similarly, the C-methylene series are distinguished by the shifts observed in the spectra of **400** and **408**. Note, for example, the shifts of C-6 and C-22 in **408** and of C-19 vs. C-19′ in **400**. The assignment

TABLE XV

^{13}C-NMR CHEMICAL SHIFTS OF VOBTUSINE AND RELATED ALKALOIDS[a,b]

Position	A	B	C	D	F	G	H	I	L	J	K	M	N
C-2	167.4	93.7	93.3	75.6	93.3	91.3	94.4[b]	94.3[d]	164.9	93.9[e]	76.0	94.1	76.9
C-3	45.7	48.7	48.0	47.8	48.7	58.9	48.8	48.2	49.4	53.4[f]	53.0	53.5[g]	53.1[h]
C-5	51.2	51.9	51.8	52.7	51.8	66.9	52.0	51.9	51.0	52.6[f]	53.0	52.8[g]	52.7[h]
C-6	45.1	31.1	30.8	37.7	51.9	44.9	31.1	31.3	43.9	31.8	39.0	31.7	39.1
C-7	54.2	55.9	55.8	51.0	55.8	54.9	56.4	56.4	54.7	55.4	51.5	55.5	51.6
C-8	—	134.2	133.0	135.6	133.0	132.3	134.2	133.1	137.5	133.8	135.6	133.5	135.9
C-9	—	114.5	114.6	114.8	113.9	112.7	113.5	113.6	121.3	121.8	120.8	121.2	131.0
C-10	—	118.1	118.8	118.3	118.4	117.9	118.5	119.4	120.3	117.4	116.2	117.4	116.3
C-11	—	110.8	111.1	110.9	112.2	112.7	110.3	110.6	127.6	127.3	127.0	127.1	126.8
C-12	—	144.9	144.9	145.1	146.8	146.8	146.0	146.1	109.2	107.3	106.3	109.0	108.1
C-13	—	137.2	136.9	137.9	135.1	135.2	136.8	136.4	142.9	148.8	149.8	148.3	149.4
C-14	26.6	25.7	24.7	25.3	25.6	22.6	25.5	25.0	52.0	53.0	52.5	53.0	52.5
C-15	79.8	80.3	81.4	82.0	80.3	78.5	80.4	81.4	56.2	56.7	56.5	56.6	56.4
C-16	93.9	31.5	31.1	29.4	31.9	31.0	32.4[c]	32.0	90.4	28.5	25.6	28.8	26.1
C-17	27.4	32.4	31.7	33.7	32.5	33.4[a]	32.6[c]	32.0	23.5	28.1	29.9	29.6	29.6
C-18	64.7	65.1	175.3	175.5	65.2	64.8	65.3	175.4	7.1	7.6	7.3	7.6	7.3
C-19	34.6	36.6	41.4	41.5	36.2	33.9[a]	36.8	41.7	26.5	28.1	28.0	28.1	27.9
C-20	46.4	44.1	43.4	44.0	43.9	39.2	44.4	43.5	37.0	36.1	35.9	36.3	36.0
C-21	68.7	63.6	63.9	65.3	63.5	72.2	63.7	63.6	70.9	66.3	67.5	66.5	67.9
C-22	—	34.1	33.5	39.0	29.4	29.3	35.1	34.9	—	33.2	38.1	34.8	39.9
C-23	—	46.1	46.0	52.7	44.1	44.1	46.5	46.3	—	44.9	48.2	44.2	49.1
OCH_3	—	55.0	55.0	54.9	56.1	56.2	55.0	55.0	—	—	—	—	—

	E											
C-2′	166.7	166.6	166.6	166.7	162.6	162.7	166.7	166.7	166.6	166.1	166.6	166.1
C-3′	—	53.7	53.8	53.8	171.2	171.5	58.0	58.3	53.7	53.4	58.0	57.7
C-5′	—	50.9	50.9	51.2	39.1	39.1	51.4	50.9	50.6	50.3	51.4	51.0
C-6′	—	44.9	44.9	45.0	44.9	45.2	44.4	44.4	44.1	44.6	46.4	44.1
C-7′	—	54.8	54.8	55.0	57.4	57.4	54.8	54.8	54.6	54.4	55.0	54.5
C-8′	137.8	137.6	137.5	137.5	137.5	138.1	137.9	137.8	137.5	137.1	137.8	137.4
C-9′	121.4	121.2	121.2	121.4	121.2	121.2	121.5	121.4	121.2	121.5	121.4	121.0
C-10′	120.5	120.4	120.4	120.5	120.8	120.8	120.5	120.5	120.5	120.2	120.5	120.2
C-11′	127.6	127.4	127.5	127.6	128.4	128.5	127.5	127.6	127.6	127.3	127.5	127.4
C-12′	109.2	109.1	109.1	109.2	109.5	109.6	109.1	109.2	109.2	108.9	109.1	108.8
C-13′	143.1	142.8	142.8	142.9	142.8	142.8	142.8	142.9	142.9	142.5	142.9	142.6
C-14′	—	39.6	39.5	40.0	49.0	48.9	38.9	38.9	40.4	40.3	39.0	38.8
C-15′	—	87.4	87.3	87.5	87.2	87.2	81.5	81.5	87.3	87.0	80.2	80.6
C-16′	92.2	94.3	94.2	94.1	92.4	92.6	94.3[b]	93.9[d]	94.1[e]	93.7	94.3	93.9
C-17′	—	27.3	27.3	27.5	27.1	27.4	28.1	28.4	27.7	27.4	27.9	27.9
C-18′	—	64.2	64.2	64.3	67.6	67.5	63.7	64.0	64.4	64.0	62.7	62.5
C-19′	—	34.8	34.8	34.8	36.2	36.3	35.1	34.9	34.9	34.6	34.8	34.6
C-20′	—	47.6	47.8	47.8	47.8	48.0	47.9	47.9	47.5	47.3	47.1	46.8
C-21′	—	68.9	68.8	68.8	67.1	66.9	69.9	69.9	69.8	69.4	70.0	69.6
C=O	168.8	168.3	168.4	168.4	167.9	168.0	168.1	168.2	168.0	167.8	168.1	167.6
OCH_3	50.8	50.9	51.0	51.0	51.0	51.0	50.9	50.9	51.0	50.6	50.9	50.5

[a] Legend: A = vandrikine (**469**); B = vobtusine (**390**); C = 18-oxovobtusine (**408**); D = 2-deoxyvobtusine (**400**); E = tabersonine (**470**); F = 3′-oxovobtusine (**416**); G = 3′-oxovobtusine *N*-oxide (**417**); H = isovobtusine (**424**); I = 18-oxovobtusine (**430**); J = voafolidine (**455**); K = voafoline (**464**); L = pachysiphine (**472**); M = isovoafolidine (**475**); N = isovoafoline (**465**). δ (ppm) from TMS.

[b] Superscripts *a–h* indicate assignments may be reversed. Data from Rolland *et al.* (*210*).

of C-6 and C-22 could be made with some certainity only when 3′-oxovobtusine (**416**) and 3′-oxovobtusine *N*-oxide (**417**) were examined. As expected, in the latter compound C-6 is dramatically deshielded on the introduction of an *N*-oxide, whereas C-22 is unaffected.

The structure of 3′-oxovobtusine (**416**) was confirmed by examination of the shift of C-14′ between this compound and vobtusine (**390**). Similarly, the *N*-oxide derivative **417** showed characteristic shifts for C-3, C-5, and C-21 on introducing this group into the upper unit.

Turning to the *Voacanga africana* alkaloids of the voafolidine type in which the 15-oxygen substituent is attached not to C-18 but to C-14, comparison was made with the alkaloid pachysiphine (**472**) in order to establish the chemical shift of the ring D carbons. Because the lower unit was visualized as being identical with that of vobtusine (**390**), chemical shifts of this portion of the molecule could be used in the voafolidine series. In this way, definitive assignments could be made for all carbon atoms except C-2 and C-16′ and C-3 and C-5

In voafolidine (**455**), C-17 is shielded with respect to the corresponding carbon in vobtusine (**390**) because of the proximity of the epoxide. Indeed, this evidence and the close correspondence of the shift of C-14 and C-15 to pachysiphine (**472**) serve to establish the stereochemistry of the epoxide as *β*. Similarly, C-23 in voafolidine (**455**) is also slightly shielded with respect to C-23 in vobtusine (**390**); in this case, the shift is due to removal of the 12-methoxy group.

The location of a 2-hydroxy group in voafolidine is substantiated both by the chemical shift of this carbon (94.1 or 93.9 ppm) and by the shift to 76.0 ppm in voafoline. These close similarities with the data of vobtusine serve to establish the complete molecular structures for voafolidine and voafoline as **455** and **464**, respectively.

Isovoafoline was previously (*205*) shown to be a stereoisomer of voafoline, although the nature and location of this stereoismerism could not be deduced. Comparison with the spectral data of voafoline proves the overall identity of the compounds and indicates that the stereoisomerism is not due to a change in stereochemistry for the epoxide group for C-21 or C-21′, or for C-17. However, the chemical shifts of carbons 3′, 14′, 15′, 16, 22, and 23 were significantly different between the two compounds, clearly indicating the nature of the stereoisomerism to be the C-14′ spiran center. Therefore, isovoafoline could be ascribed structure **465** (*210*).

The largest shift differences on changing the C-14′ stereochemistry are a shielding effect for C-15′ of about 6 ppm and a deshielding effect on C-3′ of about 4.3 ppm. As the ring F conformations **473** and **474** indicate in voafoline (**473**), C-3′ experiences 1,3-diaxial interactions with N_a and C-16, but in isovoafoline (**474**) it is C-15′ which experiences these effects. As summarized

473 Voafoline

474 Isovoafoline

in Table XVI, the shift differences between C-3′ and C-15′ in the 14*R* and 14*S* series of compounds can be used with some confidence in assigning this stereochemical center. In this way, the stereochemistry at C-14′ in 14′-isovoafolidine, 14′-isovobtusine, and 18-oxo-14′-isovobtusine could be determined.

TABLE XVI
CHEMICAL SHIFT DIFFERENCES FOR C-3′ AND C-15′ IN THE 14*R* AND 14*S* SPIRO *ASPIDOSPERMA–ASPIDOSPERMA* ALKALOIDS

14′*R*	14′*S*	$\Delta\delta$ C-3′	$\Delta\delta$ C-15′
14′-Isovobtusine	Vobtusine	4.3	−5.9
18-Oxo-14′-Isovobtusine	18-Oxovobtusine	4.5	−5.8
14′-Isovoafoline	Voafoline	4.3	−6.4
14′-Isovoafolidine	Voafolidine	4.3	−7.1

Borohydride reduction of folicangine had afforded isovoafolidine, an isomer of voafolidine of unknown configuration. The ^{13}C-NMR spectrum indicated that its stereochemistry was the same as that in voafolidine with the exception of C-14′, thus this series parallels the voafoline–isovoafoline series. The stereochemistry of the epoxide group was shown by analogy with pachysiphine to be 14β,15β; therefore, voafolidine has the complete stereochemistry **455**, and 14′-isovoafolidine is represented by structure **475** (*210*).

In the ^{13}C-NMR spectrum of folicangine an aminomethylene is further substituted by oxygen, with the presence of a new signal at 91.8 ppm and the absence of the C-3′ methylene at about 58.0 ppm.

On sodium borohydride reduction, amataine had afforded isovobtusine, which again was determined by ^{13}C-NMR to have 14′*R*-stereochemistry and the complete structure **424**. The main difference between the spectra of 14′-isovobtusine (**424**) and amataine was also the addition of a methine carbon at 91.9 ppm and the absence of a methylene carbon at about 58.0 ppm.

Similarly, the relative data of 18-oxovobtusine (**408**), 18-oxo-14′-isovobtusine (**430**), and 18-oxoamataine (**431**) indicated that the stereochemistry at position 14′ in the latter two compounds was *R* and that, like folicangine and amataine, 18-oxoamataine had an ether linkage between C-2 and C-3′.

Although several of the methylene signals could not be assigned because of the substantial shift by comparison with the parent compound (Table XVII), the chemical shift of the residual signals indicated that they belong to the same stereochemical series. In other words, folicangine, amataine, and 18-oxoamataine have the same C-3′ stereochemistry.

If the C-3′ proton is α (as in **476**), ring D′ exists in a chair conformation and it might be expected that only the shifts of C-3′, C-15′, C-14′, C-5′, and C-21′ would be affected. In fact, formation of the ether bridge also shifts the resonances of C-6′, C-7′, C-17′, C-19′ by from 3 to 6 ppm. Wenkert has suggested (*210*) that such shifts can only be explained if the C-3′ proton is β. Such a configuration, i.e., **477**, would also require inversion of the N_b' lone pair, and would thereby account for the many shift changes induced when the ether bridge is formed. On this basis, folicangine was assigned structure **478** and amataine and 18-oxoamataine structures **479** and **480**,

TABLE XVII

^{13}C-NMR Chemical Shifts of Folicangine, Amataine, and 18-Oxoamataine[a,b]

Position	478	479	480	Position	478	479	480
C-2	93.9[a]	94.6[c]	93.9[e]	C-2′	165.8	165.9	165.7
C-3	53.7[b]	49.0	48.4	C-3′	91.8	91.9	91.9
C-5	53.3[b]	53.2	52.9	C-7′	59.1	59.1	g
C-6	30.5	31.1	31.5	C-8′	136.1	137.1	g
C-7	53.7	54.4	54.4	C-9′	121.4	122.4	122.3
C-8	135.9	136.0	135.8	C-10′	121.0	120.9	120.9
C-9	122.2	114.6	114.6	C-11′	127.8	127.7	127.8
C-10	118.1	118.8	119.1	C-12′	109.1	109.0	109.1
C-11	127.8	111.1	111.6	C-13′	142.9	142.9	142.8
C-12	106.9	145.8	145.9	C-14′	38.1	38.1	38.2
C-13	147.8	136.2	136.0	C-15′	87.5	87.6	87.5
C-14	52.9	26.1	25.2	C-16′	93.0[a]	92.8[c]	92.7[e]
C-15	56.5	80.9	81.9	C-18′	67.3	67.3	67.3
C-16	32.2	32.8	32.2	C-20′	49.9	49.9	50.0
C-17	29.6	34.4[d]	33.7[f]	C-21′	70.6	70.6	70.7
C-18	7.6	65.0	175.2	C=O	168.1	168.1	168.0
C-19	28.0	35.8	40.6	OMe	51.0	50.9	51.0
C-20	35.5	44.1	43.4	NCH_2	47.5, 51.2	49.0, 50.9	48.9, 50.8
C-21	67.7	65.8	65.9	CH_2	33.9, 38.0, 38.6, 41.6	34.0[d], 38.1, 38.6 41.5	34.0[f], 37.7, 38.[illegible] 41.4
OMe	—	55.6	55.6				

[a] δ (ppm) TMS = δ ($CDCl_3$ + 76.9 ppm).

[b] Superscripts *a–f* indicate assignment may be reversed; *g* indicates signal not observed. Data fro[m] Rolland *et al.* (*210*).

respectively (*210*). These exhibit C-3′ stereochemistries reverse to those previously assigned to folicangine, 18-oxoamataine (*214*), and amataine (*213*).

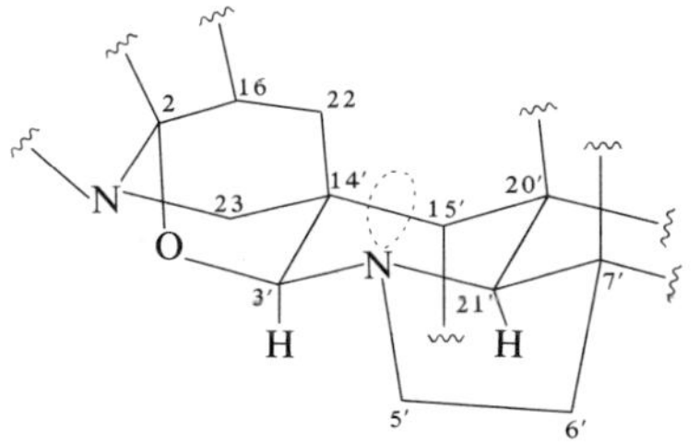

476 H-3′α

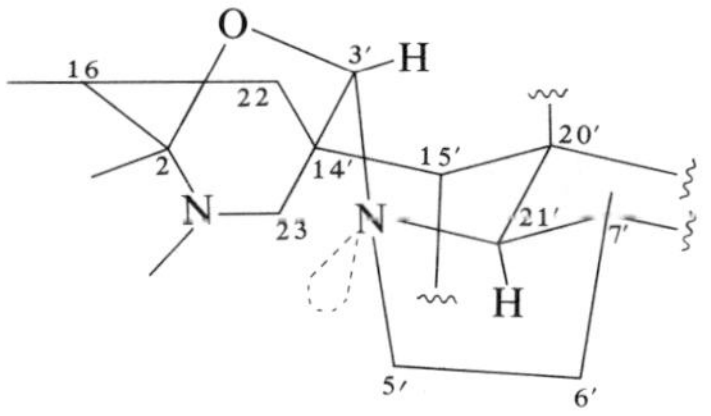

477 H-3′β

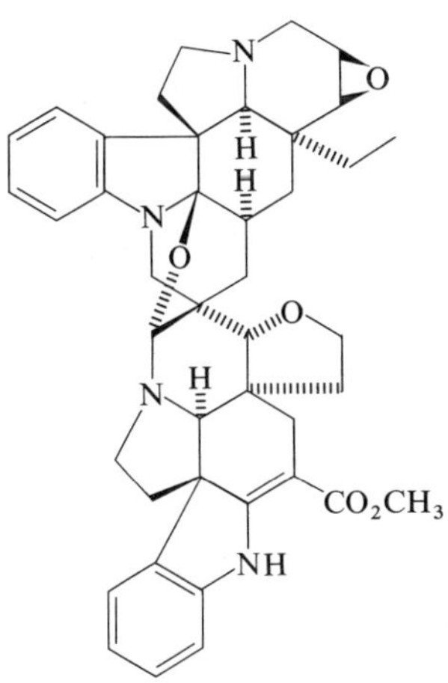

478 Folicangine

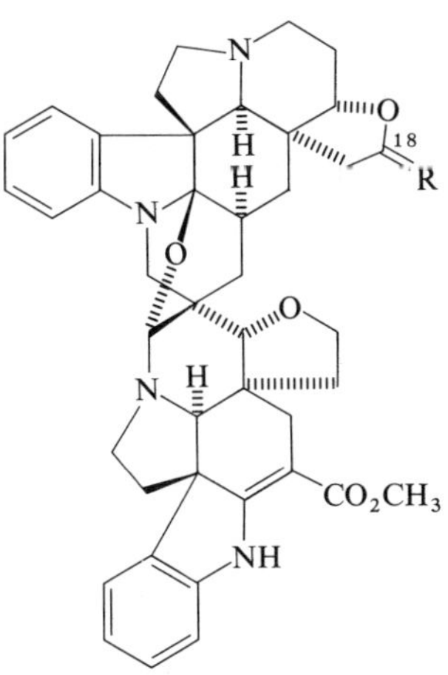

479 R = H_2 Amataine
480 R = O 18-Oxoamataine

S. Physical Data of Spiro *Aspidosperma–Aspidosperma* Alkaloids

In addition to the ^{13}C-NMR spectra of the spiro *Aspidosperma–Aspidosperma* alkaloids discussed in Section XXI,R, several other techniques have been brought to bear on the problem of structure elucidation in this series (*213*, *214*).

In almost all the ^{1}H-NMR spectra of compounds in this series (Table XVIII), a doublet ($J = 12–14$ Hz) is observed in the region 4–5.4 ppm. On several occasions, this doublet has been assigned to one of the protons on C-3. The origin of this signal has recently been re-evaluated (*213*).

TABLE XVIII
CHEMICAL SHIFT DATA FOR AN H-23 PROTON IN THE SPIRO *ASPIDOSPERMA–ASPIDOSPERMA* SERIES

Compound	δ (ppm)	*J* (Hz)	14′	Reference
Vobtusine (**390**)	5.13	14	*S*	(*197, 202, 213*)
Vobtusine *N*-oxide	5.13	14	*S*	(*213*)
Vobtusine *N,N′*-dioxide	—	—	*S*	(*213*)
Anhydrovobtusine (**391**)	4.38	12	*S*	(*213*)
Anhydrotetrahydrovobtusine	5.11	12	*S*	(*213*)
12-Demethylvobtusine (**399**)	4.83	14	—	(*202*)
2-Deoxyvobtusine (goziline) (**400**)	5.37	14	*S*	(*202, 205, 206, 213*)
2-Deoxy-3′-oxovobtusine (**418**)	4.8	13	*S*	(*203*)
3′-Oxovobtusine (**416**)	4.55	14	*S*	(*203*)
3′-Oxovobtusine *N*-oxide (**417**)	4.75	14	*S*	(*203*)
Amataine (subsessiline) (**426**)	4.75[a]	—	*R*	(*197, 207, 213*)
Dihydroamataine (isovobtusine) (**424**)	4.69	14	*R*	(*213*)
18-Oxoamataine (subsessiline lactone) (**431**)	4.78[a]	—	*R*	(*203, 214*)
Hydratoamataine (**425**)	5.07	14	*S*	(*213*)
Anhydrodihydroamataine (owerreine) (**420**)	4.55	12	*R*	(*213*)
Quimbeline (**421**)	4.95	14	*S*	(*211*)
Voafolidine (**455**)	4.4	13	*S*	(*205, 213*)
Isovoafolidine (**475**)	3.96	14	*R*	(*205, 213*)
Voafoline (**464**)	4.60	14	*S*	(*205, 213*)
Isovoafoline (**465**)	4.15	14	*R*	(*205, 213*)
Folicangine (**468**)	4.6[a]	—	*R*	(*205*)

[a] Tentatively assigned to H-3′.

Spin-decoupling (*202, 213*) and INDOR (*213*) experiments on vobtusine (**390**) established that the proton at 5.13 ppm was coupled to another proton at 3.10 ppm, indicating that each must be in the structural unit X—CH_2—C, where X may be N or O. Two possible sites are C-3′ and C-23; however, the former possibility is eliminated because 3′-oxovobtusine (**416**) and 3′-oxovobtusine *N*-oxide (**417**) also show this signal. On this basis, it was suggested (*213*) that this was one of the C-23 protons, although its stereochemistry was not assigned. In the voafolidine series, where no 12-methoxy group is present, the signal is less deshielded by about 0.7 ppm compared with vobtusine. There is also some evidence that compounds in the 14′*R* series have this proton less deshielded by about 0.5 ppm; compare, for example, vobtusine with isovobtusine, voafolidine with isovoafolidine, and voafoline with isovoafoline.

In the compounds having a C-2–C-3′ ether bridge—i.e., folicangine, amataine, and 18-oxoamataine—a singlet appears in the region where the doublet is usually observed. This proton, however, should be due to the

TABLE XIX
MOLECULAR ROTATION DATA FOR SOME SPIRO *ASPIDOSPERMA–ASPIDOSPERMA* ALKALOIDS

Compound	Stereochemistry	$[M]_D$	Compound	Stereochemistry	$[M]_D$
Vobtusine	14′*S*	−2527°	Isovobtusine	14′*R*	−3619°
2-Deoxyvobtusine	14′*S*	−2359°			
Voafolidine	14′*S*	−2064°	Isovoafolidine	14′*R*	−4417°
Voafoline	14′*S*	−2083°	Isovoafoline	14′*R*	−3810°

carbinolamine ether proton at C-3′. It is not clear at present why these derivatives and *N,N′*-dioxides of both vobtusine and 10′-nitrovobtusine do not show doublets in the region 4.5–5.1 ppm. In these cases, singlets were observed at 4.35 and 4.28 ppm, respectively, which were assigned to H-21 (*213*).

The ORD (*213*) and CD (*214*) spectra of several of the vobtusine-type alkaloids have been reported, but it is not clear at present whether these techniques can be used for the C-14′ stereochemistry. The only readily discernible pattern was in the voafolidine–isovoafolidine and voafoline–isovoafoline series. In this instance the 14′*S* series (e.g., **455**) showed double maxima at 320 ($\Delta\epsilon$ −15.3) and 331 nm (−17.0), whereas the 14′*R* series (e.g., **475**) showed only a single maximum in this region at 327 nm (−31.4) (*214*).

The more useful data for readily determining the stereochemistry at position 14′ appear to be the molecular rotation {$[M]_D$} values, as shown in Table XIX. From the limited data available, a value of about −2500° is a lower limit for the 14′*S* stereochemistry, and about −3500° is an upper limit for the 14′*R* series of compounds.

A combination of $[M]_D$ and ^{13}C-NMR shifts of positions 3′ and 15′ should, therefore, serve to define the C-14′ stereochemistry of any new compounds in this series.

XXII. Bisindole Alkaloids from *Melodinus* Species

A. SCANDOMELONINE (**483**) AND 19-EPISCANDOMELONINE (**484**)

Melodinus scandens Forst. produces a number of interesting monomeric alkaloids which were reviewed in a previous volume of this series (*162*). Additional fractionation of the alkaloids has afforded four "dimeric" alkaloids which were subsequently found to be two pairs of closely related derivatives (*219*).

Scandomelonine and episcandomelonine are isomers with molecular formula $C_{41}H_{42}N_4O_5$, having NH, carbonyl, and β-anilinoacrylate moieties; the latter unit was confirmed from the UV spectrum. The remainder of the UV spectrum was characteristic of the meloscandonine type; indeed, the two compounds displayed methyl doublets at 1.21 and 0.93 ppm, respectively, corresponding to the shifts of the C-18 protons of meloscandonine (**481**) and its 19-epimer **482** (*220, 221*). Preliminary information from the mass spectrum indicated loss of unit **438** from the molecular ion, as expected for the β-anilinoacrylate-type compound. The junction of the two halves must, therefore, involve the aliphatic (positions 3, 5, 14, and 15) carbons of the anilinoacrylate unit. The structures of these two alkaloids were deduced by analysis of their ^{13}C-NMR spectra.

Comparison of the spectra of meloscandonine (**481**) and scandomelonine indicated all the carbons to be present from the monomer with only carbons 9, 10, 11, and 12 being shifted. The major difference was the increased deshielding of C-10 (123.4 to 129.1 ppm), which had also become a singlet, thereby indicating the junction of the anilinoacrylate monomer to scandomelonine. This structural conclusion requires the assignment of all 14 aromatic and olefinic carbons, a problem solved in a novel way by Wenkert *et al.* (*219*). As quaternary carbons, C-8, C-8′, C-13, and C-13′ are readily observed as singlets in the fully decoupled spectrum. The single-frequency off-resonance-decoupled spectra could be obtained under conditions in which only one-bond and three-bond carbon–hydrogen interactions were observed. Therefore, the carbons of an unsubstituted indole nucleus will each appear as doublets of doublets, and this was observed for aromatic ring carbons of the anilinoacrylate unit. In the quinolone unit, C-12 is the highest-field aromatic carbon; under the conditions described above, it appeared as a doublet. Therefore, C-10 must be the point of attachment of the second unit.

Four aliphatic methine carbons were found to be present in the β-anilinoacrylate half, and from the one-bond coupling constants these were attached to heteroatom centers, including two carbons of an epoxide. From the molecular formula, which shows no other oxygen atoms, the meloscandonine unit must be attached at either C-3 or C-5.

Comparison with the shift data for pachysiphine (14β,15β-oxidovincadifformine) (**472**) indicated that attachment of the meloscandonine unit was to C-3 of scandomelonine (**483**). Of particular significance were the shift of C-3 and C-14 from 49.4 and 52.0 ppm, respectively, in **472** to 57.4 and 56.2 ppm, respectively, in **483**. The α-stereochemistry for this substitution at C-3 was apparent from the relatively small shift of the resonances of C-5 and C-6 in **483** compared with **472** and from the moderate γ-shift of C-21 from 70.9 in **472** to 61.5 ppm in **483**.

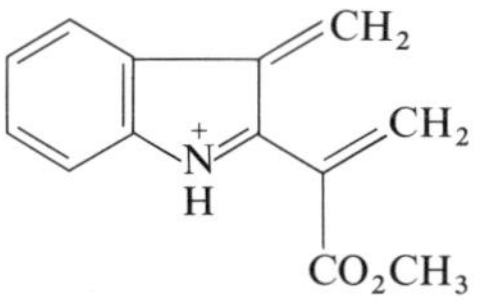

438 *m/e* 214

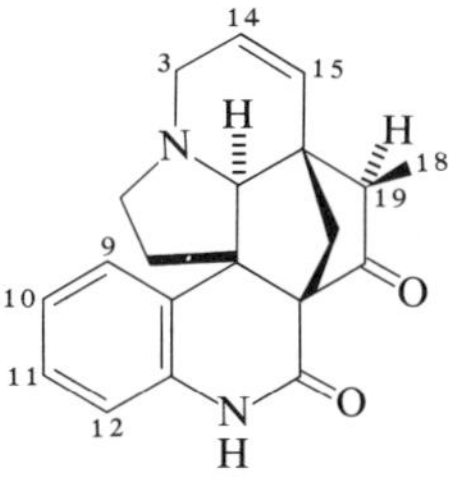

481 Meloscandonine
482 19-Epimeloscandonine

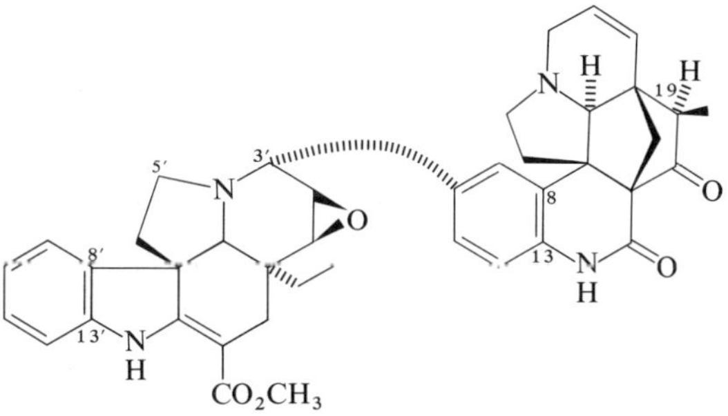

483 Scandomelonine
484 19-Episcandomelonine (19α-methyl)

Episcandomelonine was found to differ from scandomelonine (**483**) only in the stereochemistry at C-19 in the meloscandonine unit. Significant in the ^{13}C-NMR spectrum compared with that of **483** were the shifts of C-17, C-18, and C-21. Thus, episcandomelonine was assigned structure **484**, with a 19α-methyl group (*219*).

B. Scandomeline (486) and 19-Episcandomeline (487)

Two other closely related alkaloids from *Melodinus scandens*, scandomeline and episcandomeline, were also studied at this time by Wenkert *et al.* (*219*).

The molecular formula, $C_{42}H_{46}N_4O_6$, for each alkaloid was established by high-resolution mass spectrometry. The IR spectra indicated the presence of OH (ν_{max} 3540 cm^{-1}), NH (ν_{max} 3340 cm^{-1}), saturated ester (ν_{max} 1725 cm^{-1}), and vinylogous urethane (ν_{max} 1665 and 1610 cm^{-1}) functionalities. The UV spectrum (λ_{max} 214, 257, 300, and 325 nm) confirmed this last functionality and indicated the presence of an *o*-toluidine unit in the second moiety. The difference between scandomeline and episcandomeline was deduced from the ^{1}H-NMR spectrum to be the configuration of the C-19 methyl group.

The ^{13}C-NMR spectra established that the alkaloids contain the 3′α-pachysiphinyl unit attached to C-10 of the second unit. The latter unit proved to have a skeleton not found previously in this series of alkaloids.

Some similarities with scandine (**485**) were observed, particularly in the shift of the C, D and E ring carbons. A carbonyl of the carbomethoxy group was also evident, but no other carbonyl carbons were observed. The most significant new peak was a quaternary carbon at 88.1 ppm, indicating attachment to both O and N. The vinyl group of **485** was missing, being

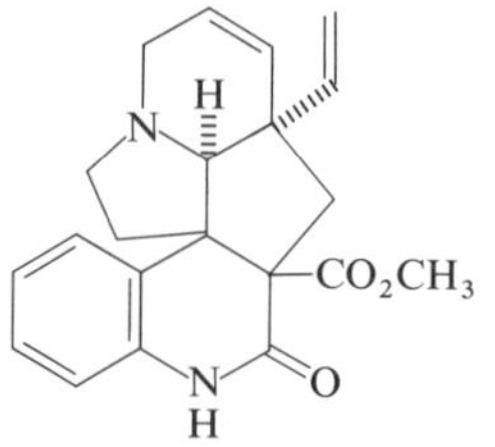

485

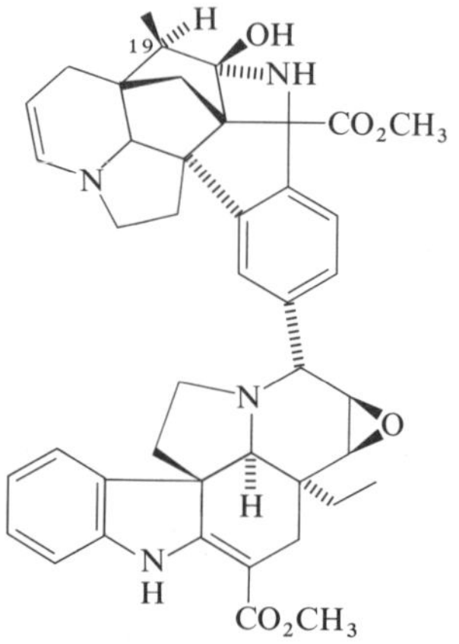

486 Scandomeline
487 19-Episcandomeline (19α-methyl)

replaced by a methyl and a methine. On this basis, scandomeline was assigned structure **486**, and episcandomeline structure **487**. The shift of C-21 (γ to C-18) was particularly diagnostic of the C-19 stereochemistry.

Wenkert and co-workers have also commented on the biogenesis of these alkaloids (*219*). The initial alkaloid important in the biogenesis is 18,19-dehydrotabersonine (**488**), which can undergo electrophilic hydroxylation at C-16. Rearrangement gives a scandine intermediate (**485**) which undergoes further reductive rearrangement to **489**, a unit which thus far has only been found in the dimeric alkaloids scandomeline (**486**) and episcandomeline (**487**). Rearrangement of unit **489** affords the meloscandonine (**481**) skeleton (Scheme 25).

In vitro evidence for the latter rearrangement was obtained when scandomeline (**486**), on heating with acetic anhydride at 100°, gave a 15% yield of scandomelonine (**483**).

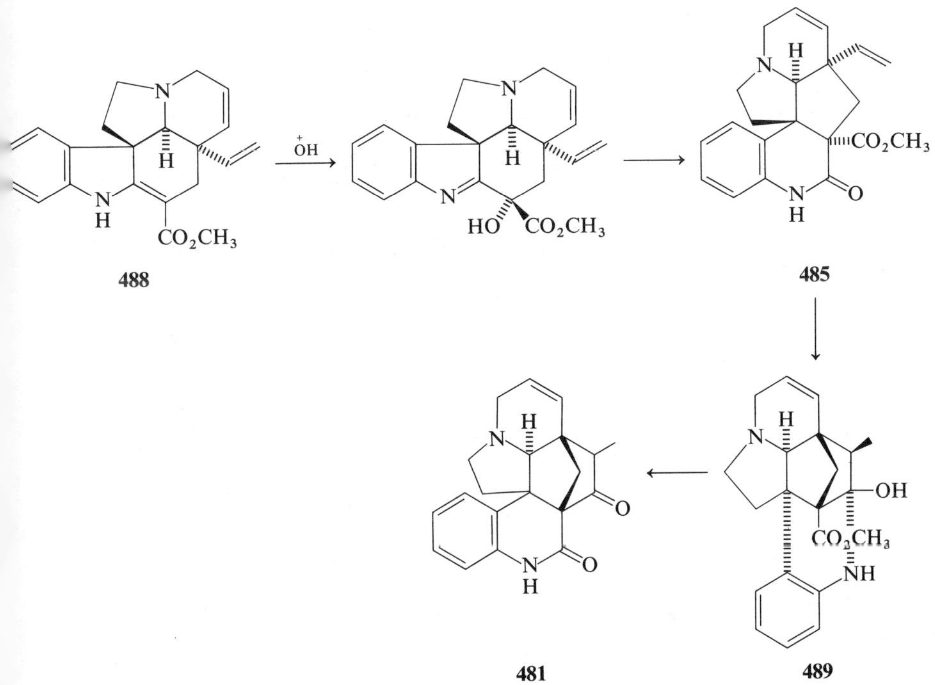

488 485 481 489

SCHEME 25.

C. METHYLENE BIS-1,1′-MELONINE

From the twigs and leaves of *Melodinus celastroides* several monomeric alkaloids were obtained together with a new dimeric alkaloid methylene bis-1,1′-melonine (*223*). The structure was not given in the manuscript.

The alkaloid was obtained as a bishydrochloride having $[\alpha]_D$ +71° and λ_{max} 248 and 298 nm. In the NMR spectrum, aromatic protons were observed between 6.5 and 7.2 ppm and a triplet at 0.87 ppm. The methylene group joining the two monomers appeared as an AB system ($J = 11$ Hz) at 5.08 and 5.15 ppm (*223*).

XXIII. *Aspidosperma*–Eburnea Type

A. CRIOPHYLLINE (**492**)

Criophylline was obtained as a crystalline solid, mp 278°–280°, from the leaves of *Crioceras dipladeniiflorus* (Stapf.) K. Schum. (Apocynaceae) (*224*, *225*), and a structure was deduced mainly on the basis of its ^{13}C-NMR spectrum.

The UV spectrum of criophylline indicated the presence of two chromophores, a dihydroindole, and a β-anilinoacrylate. In the ^{1}H-NMR spectrum only one carbomethoxy methyl was observed, which was thought to be part of the anilinoacrylate system. The molecular ion at 646 mu analyzed for $C_{40}H_{46}N_4O_4$ and gave a fragment at *m/e* 432 ($C_{27}H_{34}N_3O_3$). This ion would arise by loss of unit **438** from the anilinoacrylate half of the molecule (*162*).

The noise- and single-frequency-decoupled ^{13}C-NMR spectra in the region below 80 ppm indicated three methyl groups, ten methylenes, eight methines, and four quaternary carbons. The three methyl groups could be assigned to two ethyl groups (two triplets in the ^{1}H-NMR spectrum) and the carbomethoxy methyl.

Above 80 ppm, eight quaternary and seven methine carbons were observed. In the absence of olefinic protons, the seven methine protons must be aromatic and there must be a linkage between the aromatic ring of one unit and the aliphatic system of the second unit.

Because of the large number of methine carbons below 80 ppm, it appeared that the two oxygens remaining to be accounted for were epoxide functionalities. Comparison with the ^{13}C-NMR spectra of the alkaloids andrangine (**490**) and hazuntinine (**491**) indicated that the bisindole alkaloid criophylline was derived by a unification of **490** and **491**.

All the saturated carbons of andrangine appeared at virtually the same chemical shift in criophylline. Thus, it is the aromatic unit of andrangine

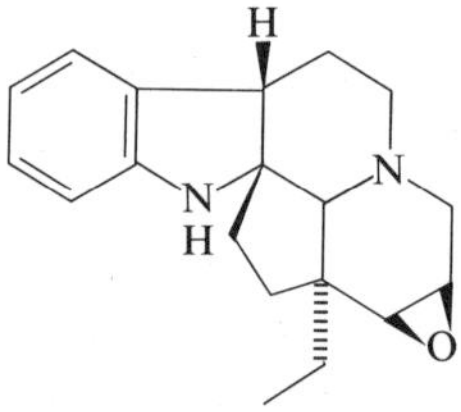

490 Andrangine

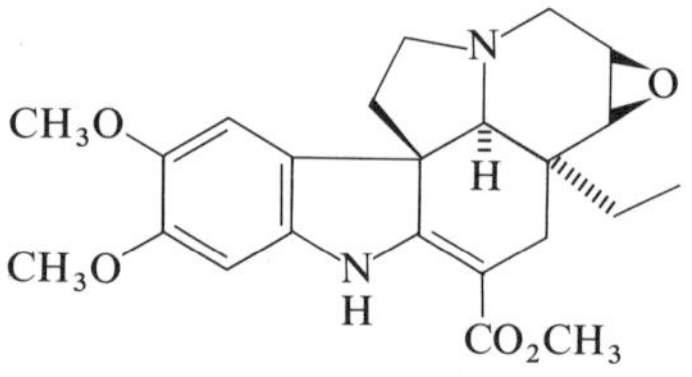

491 Hazuntinine

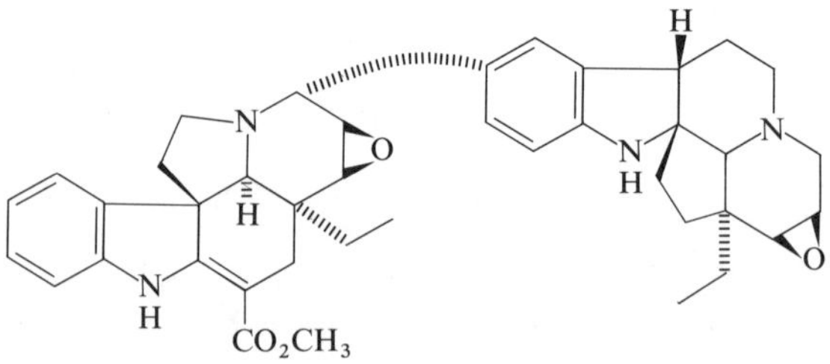

492 Criophylline

which is somehow linked to the aliphatic portion of an hazuntinine-type compound. However, in the single-frequency-decoupled spectrum of criophylline only three methylene groups adjacent to nitrogen were observed. Thus, the aromatic ring of andrangine must be attached at either C-3′ or C-5′ of the hazuntinine-type compound. From the marked shift of the C-21′ resonance from 70.8 ppm in hazuntinine to 62.0 ppm in criophylline, there must be a strong steric interaction between the C-3′ (or C-5′) –aromatic system bond and the nitrogen–C-21′ bond. This is possible if the group is substituted at C-5′ or C-3′ with an α-stereochemistry.

The choice between substitution at C-3α or at C-5α was made tentatively on the basis of reconciliation of a number of chemical shifts, particularly the β effect (51.8 ppm in monomer, 56.8 ppm in dimer) observed for the C-14′ resonance and the very small shielding (1 ppm) observed for the C-6′ resonance. Criophylline therefore has structure **492** (*224*).

B. PAUCIVENINE (493)

In addition to several monomeric alkaloids, the leaves of *Melodinus balansae* Baillon var. *paucivenosus* (S. Moore) Boiteau have also yielded a new bisindole alkaloid, paucivenine (*226*). The new compound was obtained as an amorphous gum showing a molecular ion at *m/e* 616, in agreement with the molecular formula $C_{40}H_{48}N_4O_2$. Fragmentation in the mass spectrum occurred principally to give ions at *m/e* 336 and 280. The observation of ions at $M^+ - 70$ and at *m/e* 85 and 70 suggested that half of the molecule had the eburnane skeleton, and ions at *m/e* 109, 121 and 122 were regarded as being typical of an *Aspidosperma* nucleus having a 14,15-double bond.

The UV spectrum indicated the presence of indole and indoline nuclei, and consequently the formation of the ion at *m/e* 336, postulated as tabersonine (**470**), should involve hydrogen migration to the eburnea unit. A 2,16-dihydro-β-anilinoacrylate derivative typically also shows loss of C-16 and C-17 on fragmentation, but no ion at *m/e* 252 (338–86) was observed. These data indicate that C-16 or C-17 is probably involved in the linkage between the two units.

The $M^+ - 70$ ion implies loss of the methylenes at positions 3, 5, 14, and 15, and because an indolic UV spectrum was observed for this half of the molecule, it was suggested that C-16 or C-17 of this unit was involved in the linkage.

In the formation of the ions at *m/e* 336 and 280, a hydrogen radical is transferred from the *Apidosperma* to the eburnea unit. Considering the probable nature of the ion at *m/e* 336, it was thought that the migrating hydrogen was derived from C-2. On this basis, paucivenine was assigned

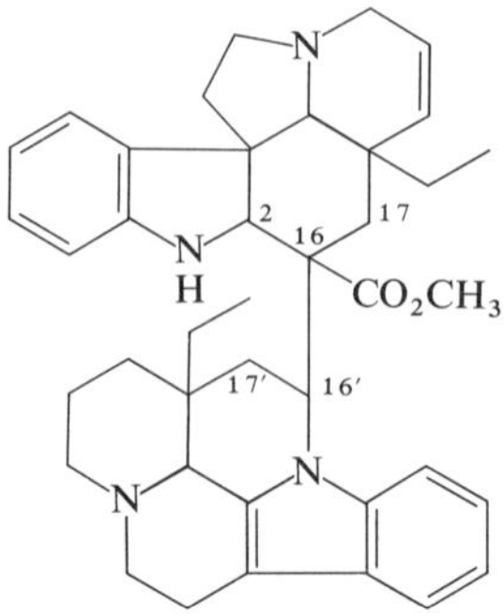

493 Paucivenine

structure **493**, in which the linkage of the two monomer units is between C-16 and C-16′ (*226*). Paucity of material prevented the verification of this linkage or determination of the stereochemistry.

C. Partial Synthesis of Novel Bisindole Derivatives

In an effort to investigate the potential of new optically active synthetic products, Takano and co-workers have studied the coupling of (−)-vindoline (**378**) and (+)-eburnamenine (**494**) (*227*).

Refluxing an equimolar mixture of **378** and **494** in 1.5% methanolic HCl for 3 hr gave a mixture of the dimer **495** (33% yield) and **496** (44% yield). Dimer **495**, mp 237°–238°, $[\alpha]_D$ −196.7°, showed a molecular ion at *m/e* 734 and fragment ions at *m/e* 467, 282, 135, 122, and 121 typical of the vindoline unit. The characteristic NMR signals of a vindoline unit were observed at 0.86 ppm (C-18), 2.08 ppm ($-OCOCH_3$), 2.80 ppm ($N-CH_3$), 3.80 ppm ($-CO_2CH_3$), 3.92 ppm ($-OCH_3$), 5.25 ppm (H-14), 5.40 ppm (H-17), and 5.85 ppm (H-15).

Similarly, dimer **496**, mp 187°–188°, $[\alpha]_D$ −3.5°, also showed a molecular ion at *m/e* 734 and fragmentation indicating the presence of a vindoline unit. This was confirmed by the NMR spectrum, which showed signals at 0.89, 2.09, 2.80, 3.83, 3.98, 5.20, 5.40, and 5.76 ppm corresponding to those described above.

That eburnamine was a component of each dimer was established by reductive cleavage (Sn in refluxing hydrochloric acid). The product from dimer **495** was identical, including CD curve, with natural dihydroeburnamenine (**497**), whereas the compound from **496** was antipodal and must therefore have structure **498**.

The point of attachment of the two units was established from the NMR spectra of the two dimers. Each dimer showed two singlets for the C-9′ and C-12′ protons (6.17 and 6.65 ppm in **495**, 6.24 and 6.70 ppm in **496**) which

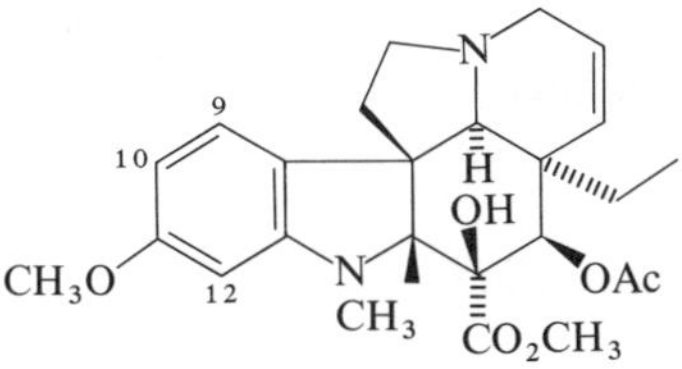

378 (−)-Vindoline

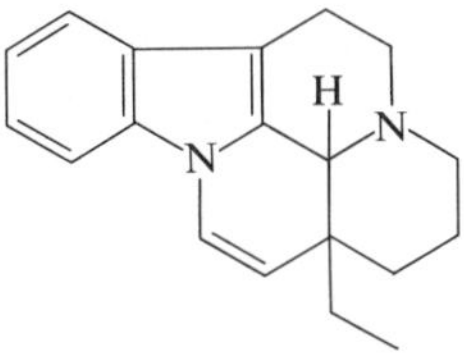

494 (±)-Eburnamenine

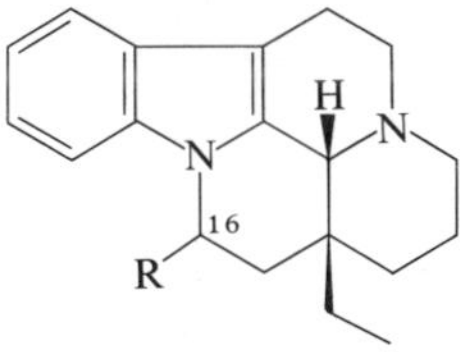

495 R = α-10-Vindolinyl
497 R = H

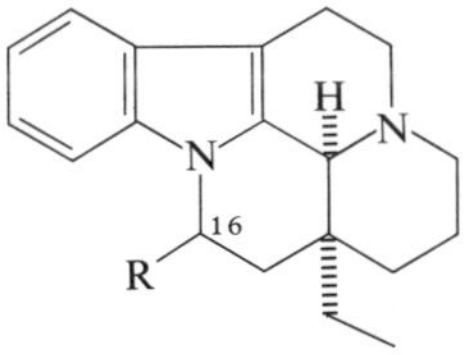

496 R = β-10-Vindolinyl
498 R = H

indicated that coupling has taken place at the C-10′ position of vindoline. In addition, the C-16 proton of the eburnamine unit appeared in each spectrum (5.50 ppm in **495**, 5.60 ppm in **496**) as a doublet of doublets ($J = 5$ and 11 Hz). In each isomer, then, H-16 is axial, and the dimers have structures **495** and **496** (*227*).

D. PLEIOMUTINE (499)

Although its name might suggest a relationship to pleiocraline, pleiocorine, or pleiomutinine, pleiomutine (**499**), as discussed in previous volumes of this

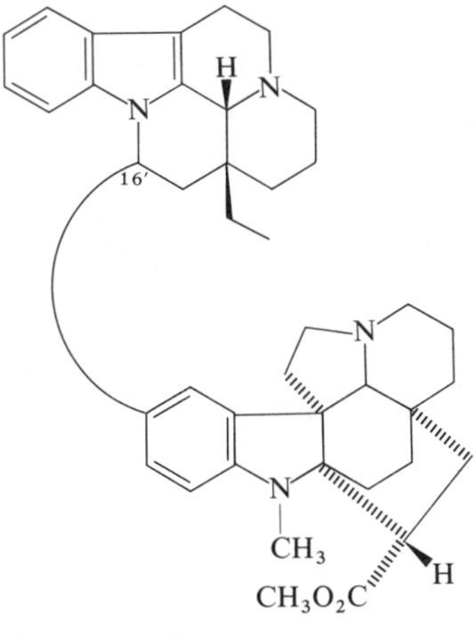

499 Pleiomutine

treatise (*12*), is actually derived from an eburnamine unit and a pleiocarpinine unit. The stereochemistry at C-16′ remains unknown.

XXIV. *Aspidosperma*–Canthinone Type

The previous work (*228–230*) on haplophytine (**500**) has been summarized in earlier volumes of this series (*162*, *231*).

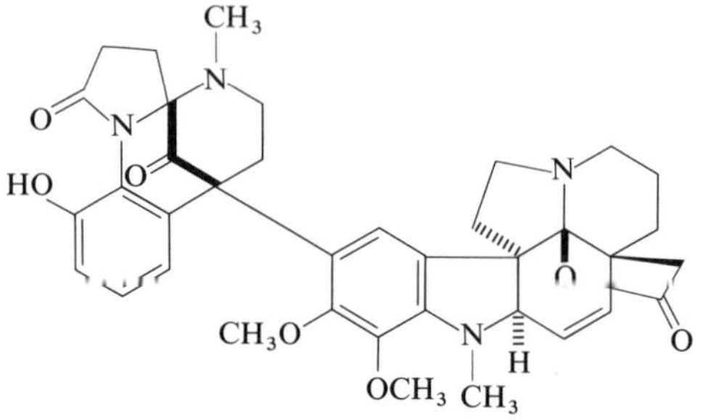

500 Haplophytine

XXV. *Aspidosperma*–Cleavamine Type

A. Introduction

At the time the last summary of the *Aspidosperma*-cleavamine alkaloids was compiled (Volume XI, Chapter 9), the main features of the structure and stereochemistry of vinblastine* and vincristine (leurocristine)* had been established, but the structures of leurosine, leurosidine, vincathicine, and catharine were still obscure. In the intervening period, these remaining structures have been clarified, together with the structures of some more recently isolated alkaloids of this group.

However, in view of the clinical application of vinblastine and vincristine in the treatment of leukemia and Hodgkin's disease, the most intensive activity in this area has been devoted naturally to the total synthesis of these alkaloids, and to the synthesis of structural analogs in the search for compounds having a higher activity: toxicity ratio and for compounds possessing useful clinical activity in the treatment of other human neoplasms.

* The name vinblastine has largely superseded the original name vincaleukoblastine, and will be used throughout this chapter. Similarly, vincristine, the name approved by the AMA, will be used in preference to the synonym, leurocristine.

In the following discussion, structural and stereochemical aspects of the alkaloids will be treated first, followed by transformations and interconversions within the dimeric alkaloid series. The synthesis of the alkaloids and related compounds will be discussed subsequently.

B. Vinblastine (502) and Vincristine (501)

The X-ray crystal structure determination of leurocristine (vincristine) methiodide (*232*) defined the structure and absolute stereochemistry as shown in **501**; vinblastine must then be **502**, in view of the known relationship between these alkaloids.

As an aid to the determination of the structures of other bisindole bases in this series, the ^{13}C-NMR spectra of vinblastine and its relatives have been thoroughly examined (*233–235*), and complete assignments for all the carbon atoms in the molecules have been made (Table XX). The NMR data for vinblastine itself are considered to be consistent with the conformation shown in structure **502** for the piperidine ring in the velbanamine component, because the chemical shifts for C-18′–C-21′ are closely similar to those of the analogous nuclei in 1,3-diethyl-3-piperidinol, a velbanamine model in which the 3-hydroxyl group is axially disposed (*234*). An axially placed hydroxyl group would also permit hydrogen bonding between the hydroxyl hydrogen atom and the lone electrons on $N_{b'}$; such a phenomenon is considered to account for the somewhat reduced basicity of $N_{b'}$ in the velbanamine component of vinblastine (*236*).

The absolute stereochemistry at C-16′ in the vinblastine group of alkaloids can also be deduced from the ORD and CD spectra (*237*, *238*). This was rendered possible by the synthesis (*237*, *239*) and structure and stereochemistry determinations (*237*, *240*) of several C-16′ compounds of the epi series, i.e., having the unnatural configuration at C-16′. Because this chiral center controls the relative geometry of the indole and dihydroindole components in the molecule, the optical rotation arising from dipolar coupling between electronic transitions in these two chromophores can in turn be correlated with the absolute stereochemistry at C-16′. Such dipole coupling will inevitably be reflected in the Cotton effects exhibited by the isomers. The long-wavelength transitions (280–300 nm) gave rise to complex CD spectra that were too difficult to interpret; however, at lower wavelengths (200–230 nm) the two C-16′ epimeric series gave rise to split Cotton effects of opposite sign, one close to 210 nm and the other close to 224 nm. Configurational changes elsewhere in the molecule, e.g., at C-20′, proved relatively insignificant. Hence the configuration at C-16′ can be deduced from the sign of the Cotton effects at low wavelengths, and the complete stereochemistry

TABLE XX

^{13}C-NMR DATA OF THE VINBLASTINE GROUP OF ALKALOIDS[a,b]

Carbon	Vinblastine (**502**) (*235*)	Deacetyl-vinblastine (*235*)	De-*N*-methyl-vinblastine (**516**) (*235*)	Deacetoxy vinblastine (**529**) (*235*)	Leurocolombine (**549**) (*235*)	Vincadioline (**553**) (*235*)	Leurosidine (**506**) (*234*)	Leurosine (**503**) (*234*)	Vincathicine (**543**) (*274*)
2	83.3	82.8	74.2	81.8	83.3	83.4	83.1	83.1	83.0
3	50.2	50.4*	50.5	50.1	50.5	50.4	50.2	50.2	50.6[c]
5	50.2	49.8*	51.1	50.1	50.5	50.4	50.2	50.2	50.9[c]
6	44.6	44.7	42.9	44.7	44.4	44.6	44.5	44.5	43.6
7	53.2	53.2	53.2	53.8	53.2	53.3	53.1	53.1	53.0
8	122.6	122.8	122.7	121.8	124.0	123.0	123.0	123.0	122.8
9	123.5	123.9*	124.2*	123.7*	123.7*	123.6*	123.4	123.4	123.6
10	121.1	120.9	120.7	120.4	120.0	120.6	120.4	120.4	120.6
11	158.0	158.0	157.9	157.8	158.7	158.1	157.6	157.6	159.1
12	94.2	93.9	93.9	93.5	94.6	94.2	94.0	94.0	94.3
13	152.5	152.5	149.1	152.8	153.7	152.7	152.8	152.8	152.8
14	124.4	124.2*	124.5*	124.0*	124.5*	124.5*	124.3	124.3	124.6
15	129.9	130.0	129.9	135.9	129.9	130.0	129.7	129.7	130.5
16	79.7	80.7	79.7	77.1	79.5	79.6	79.5	79.5	79.6
17	76.4	74.1	76.5	38.1	76.3	76.5	76.2	76.2	76.3
18	8.3	8.6	8.3	8.6	8.4	8.4	8.3	8.3	7.6
19	30.8	32.9	31.2	34.3	30.9	30.9	30.7	30.7	30.6
20	42.7	42.4	43.4	37.2	42.7	42.7	42.6	42.6	42.8
21	65.5	66.4	66.7	65.5	65.8	65.7	65.5	65.5	65.7
$\underline{C}O_2CH_3$	170.8	173.1	170.8	173.6	170.9	170.9	170.7	170.7	170.7
$CO_2\underline{C}H_3$	52.1	52.3	52.7	52.4	52.3	52.2	51.9	52.1	52.2

C_{11}-O$\underline{C}H_3$	55.8	55.8	55.7	55.7	55.4	55.8	55.7	55.7	55.4
$N_a\underline{C}H_3$	38.3	38.6	—	38.1	38.2	38.3	38.2	38.2	38.4
O$\underline{C}$OCH_3	171.6	—	172.7	—	171.6	171.7	171.4	171.4	171.7
OCO$\underline{C}H_3$	21.1	—	21.0	—	21.1	21.1	21.0	21.0	21.1
2′	131.4	131.3	131.7	131.6	130.4	131.6	130.2	130.7	187.0
3′	48.0	48.1	48.4	48.4	56.0	43.2	43.9	42.3	52.6[d]
5′	55.8	55.8	55.7	55.7	56.0	55.6	53.9[a]	49.6	55.0[d]
6′	28.2	28.7	28.8	28.8	27.3	28.5	21.4	24.6	32.2[e]
7′	117.0	117.0	116.9	116.8	117.2	116.9	116.8	116.7	57.1[f]
8′	129.5	129.4	129.5	129.5	129.3	129.4	128.9	129.1	144.2
9′	118.4	118.4	118.5	118.4	118.5	118.5	117.9	118.1	124.6[g]
10′	122.1	122.2	122.2	122.1	122.4	122.3	122.0	122.2	127.7[g]
11′	118.7	118.7	118.8	118.7	119.0	118.9	118.6	118.4	124.1[g]
12′	110.4	110.4	110.3	110.3	110.6	110.5	110.2	110.3	121.3[g]
13′	135.0	134.9	135.0	134.9	135.1	134.9	134.5	134.6	154.0
14′	30.1	30.2	30.4	30.3	71.3	39.2	29.8	33.5	28.9
15′	41.4	41.4	41.6	41.6	50.5	75.2	40.4[b]	60.3	49.2
16′	55.8	55.8	55.9	55.7	54.3	55.8	55.4	55.3	63.1[f]
17′	34.4	34.3	34.5	34.5*	40.7	32.8*	35.1[b]	30.7	38.7[e]
18′	6.9	6.9	6.9	6.9	6.9	6.2	7.1	8.6	7.0
19′	34.4	34.3	34.3	34.7*	34.0	29.2*	38.5[b]	28.0	34.4[e]
20′	69.4	69.5	69.8	69.7	69.9	71.3	71.8	59.9	75.8
21′	64.2	64.3	64.5	64.5	63.6	60.3	55.5[a]	54.0	63.1[d]
$\underline{C}O_2CH_3$	174.9	175.1	175.0	175.2	174.4	174.8	173.9	174.1	174.8
$CO_2\underline{C}H_3$	52.3	52.3	52.4	52.4	52.6	52.4	52.1	52.3	52.6

[a] Superscripts *a*–*g* indicate signals may be interchanged.

[b] Asterisk indicates tentative assignment.

of a partially synthetic bisindole alkaloid of this group can be deduced without recourse to a complete X-ray structure determination, as was formerly required. The alkaloids included in this study were vincristine (**501**), vinblastine (**502**), leurosine (**503**), isoleurosine (**504**), pleurosine (**505**), leurosidine (**506**), anhydrovinblastine (**507**), deoxyvinblastine B (**508**), and the synthetic 16′-epi compounds (**509**–**515**). In addition, the chiroptical properties of some other pairs of 16′-epimers, prepared from allocatharanthine or coronaridine derivatives, and dihydrotabersonine derivatives or vindorosine were studied.

ar-Tritiated vinblastine (i.e., [9, 12, 9′, 10′, 11′, 12′-$^{3}H_{6}$]vinblastine) has been prepared and analyzed by means of tritium NMR spectroscopy; this provides a rapid, nondestructive, and direct method for the analysis of tritium on a very small scale, and should be applicable to the analysis of vinblastine recovered from animal tissues in biological experiments. The aromatic tritium labels are readily identified, and are normally stable; trivial loss during biological experiments can therefore be discounted. Equally, loss of a label by metabolic processes can readily be deduced (*241*).

The separation of vinblastine and vincristine from *Catharanthus roseus* continues to receive attention, and several procedures have been reported (mainly in the patent literature) for the isolation and separation of these alkaloids (*242–247*). *C. roseus* extracts have been found to contain *N*-demethylvinblastine (**516**), and the yield of vincristine can therefore be increased by formylating the alkaloid mixture before separation and purification (*243–246*).

High-performance liquid chromatography is recommended as a rapid, reliable, reproducible, and sensitive method for the quantitative separation and determination of vinblastine and its congeners, whether monomeric or dimeric. Retention times of the alkaloids differed widely; therefore, good separations were possible (*248*). Detection of the alkaloids was achieved by monitoring the chromatograms with a variable-wavelength UV detector, the wavelength used in most cases being 298 nm.

The use of vincristine in the treatment of leukemia has resulted in the development of methods for its preparation by the oxidation of vinblastine (*249–252*). The preferred reagent appears to be chromic acid at low temperatures (−60°). Some demethylation of the N_a-methyl group also occurs; hence, formylation of the reaction mixture increases the yield of vincristine obtainable. The reagent is not universally satisfactory, however, and is reported not to succeed with several relatives of vinblastine, e.g., 15′,20′-anhydrovinblastine (**507**) and 15′,20′-anhydro-16-decarbomethoxy-16-*N*-methylcarboxamidovinblastine (**517**); these may be oxidized to the corresponding

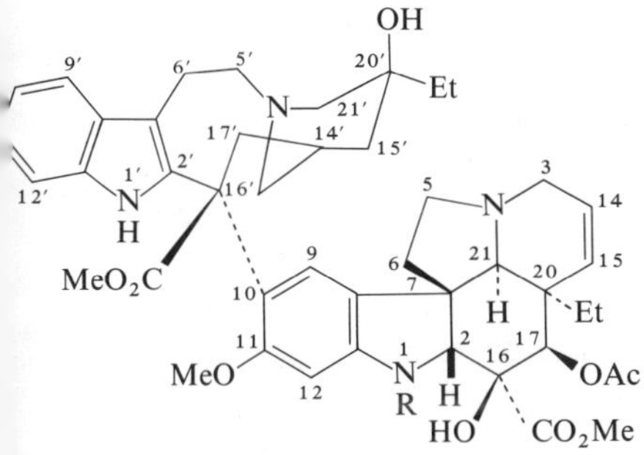

501 R = CHO Vincristine
502 R = Me Vinblastine
516 R = H Demethylvinblastine

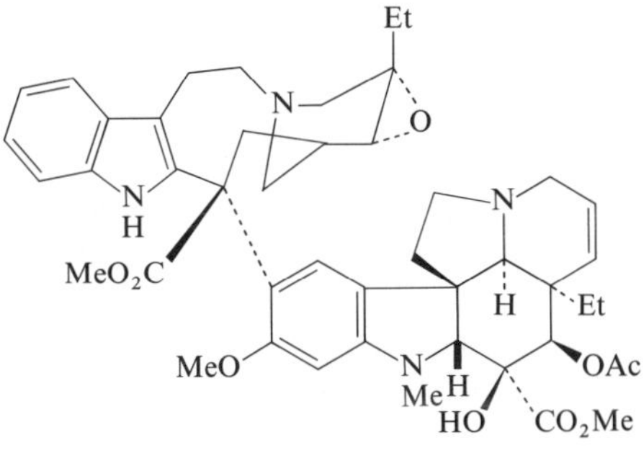

503 Leurosine
505 Pleurosine ($N_{b'}$-oxide)

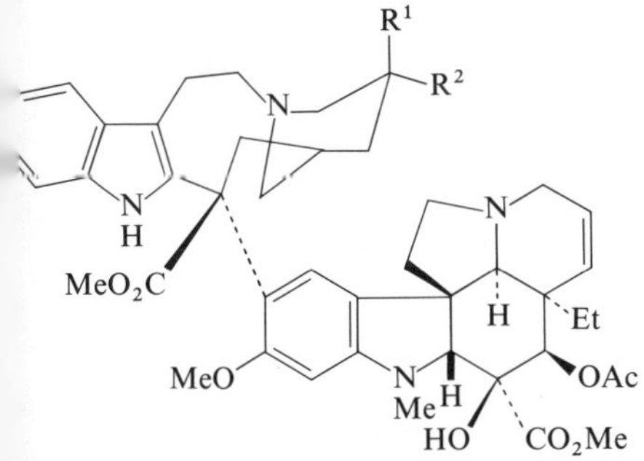

506 R^1 = Et, R^2 = OH Leurosidine
504 R^1 = H, R^2 = Et Isoleurosine
508 R^1 = Et, R^2 = H Deoxyvinblastine B

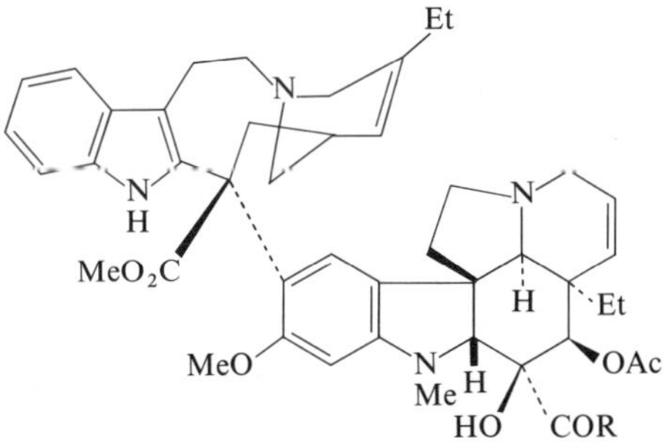

507 R = OMe Anhydrovinblastine
517 R = NHMe

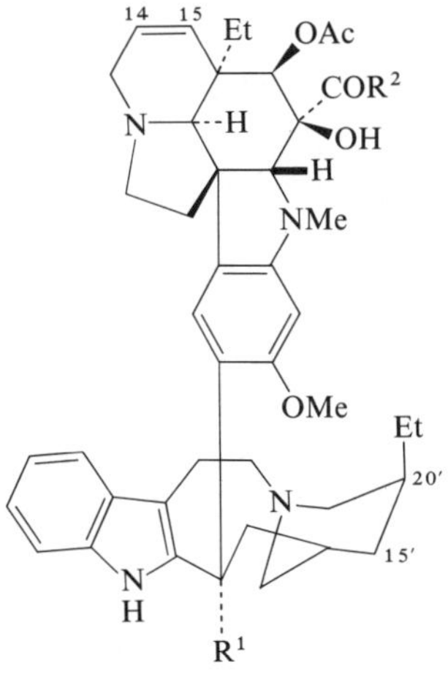

	R^1	R^2	
509	H	OMe	
510	H	OMe	14,15-dihydro
511	CO_2Me	OMe	
512	CO_2Me	OMe	20′-epimer
513	CO_2Me	OMe	14,15-dihydro
514	H	$NHNH_2$	
515	CO_2Me	OMe	15′,20′-dehydro

N_a-formyl compounds in better yield by a large excess of Jones' reagent at −78° (*253, 254*).

The use of microorganisms for the removal of the N_a-methyl group of vinblastine has also been investigated; thus, incubation of vinblastine with *Streptomyces albogriseolus* affords N_a-demethylvinblastine (*255*). The reaction, however, is not specific, and competing reactions also occur. Oxidation of the velbanamine component to the related hydroxyindolenine, followed by internal ether formation, is one such reaction, the product being the ether **518** (*256*). In the presence of a different *Streptomyces*, *S. panipalus* A36120, hydroxylation of the aromatic ring in the velbanamine component merely occurred, and the product was 10′-hydroxyvinblastine (*256*).

Included among the numerous transformation products of vinblastine and vincristine that have been prepared for biological testing are a number of amides formed by the reaction of the appropriate alkaloid with an amine in anhydrous methanol (*257*). The amines used in this study were ammonia, methylamine, ethylamine, dimethylamine, hydrazine, 2-aminoethanol, isopropylamine, 2-cyanoethylamine, and 2-cyanoethyldimethylamine; generally, amide formation was accompanied by some deacetylation.

Other derivatives prepared include several anhydro compounds resulting from the removal of the 20′-hydroxyl group (*258*). This work was stimulated by the observation that, whereas modification of the functional groups in the vindoline component led to unpredictable changes in pharmacological activity, changes in the velbanamine component led to products with a lower oncolytic activity, but also a decrease in toxicity. The removal of the C-20′ hydroxyl group was achieved by means of sulfuric acid, a crude reagent which nevertheless did not affect the various functional groups in the molecule except for the C-17 acetoxy group, which suffered deacetylation; reacetylation then gave the desired anhydro derivative. As might be expected, the dehydration reaction was not regiospecific; vinblastine (**502**), for example, gave a mixture of 15′,20′-anhydrovinblastine (**507**) with the two geometrically isomeric 19′,20′-anhydrovinblastines (**519** and **520**) in almost equimolecular amounts. Vincristine and deacetylvinblastine amide behaved similarly, but in contrast 16′-decarbomethoxyvinblastine and 16′-epi-16′-decarbomethoxy-17-deacetylvinblastine gave predominantly the 15′,20′-anhydro compound. Of the 15 anhydro derivatives obtained and tested biologically, anhydrodeacetylvincristine (**521**) proved to have a considerable potency against one tumor (B-16 melanoma) against which vincristine had little effect, and it also exhibited decreased toxicity. This is clearly an area that will be the subject of further investigation.

15′,20′-Anhydrovinblastine (**507**) has also been used as a starting material for the preparation of novel vinblastine derivatives (*259, 260*). The functionalization of the 15′,20′-double bond in this molecule by oxidative processes is considerably restricted by the fact that oxidation also occurs elsewhere; peracids and positive halogen, for example, attack $N_{b'}$ in the velbanamine component, or the methylene group (C-3′) adjacent to $N_{b'}$. Thus, oxidation of vinblastine by iodine in the presence of sodium bicarbonate gives the 3′-lactam **522** (*260*). However, oxidation of **507** by means of *tert*-butyl hydroperoxide and trifluoroacetic acid gave an epoxide, identified as leurosine (**503**) (see Section XXV,F), and oxidation of 15′,20′-anhydrovinblastine-$N_{b'}$-oxide (itself prepared from **507** and *m*-chloroperbenzoic acid) by means of osmium tetroxide in pyridine–tetrahydrofuran gave a diol $N_{b'}$-oxide which could be reduced to the corresponding diol, 15′-hydroxyvinblastine (**523**), by hydrogen sulfide. The stereochemistry at positions 15′ and 20′ in **523** was tentatively assigned as shown, because attack at the double bond in **507** was believed to be facile only at the β-face, i.e., opposite the bulky vindoline component. There has been some discussion of this point, and in other cases, e.g., the oxidation of **507** itself, attack certainly proceeds at the α-face; hence the stereochemistry assigned to **523** has not by any means been rigidly established. This diol (**523**) has the oxygenation pattern postulated for the alkaloid vincadioline (see Section XXV,H), but its identity (or otherwise) with vincadioline has not yet been demonstrated (*259*).

At low temperatures (−3°) in tetrahydrofuran the use of 1 equivalent of osmium tetroxide affords some **523** (after reductive work-up), but the major product is the 3′-lactam **524**, in which the 15′,20′-double bond has escaped

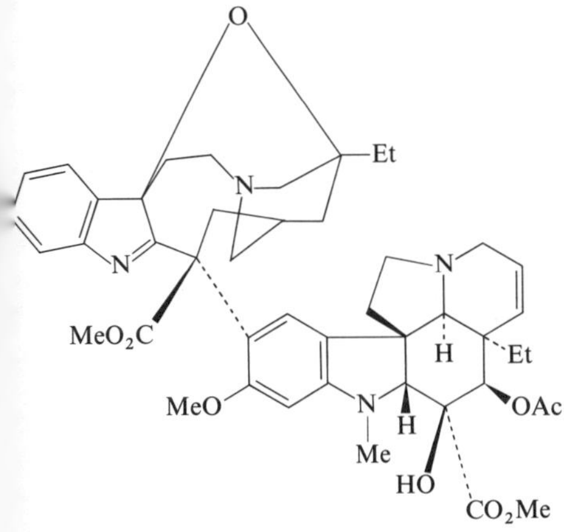

O Et N N MeO₂C MeO N Me H HO N H Et OAc CO₂Me

518

R¹ R² N N H MeO₂C MeO N Me H HO N H Et OAc CO₂Me

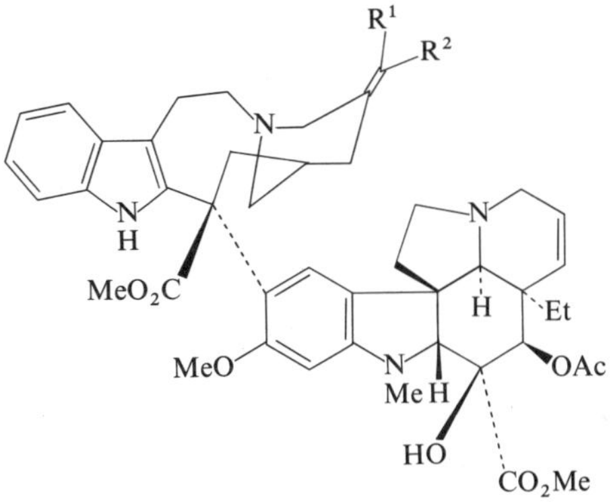

519 $R^1 = H, R^2 = Me$
520 $R^1 = Me, R^2 = H$

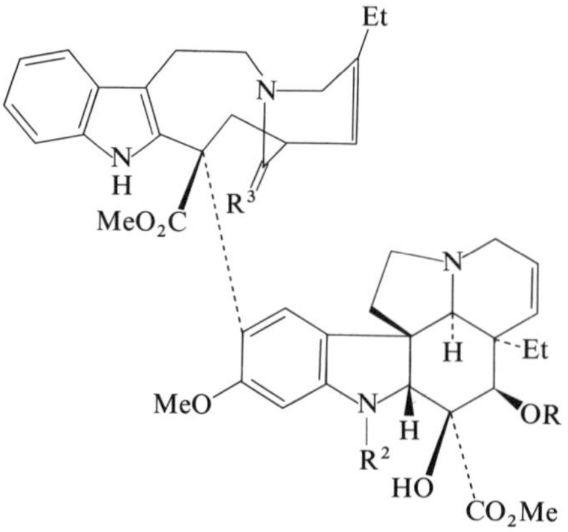

	R^1	R^2	R^3
521	H	CHO	H_2
524	Ac	Me	O

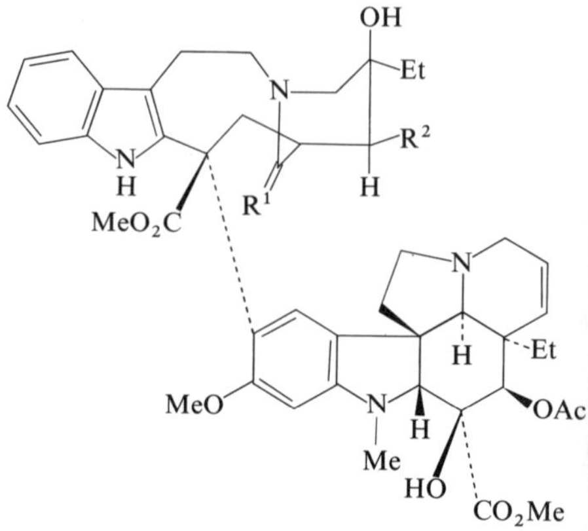

	R^1	R^2
522	O	H
523	H_2	OH
525	O	OH

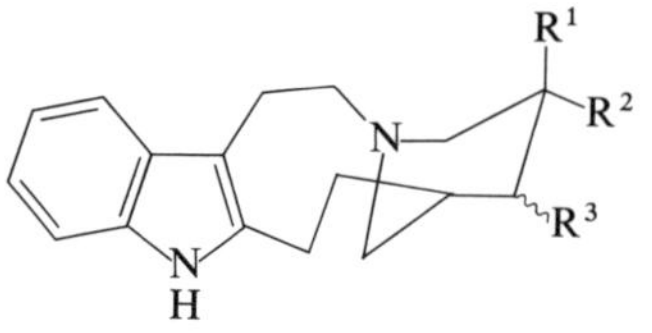

	R^1	R^2	R^3
526	H	Et	OH
528	Et	OH	H

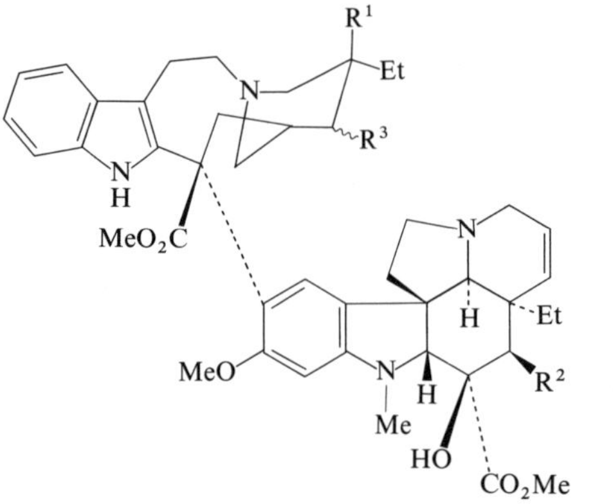

	R^1	R^2	R^3
527	H	OAc	OH
529	OH	H	H

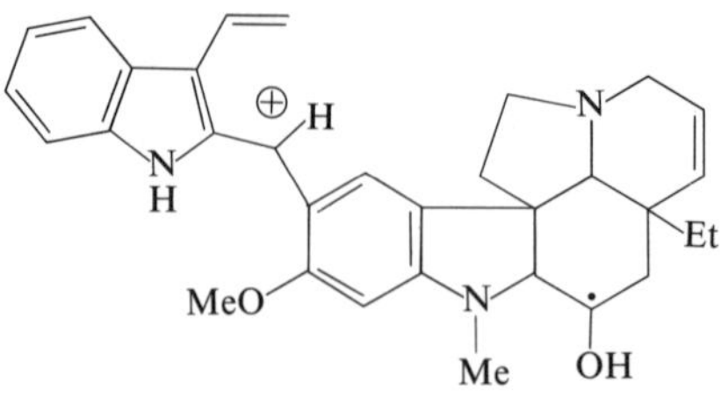

530

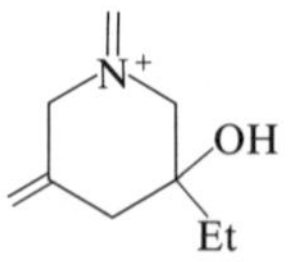

531

oxidation (*260*). When 2 equivalents of osmium tetroxide were used, both oxidation reactions ensued, and the product was 3′-oxo-15′hydroxyvinblastine (**525**). Again, the stereochemistry assigned to positions 15′ and 20′ is tentative.

C. LEUROSIDINE (VINROSIDINE) (**506**)

The elemental composition, spectroscopic properties, and functional group determination of leurosidine make it quite clear that it is an isomer of vinblastine. Furthermore, the reductive cleavage of leurosidine (Sn–$SnCl_2$–HCl) with production of deacetylvindoline renders it almost certain that the source of the isomerism resides in the indole component of the molecule, a conclusion that is reinforced by the isolation from the same reaction of cleavamine and a hydroxydihydrocleavamine, vinrosamine, which is not identical with velbanamine (*236*). Because the second pK_a of leurosidine is 1.4 units higher than the corresponding pK_a of vinblastine, it was deduced that the hydroxyl group of the indole component is in a different environment, one in which hydrogen bonding with the lone electrons on $N_{b'}$ is not permitted. The ease of acetylation of leurosidine compared with vinblastine suggested that the hydroxyl group is secondary, and this was supported by the apparent oxidation of leurosidine by the Pfitzner–Moffat reagent. These data, together with a consideration of the properties of vinrosamine (initially regarded as **526**), led to the conclusion that leurosidine (**527**) is an isomer of vinblastine in which the hydroxyl group is located at C-15′, probably in the α (axial) configuration (*236*).

A close examination of the ^{13}C-NMR spectrum of leurosidine, however, rendered structure **527** untenable. The spectrum showed a close similarity to that of vinblastine, including a low-field signal due to C-20′, and there was certainly no signal that could be attributed to C-15′ carrying an oxygen substituent. The data could only be satisfactorily interpreted by postulating that leurosidine (**506**) is the C-20′ epimer of vinblastine (*234*), a conclusion which is substantiated by the reported epimerization of vinblastine to give leurosidine (*261*); vinrosamine, accordingly, must have structure **528**. Evidently, the greater ease of acetylation of leurosidine compared with vinblastine is due, at least in part, to the impossibility of hydrogen bonding of the C-20′ hydroxyl group with the lone electrons on $N_{b'}$. The reported oxidation of leurosidine by the Pfitzner–Moffat method is obviously misleading, and the reaction requires further investigation; it is significant that the product of the oxidation appears not to have been thoroughly characterized.

Transformation products prepared from leurosidine for biological testing include a series of amides resulting from reaction with various primary and secondary amines (*257*). As with vinblastine, the oxidation of leurosidine by means of iodine and sodium bicarbonate affords a convenient preparation of the 3′-lactam (*260*).

D. Deacetoxyvinblastine (**529**)

In the course of purifying substantial quantities of vinblastine, Neuss *et al.* (*262*) isolated a new alkaloid, deacetoxyvinblastine (**529**), $C_{44}H_{56}N_4O_7$, mp 183°–190°, $[\alpha]_D^{26°}$ + 95.3° ($CHCl_3$). The main features of its structure, which immediately place it in the vinblastine group, are readily apparent from its spectrographic properties, particularly its NMR and mass spectra. Thus, there is no signal owing to an acetate methyl group in the NMR spectrum, and on acetylation a monoacetate is formed, whose methyl group gives rise to an NMR signal almost exactly coincident with that arising from the C-16 acetate group in 16-acetylvinblastine. The C-17 acetate methyl in vinblastine and its derivatives absorbs at somewhat lower field, and can be clearly differentiated from the C-16 acetate. It should be noted that in both alkaloids the C-20′ alcohol function escapes esterification under normal conditions.

The absence of a function at C-17 is clearly indicated by the mass spectra. The alkaloid itself gives a molecular ion at *m/e* 752, corresponding to $C_{44}H_{56}N_4O_7$, and this is confirmed by a molecular ion at *m/e* 694 in the spectrum of the 16′-decarbomethoxy-16-hydrazide, where intermolecular transmethylation of N_b by an ester methyl group, so common in this series, cannot occur. The hydrazide also exhibits important peaks at *m/e* 493 and 154, owing to the fragments **530** and **531**, the former of which emphasizes the absence of a substituent at C-17 (cf. the peak at *m/e* 509 in the spectrum of the corresponding deacetylvinblastine derivative), and the latter the presence of a hydroxyl group at C-20′.

E. Isoleurosine (Deoxyvinblastine A)

In the early extractions of the alkaloid leurosine, it was always observed to be contaminated with substantial amounts of another alkaloid, that was initially believed to be isomeric with leurosine and therefore was named isoleurosine. However, high-resolution mass spectroscopy proves that isoleurosine is $C_{46}H_{58}N_4O_8$ and, in fact, contains two more hydrogen atoms and one less oxygen atom (*263*), than leurosine (**503**). Its structure (**504**) becomes evident from reductive cleavage of the corresponding 16′-decarbomethoxy-17-*O*-deacetyl hydrazide (**532**) in acid solution, which yields deacetylvindoline hydrazide (**533**) and 20α-dihydrocleavamine (**534**). Because the NMR spectrum of isoleurosine exhibits the characteristic singlets for

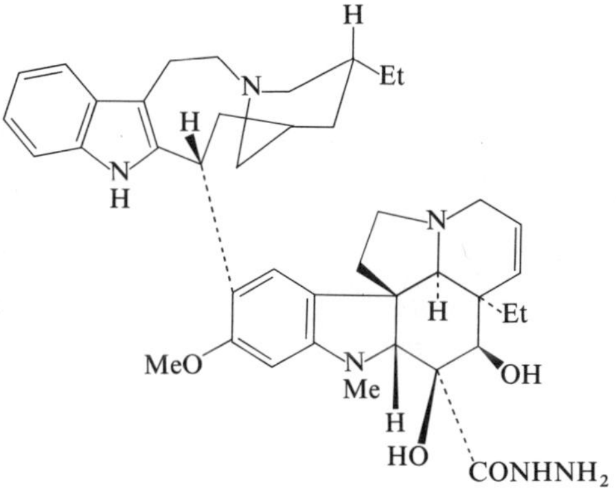

532

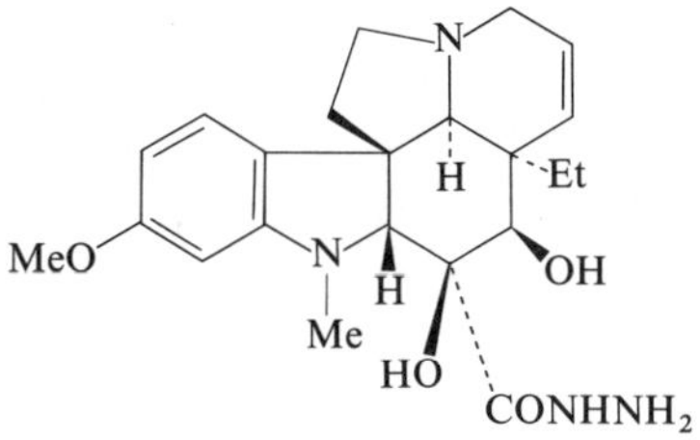

533

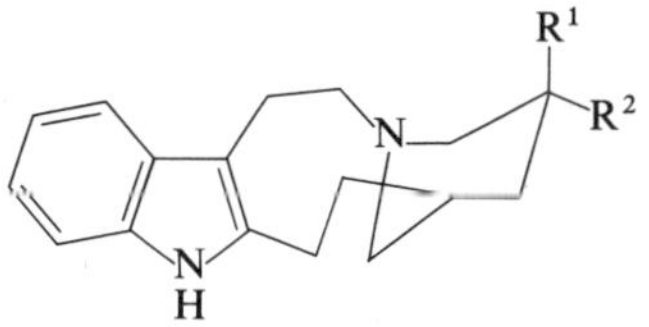

	R^1	R^2
534 20α-Dihydrocleavamine	H	Et
535 20β-Dihydrocleavamine	Et	H

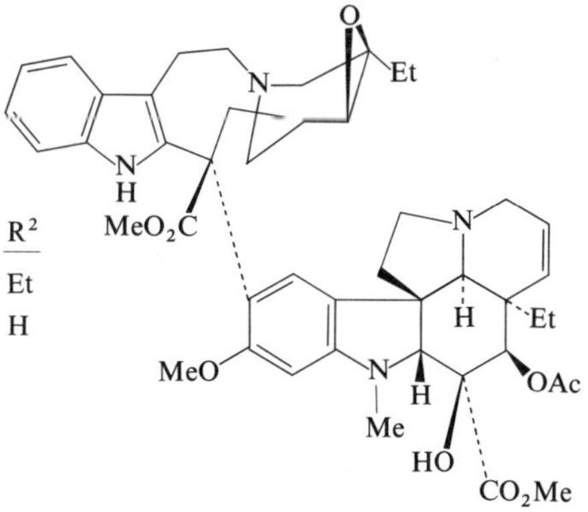

537

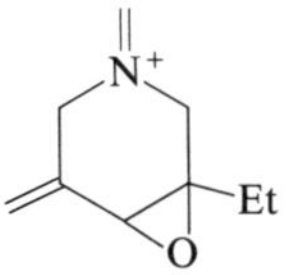

536

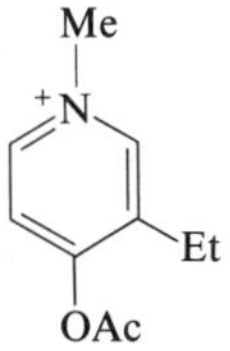

538

aromatic protons at positions 9 and 12, the vindoline unit must be attached to C-10 as in the other alkaloids, and isoleurosine (**504**) is simply 20′-deoxyvinblastine (*263*).

F. Leurosine (503), Pleurosine (505), Catharine (542), and Vincathicine (543)

Leurosine, $C_{46}H_{56}N_4O_9$, contains two hydrogen atoms fewer than vinblastine, and on treatment with Raney nickel in refluxing alcohol is converted into a deoxyvinblastine, isomeric with isoleurosine (deoxyvinblastine

A), and designated deoxyvinblastine B (**508**); some isoleurosine is also formed (*263*). Reductive acid cleavage of deoxyvinblastine B or its 16′-decarbomethoxy hydrazide afforded the appropriate vindoline derivative, together with 20β-dihydrocleavamine (**535**). Hence deoxyvinblastine B differs from deoxyvinblastine A (isoleurosine) only in the configuration at C-20′, and leurosine contains an oxygen atom in place of two hydrogen atoms in these isomers. This result is confirmed by the reductive acid fission of leurosine or its hydrazide, which gives deacetylvindoline or its hydrazide, together with cleavamine. The NMR spectrum (singlets owing to protons at positions 9 and 12) indicates that here also the vindoline component is substituted at C-10.

The position and function of the remaining oxygen atom remain to be determined. From the results of acid fission, it cannot be present in the vindoline component, but the isolation of cleavamine or 20β-dihydrocleavamine indicates that it must be situated in the indole component, and that C-15′ and/or C-20′ may be implicated. This is confirmed by the prominent ion at *m/e* 152 in the mass spectra of leurosine and its hydrazide, compared with one at *m/e* 138 in the spectra of compounds (e.g., isoleurosine–deoxyvinblastine B derivatives, and dihydrocleavamine) lacking an oxygen atom in the indole component. This remaining oxygen atom appears not to be present as a hydroxyl group, for leurosine only gives a monoacetate, and it must therefore be contained in an ether bridge; this also accounts for the number of hydrogen atoms in the molecule (*263*).

If it is assumed that one terminus of this oxide bridge is at C-20′, as in vinblastine, the other terminus must be such that complete removal of the oxygen is possible under mild hydrogenolytic conditions; leurosine must then be a benzylic or carbinolamine ether, or an epoxide. That leurosine is an epoxide later became apparent from a detailed examination of its mass and NMR spectra. In the mass spectrometer a double-benzylic fission would be expected to occur in the indole portion of the molecule to give an ion (**536**) at *m/e* 152, as observed. The presence of a 15′,20′-epoxide function also explains the NMR signal at δ3.1 (d, $J = 4.1$ Hz), appropriate to the presence of an epoxymethine proton at C-15′ coupled with one hydrogen at C-14′. The chemical shifts and multiplicities of these signals are consistent only with an epoxide function in this position.

In view of the coupling constant ($J_{14',15'}$) the configuration at C-20′ was tentatively regarded as the same as in vinblastine; thus leurosine was formulated as **537**. Some support for the 15′,20′-epoxide group also comes from the acetolysis of leurosine, which affords a vicinal hydroxyacetate. Although not purified and characterized, this product gave a prominent peak in the mass spectrum at *m/e* 180, attributed to the ion **538**; this could be readily generated from the expected hydroxyacetate (*15′*-acetoxyvinblastine?) by fission of the 5′,6′- and 14′,17′-bonds, followed by dehydration and aromatization (*264*).

Confirmation of the 15′,20′-epoxide linkage in leurosine was also forthcoming from its ^{13}C-NMR spectrum. An initial study (*233*) revealed that C-20′ absorbs at anomalously high field and must be part of a small, strained ring. A later, refined study showed that the second oxycarbon was also at anomalously high field and can only be satisfactorily explained by its presence in an epoxide function. This is further corroborated by the very high coupling constant for the C-15′–H-15′ nuclei, which is again only compatible with an epoxide (*234*).

In a repetition of the reaction of leurosine with acetic acid, Neuss and his collaborators obtained not the expected vicinal hydroxyacetate but an isomer of leurosine, which could also be prepared, and more efficiently, by treatment of leurosine with sulfuric acid (*234*). This product is an indole–indoline in which the vindoline component is intact; the epoxide function has clearly been severed, and the molecule now contains an acrylic ester unit, but it retains the ethyldialkylmethanol group characteristic of vinblastine. Structure **539**, derived for this product from the combined spectrographic evidence, arises by internal nucleophilic attack on the protonated epoxide by the β-position (C-7′) of the indole ring, followed by a 1,2-shift of the newly formed bond to position 2′ of the indole ring, and rearomatization by loss of a proton from C-17′ with fission of the 2′,16′-bond. If the configurations of C-16′ and C-14′ are assumed to be the same as in vinblastine, this reaction would only be expected to occur readily if the epoxide group had the α-configuration; leurosine, therefore, has the structure and stereochemistry shown in **503**.

Pleurosine, another constituent of *Catharanthus roseus*, has an IR spectrum almost identical with that of leurosine, but contains an additional oxygen atom. The pK_a value of its more basic $N_{b'}$ is two units lower than the corresponding pK_a of leurosine. However, reduction of pleurosine by means of zinc and acetic acid affords leurosine quantitatively; pleurosine (**505**), therefore is the $N_{b'}$-oxide of leurosine (*263*).

Leurosine is an epoxide of 15′,20′-anhydrovinblastine, and attempts have naturally been made to prepare it by the oxidation of the latter. Because oxidizing agents such as peracids and positive halogen were reported (see above) to attack other positions in the molecule, the use of *tert*-butyl hydroperoxide in the presence of trifluoroacetic acid and tetrahydrofuran was investigated, a reaction which in fact gives a moderately good yield (51%) of leurosine (**503**) directly (*259*, *265*). An even better yield (62%) of leurosine was subsequently obtained by the oxidation of anhydrovinblastine hydrochloride with mercuric acetate, followed by quenching of the reaction mixture with aqueous sodium borohydride (*265*).

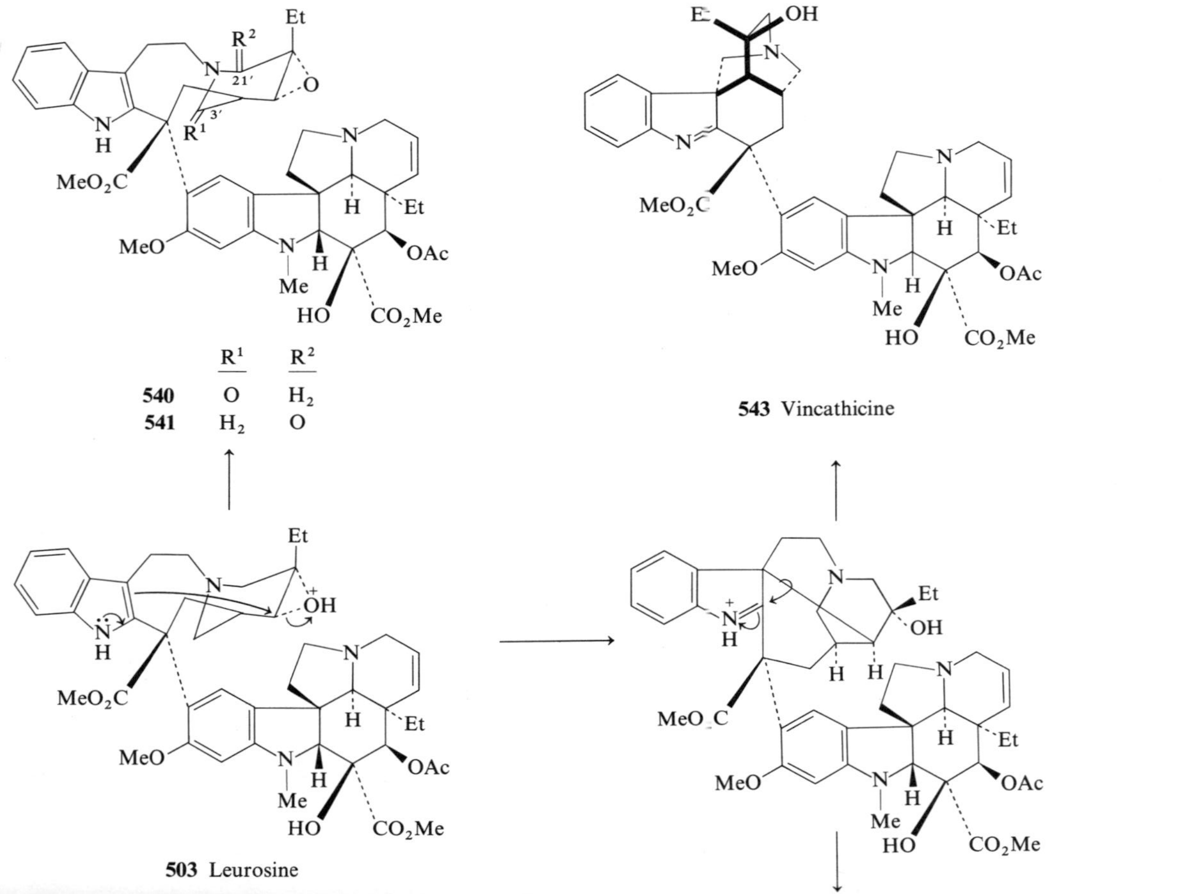

	R^1	R^2
540	O	H_2
541	H_2	O

543 Vincathicine

503 Leurosine

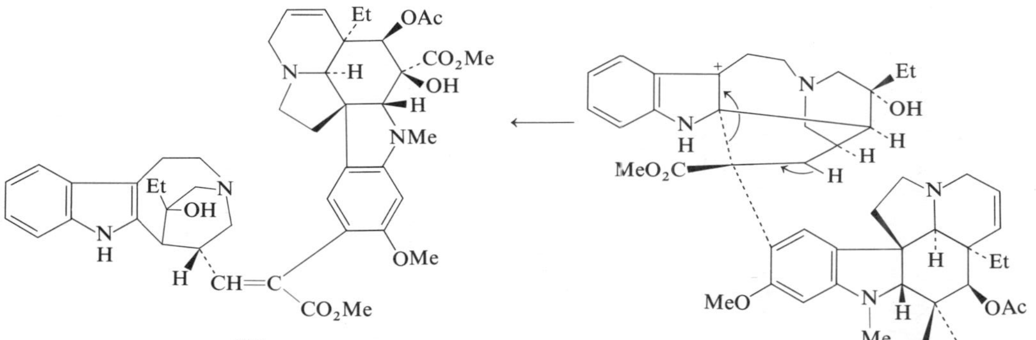

539

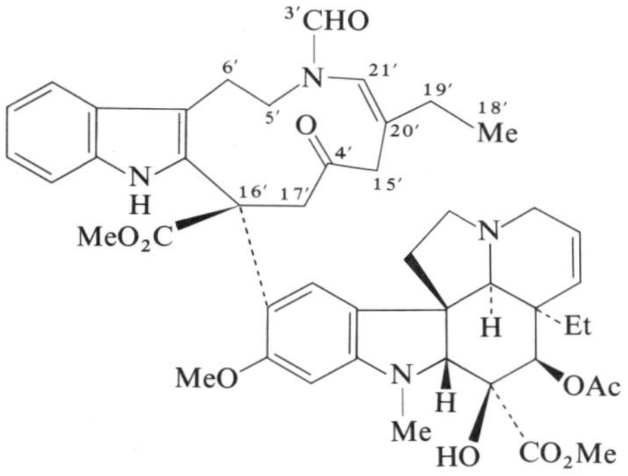

542 Catharine

Other workers (*266*) have attempted the same conversion, and have found that the oxidation of 15′,20′-anhydrovinblastine (**507**) with *p*-nitroperbenzoic acid in HMPA gives leurosine $N_{b'}$-oxide (pleurosine), which can be reduced (Zn–AcOH) to give leurosine in an overall yield of 27%. Some anhydrovinblastine (34%) can also be recovered from the reaction mixture after reduction.

These conversions, it should be noted, do not unambiguously settle the configuration of the epoxide function in leurosine; however, although a difference of opinion has been expressed, the weight of the evidence suggests (*266*) that the less-hindered side of the double bond in anhydrovinblastine is the α-face. Hence, leurosine would be expected to be the α-epoxide (**503**).

This oxidation of anhydrovinblastine appears to be a facile process, and can be achieved by reagents other than those mentioned above, e.g., lead tetraacetate (*267*). Even these reagents may not be necessary, as anhydrovinblastine is rapidly oxidized to leurosine in air if it is not stored in an inert atmosphere, and the oxidation is even faster in solution; as much as 40% of leurosine can be obtained after only 72 hr at room temperature. On the basis of this evidence, it has been suggested that leurosine is not a *bona fide* alkaloid, but an artifact; certainly, anhydrovinblastine itself has not been isolated from any *Catharanthus* species to date, and so if any is present in the plant it may well suffer oxidation to leurosine during the isolation procedure (*267*).

Other evidence, however, seems to indicate that leurosine is present in intact *C. roseus* plants (*268*). Thus, anhydrovinblastine (**507**) sulfate is stable for prolonged periods (up to 100 hr) in buffered aqueous solution in the absence of added oxidant. A slow conversion of anhydrovinblastine into leurosine (**503**) was observed in the presence of 1.3 mol equivalent of hydrogen peroxide, but the rate was increased by a factor of 12 following addition of horseradish peroxidase, and a yield of 65% of leurosine could be obtained after only 1.5 hr. Similarly, the conversion of anhydrovinblastine into leurosine was also noted in the presence of cell-free extracts of *C. roseus* and added hydrogen peroxide; here the rate of formation of leurosine was approximately five times that observed in the absence of the *C. roseus* enzymes. This result is less well defined, presumably owing to the complexity and the number of enzymes probably present in the *C. roseus* extracts, but it does suggest that leurosine is a natural product, formed from anhydrovinblastine by a peroxidase-type enzyme.

As noted above, the oxidation of these alkaloids by means of iodine and sodium bicarbonate gives the corresponding 3′-lactam; leurosine, accordingly, gives the lactam **540** (*260*). The use of oxygen and trifluoroacetic acid gives a different result, and the product from leurosine, initially believed to be the 21′-lactam **541**, was subsequently shown (*269*) to be identical with the alkaloid catharine (**542**), a minor, oncolytically inactive alkaloid of *C. roseus* (*192*, *270*) whose structure has recently been elucidated by the X-ray crystal structure determination of its acetone solvate (*271*, *272*). Catharine (**542**) can also be obtained by the overoxidation of 15,20′-anhydrovinblastine, also by means of oxygen and trifluoroacetic acid (*260*, *265*). A radical mechanism appears to be involved in this reaction, for which the presence of acid is not essential; leurosine, for example, can be converted into catharine in 48% yield by the action of *tert*-butyl hydroperoxide in dichloromethane (*269*). In view of the ease of this oxidation catharine may also prove to be an artifact.

In trifluoroacetic acid solution catharine suffers fission with release of vindoline, as expected; the other component, however, undergoes a double cyclization with formation of the pentacyclic intermediate **542c** (*272*). Evidently, the double bond in the enamide (**542a**) initially generated is sufficiently nucleophilic to attack C-16 with formation of the 16,20-bond; the *N*-formylimmonium ion (**542b**) so produced is, not surprisingly, sufficiently electrophilic to allow formation of the 7,21-bond. The product isolated from this reaction, following reduction by means of sodium cyanoborohydride, is therefore the pentacyclic compound (**542d**). The consequences of removal of the *N*-formyl group in **542d** are also of interest. In the product **542e**, N_b is sufficiently close to C-14 to allow cyclization to the carbinolamine form **542f**, the structure of which was established unequivocally by the X-ray method. The hexacyclic form (**542f**) also appears to be preferred in solution, as the molecule does not contain a ketonic carbonyl group (^{13}C-NMR spectrum). In basic solution, however, **542f** must be in equilibrium with **542e**, because the four protons on C-15 and C-17 can be exchanged for deuterium. The secondary alcohol related to **542e** is the product of sodium borohydride reduction (*272*).

The amorphous alkaloid vincathicine, $C_{46}H_{56}N_4O_9$, was initially isolated as its crystalline sulfate, and suspected to contain an oxindole chromophore (*273*). No further work was reported until it was observed that vincathicine can be obtained by the treatment of leurosine (**503**) with acid, and its structure may therefore be closely related to one of the compounds involved in the rearrangement to compound **539**, described above. In fact, vincathicine proves to be the indolenine derivative (**543**), formed simply by nucleophilic opening of the epoxide function by the β-carbon (C-7′) of the indole ring, followed by deprotonation (*234*, *274*, *275*).

In consonance with structure **543**, the ^{1}H- and ^{13}C-NMR spectra indicate clearly the presence of a 10-vindolinyl component in vincathicine. Of the remaining 21 carbon absorptions, the two owing to the ester group are easily assigned. Seven absorptions correspond to sp^2 hybridized carbon atoms, instead of the eight required for a substituted indole derivative, and one of

Catharine (**542**)

TFA

CHO
N
O
Et
N
H
MeO$_2$C
H
N
H
Et
MeO$^+$
N
Me
H
OAc
HO
CO$_2$Me

Vindoline
+
CHO
N
Et
20
N$^+$
H
16
O
CO$_2$Me

542a

CHO
N
Et
N$^+$
H
MeO$_2$C
O

CHO
7
N$^+$
21
N
H
Et
MeO$_2$C
O

542c

542b

$NaBH_3CN$

CHO
N
Et
N
H
MeO$_2$C
O

NH
Et
N
H
14
MeO$_2$C
O

542d

542e

N
Et
14
OH
20
15
N
H
16
17
MeO$_2$C

542f

these falls in the carbonyl region, but not close to the position expected of an oxindole carbonyl carbon. Taken in conjunction with the absence of an indole NH and the reduction of vincathicine by sodium borohydride to a dihydro derivative containing an additional exchangeable proton, it is clear that vincathicine is an indolenine derivative.

Thus C-7′ has become sp^3 hybridized, and far-reaching changes also appear to have occurred in the nonaromatic part of the indole component. In view of the production of vincathicine by the acid treatment of leurosine, two structures are possible for the initial product, depending on whether a 7′,15′- or a 7′,20′-bond is formed; either structure (following deprotonation) would explain the multiplicities of the carbon resonances of the indolenine component of vincathicine (i.e., three nonprotonated carbon atoms apart from the sp^2-hybridized carbons, two methines, six methylenes, and two methyl groups), but the absorptions owing to C-15′ and C-20′ are distinctly in favor of structure **543** in which a 7′,15′-bond has been formed.

G. VINAMIDINE (CATHARININE) (545)

Vinamidine, $C_{46}H_{56}N_4O_{10}$, is one of three bases recently isolated from *Catharanthus roseus* (*276*). Aside from the presence of a vindoline component, diagnosed by the usual application of ^{1}H- and ^{13}C-NMR spectroscopy, the molecule contains a C_{21} indole unit which includes carbomethoxy, *N*-formyl, and ketone carbonyl groups. Initially, a possible relationship with catharine was assumed, but because vinamidine is not identical with dihydrocatharine, structure **544** was tentatively proposed. The positioning of the carbonyl group at C-15′ was also believed to account for the chemical shift of C-18′ which is similar to that of C-18′ in vinblastine, the proposal being that the absence of a C-20′ oxygen substituent in vinamidine was offset by the effect of the C-15′ carbonyl group.

Subsequently, catharinine, $[\alpha]_D$ $-32°$ ($CHCl_3$), an amorphous base isolated from *C. longifolius* and *C. ovalis* and so named because of a suspected structural affinity with catharine, was shown to be identical with vinamidine, and to have structure **545** (*277, 278*).

The UV spectrum of catharinine is similar to that of catharine, and is unchanged in acid solution, consistent with the presence of a nonbasic $N_{b'}$. Its structural similarity to catharine suggested that it could be a C-20′ epimer of dihydrocatharine; however, dihydrocatharine is reduced by sodium borohydride to a hydroxy ester which easily lactonizes. In contrast, catharinine is reduced to catharininol which shows no tendency to lactonize. In any event, the differences in the CD curves of catharinine and dihydrocatharine are too great to be explained simply by a difference in configuration at C-20′.

Catharine (**542**) may well be formed by a biogenetic cleavage of the piperidine ring in leurosine, and an alternative fragmentation may lead to catharinine. Three such possibilities can be envisaged (see Scheme 26). Of these, possibility c is not favored because the CD spectrum of catharinine is similar to those of the vinblastine group of alkaloids, with the configuration 16′*S* and 14′*R*; in c, C-14′ is no longer an asymmetric center. However, the ^{13}C-NMR spectrum, although not easy to interpret precisely owing to a superposition of some signals, is consistent with possibility a.

A decision among the various possibilities was finally made from the results of reductive acid cleavage. Cathariminol affords vindoline, deacetylvindoline, and an indole derivative which shows a peak, albeit rather weak, corresponding to loss of an ethyl group from the molecular ion; this result

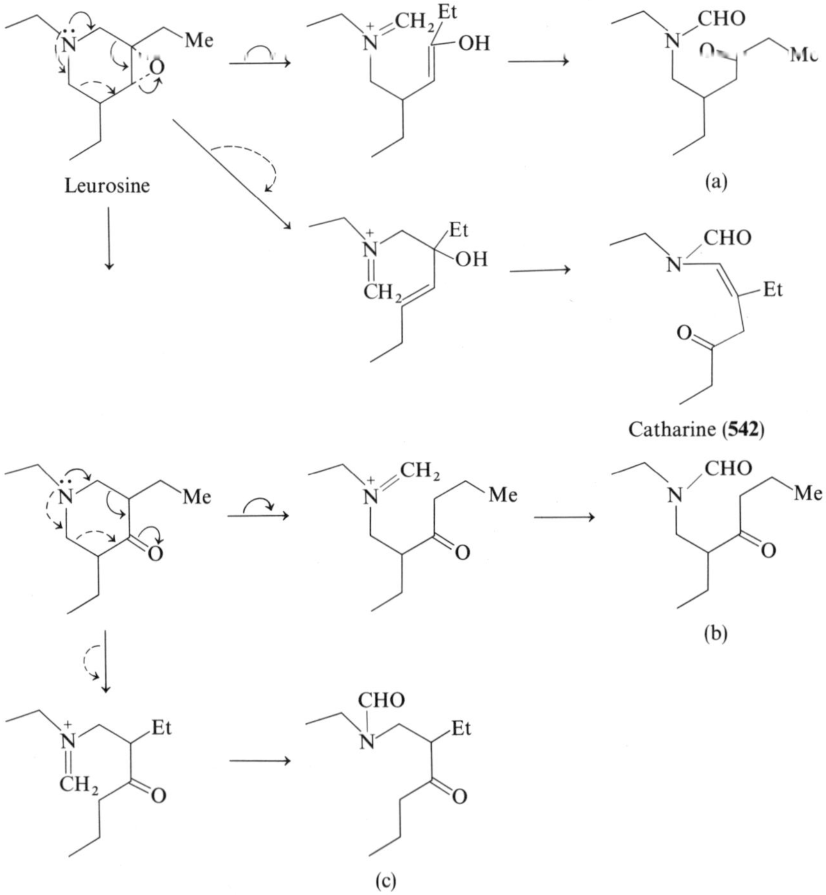

Leurosine

(a)

Catharine (**542**)

(b)

(c)

SCHEME 26.

clearly disfavors possibility b. Similarly, catharinine gives vindoline, deacetylvindoline, and an indole derivative (**546**) together with its 16′-decarbomethoxy derivative (**547**). Both **547** and catharinine give, on electron impact, an ion at $M^+ - 72$, owing to loss of a fragment C_4H_8O by a McLafferty rearrangement (i.e., **547** → **548**); such a fragmentation is not observed in the mass spectrum of catharininol. These data are consistent only with possibility a (Scheme 26), and hence with structure **545** for catharinine. Final proof of this structure was obtained by the X-ray crystal structure determination of the indole derivative **546**. This product has the 16*S*-configuration, and is clearly the thermodynamically preferred isomer. Catharinine itself has the 16′*S*-configuration, as shown in **545**, according to its ORD and CD spectra (*277*).

H. Leurocolombine (549), Pseudovinblastine Diol (550), and Vincadioline (553)

Leurocolombine, pseudovinblastine diol, and vincadioline constitute a trio of alkaloids which contain an additional hydroxyl group in the velbanamine component of the molecule. The structure of leurocolombine (**549**) was deduced almost entirely from its mass and NMR spectra, which indicated clearly the presence of a vindoline unit substituted at position 10. The signals arising from the vindoline carbon atoms having been identified, the remaining absorptions in the ^{13}C-NMR spectrum of leurocolombine were then assigned by the inversion–recovery method in conjunction with off-resonance decoupling. The unsubstituted indole ring was readily identified; of the aliphatic carbon atoms, three gave rise to singlets and were therefore non-protonated. One of these was the carbon atom of the ester carbonyl group, another corresponded well to C-20′ of vinblastine, and the third was at low field, appropriate to a carbon singly bonded to oxygen. Because the methine doublet corresponding to C-14′ in vinblastine was absent in the spectrum, the additional hydroxyl group present in leurocolombine must be situated here, and the complete structure for leurocolombine is **549**.

The siting of this second hydroxyl group at C-14′ is also consistent with the mass spectrum, since the expected fragment at *m/e* 170 was responsible for only a small peak; however, there was a peak at *m/e* 152 produced by the dehydration of this fragment. This may be contrasted with the mass spectra of vinblastine and other alkaloids containing hydrogen at C-14′, which by double-benzylic fission give an ion at *m/e* 154 which is responsible for the base peak in the spectrum.

The configuration at C-20′ in leurocolombine is considered to be the same as in vinblastine (**502**), in view of the similarity of the chemical shifts of C-18′ in **502** and **549**, but the configuration at C-14′ is unknown (*276*).

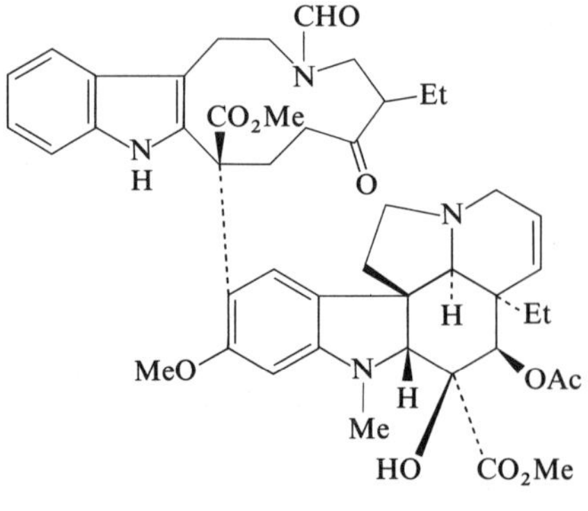

544

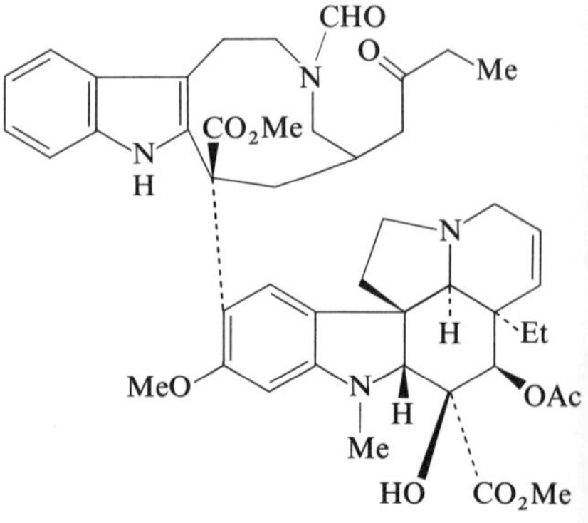

545 Catharinine (vinamidine)

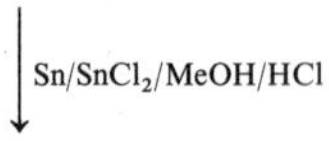

Sn/$SnCl_2$/MeOH/HCl

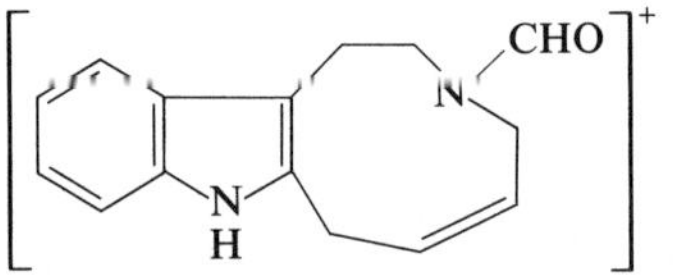

548

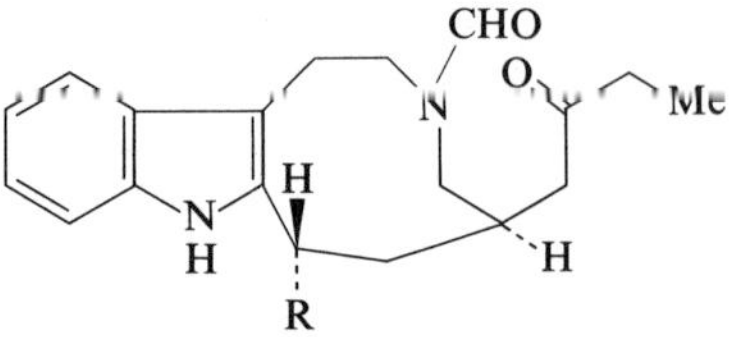

546 R = CO_2Me
547 R = H

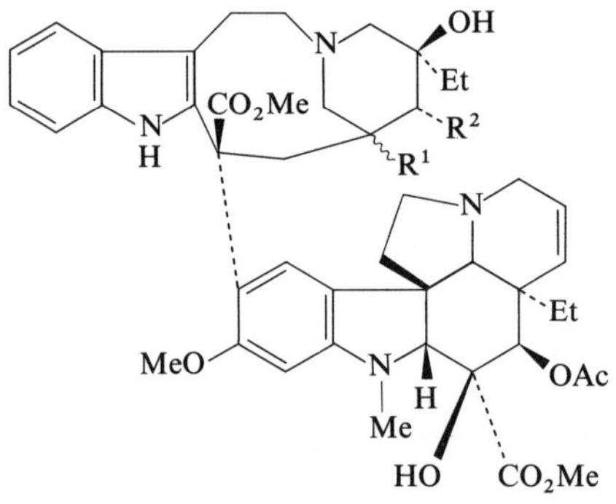

549 R^1 = OH, R^2 = H Leurocolombine
553 R^1 = H, R^2 = OH Vincadioline

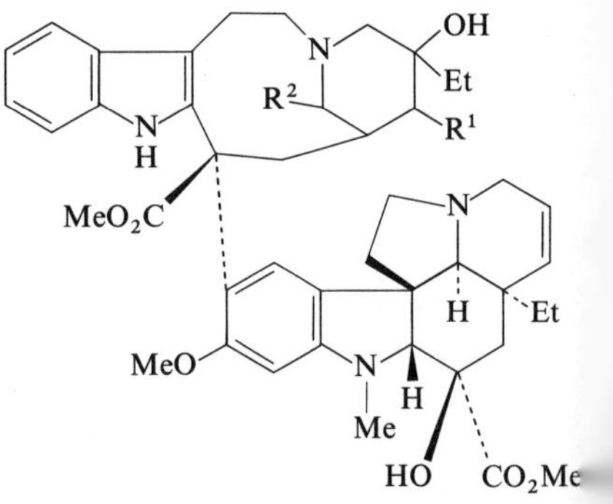

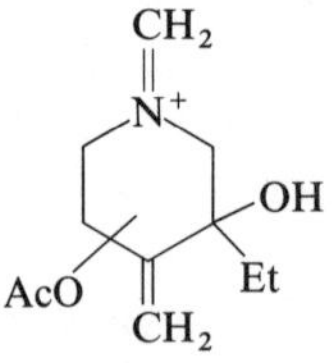

m/e 212

551

Pseudovinblastine diol

	R^1	R^2
552	H	OH
550	HO	H

Pseudovinblastine diol (**550**), $C_{44}H_{56}N_4O_8$, has the molecular formula of a deacetylvinblastine but differs from all the other bases in this series in that the dihydroindole component is not vindoline, but deacetoxyvindoline. Accordingly, the characteristic vindoline fragments at *m/e* 469 and 282 are observed in the pseudovinblastine diol mass spectrum 58 units lower, and the ^{1}H-NMR spectrum lacks the absorptions owing to an acetate methyl group and a low-field C-17 proton. The characteristic "velbanamine" fragments at *m/e* 154 and 355 are shifted in the mass spectrum of pseudovinblastine diol to *m/e* 170 and 371, and to *m/e* 212 and 413 in the spectrum of its acetate. The peak at *m/e* 212 (**551**) is particularly significant, for it offers direct proof of the presence of an additional hydroxyl group in the aliphatic part of the indole unit. Because the signals owing to the C-15′ proton and the C-15′ acetate methyl group in the NMR spectrum of pseudovinblastine diol acetate are closely similar to the corresponding absorptions in the spectrum of vincadioline acetate, pseudovinblastine diol (**550**) is also regarded as having a C-15′hydroxyl group; however, based on the evidence available at present, the alternative, carbinolamine structure **552**, cannot be conclusively excluded (*276*).

With regard to vincadioline, only proposed structure **553** can be quoted, as no details concerning the elucidation of its structure have yet been revealed (*279*).

I. Vincovaline (554)

Vincovaline, $[\alpha]_D$ $-118°$ ($CHCl_3$), a constituent of *Catharanthus ovalis*, is isomeric with vinblastine and appears to be the first base in this series in which vindoline is coupled to an indole component of the coronaridine group (*280*). The mass spectrum of vincovaline is very similar to that of vinblastine, and there are also similarities in the ^{1}H-NMR spectrum, which discloses the attachment of the indole portion to position 10 of the vindoline molecule; together, the spectra eliminate positions 18′, 19′, 3′, 5′, 14′, and 21′ as possible sites for the hydroxyl group. One significant difference from the vinblastine spectrum is the much lower chemical shift of one of the C-18 methyl groups in the vincovaline spectrum, that indicates a somewhat different environment for this group. Because acetylation only gives a monoacetate, which is presumably the acetate of the vindoline C-16 hydroxyl group, the remaining hydroxyl group in vincovaline is assumed to be attached to C-20′; its resistance to acetylation suggests that it may have the same configuration (β) as in vinblastine.

Vincovaline thus appears to be a stereoisomer of vinblastine, and from a comparison of its CD spectrum with those of various synthetic bisindole bases epimeric at positions 14′ and 16′, it was deduced to have the 16′*R*,14′*S*-

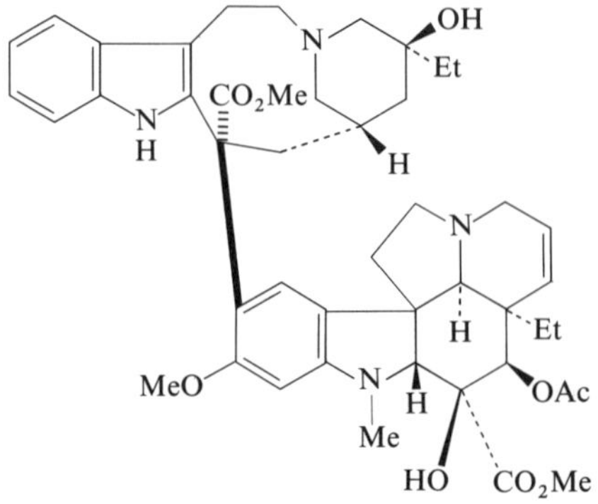

554 Vincovaline

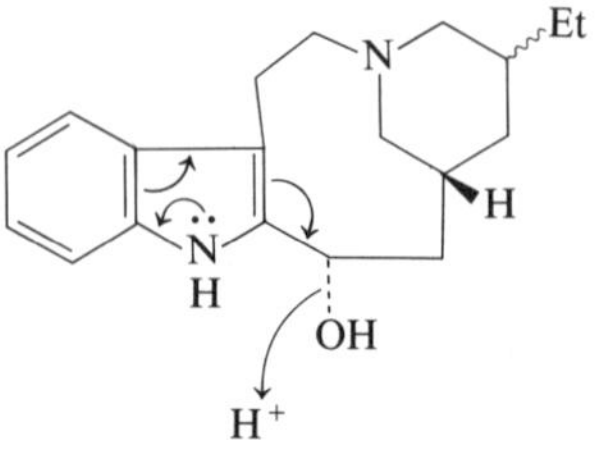

555

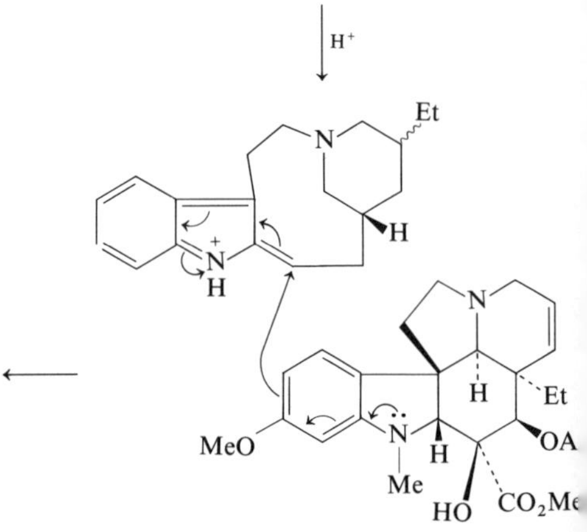

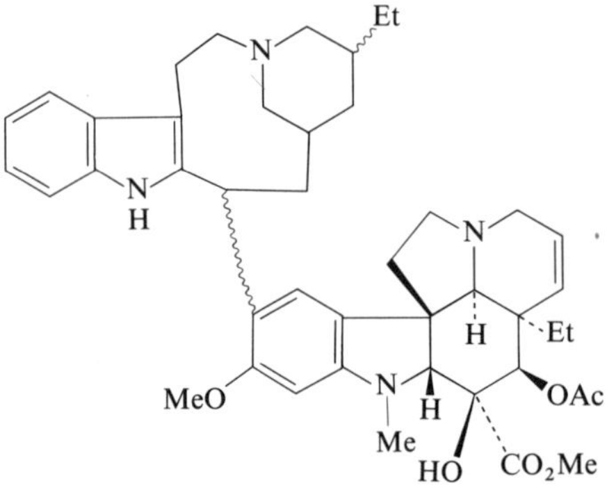

556

378

configuration depicted in **554**, in which the indole component has the same stereochemistry as 20-epipandoline at positions 14′ and 20′ (*280*).

A co-occurring base, vincovalinine, $C_{44}H_{54}N_4O_7$, is 16′-decarbomethoxyleurosine, and a third base, vincovalicine, $C_{46}H_{54}N_4O_{10}$, appears to have as dihydroindole component N_a-formyl-N_a-demethylvindoline (as in vincristine); however, insufficient material has so far been obtained to allow a full structural elucidation (*280*).

J. Synthesis of the Vinblastine Group of Alkaloids

During the last few years intensive efforts have been made to synthesize vinblastine and its relatives from monomeric components, the majority of the very considerable amount that has been achieved coming from the laboratories of Potier and Kutney, with some contributions also by Atta-ur-Rahman.

The obvious strategy was to condense an appropriate indole component at the reactive position 10 of the vindoline nucleus. Less obvious was the nature of the indole component, although the required substitution, at a benzylic position to an indole α-carbon atom, seemed to offer a plausible mode of condensation, provided that an effective leaving group was situated at this position. In the initial effort a simple application of this concept was attempted, but all subsequent attempts have been based on either the chloroindolenine approach or the modified Polonovski reaction. Coupling of the two monomeric components by either of these reactions is feasible, but this in principle can lead to either the natural or the unnatural stereochemistry at C-16′, and a considerable effort was required before appropriate conditions for ensuring the desired stereochemistry at this position were satisfactorily delineated.

The first synthesis of a bisindole base was reported by Harley-Mason and Atta-ur-Rahman (*281*), who condensed 16-hydroxydihydrocleavamine (**555**) with vindoline (**378**) in the presence of cold 1% methanolic HCl. The product was obtained as a separable mixture of two diastereoisomers, which had the molecular formula of a 16′-decarbomethoxy-20′-deoxyvinblastine (**556**); the detailed stereochemistry of this product was not elucidated, and the work appears not to have been pursued further. However, it signaled the first synthetic approach to the vinblastine series, and it established the feasibility of forming the vital 10,16′-bond.

The first application of the chloroindolenine approach was recorded the following year (*263*), when the chloroindolenine (**557**) from 20β-dihydrocleavamine was condensed in acid solution with deacetylvindoline hydrazide to give the dimeric hydrazide **558**, having the unnatural configuration at C-16′, because the chemical shift of the C-16′ proton differed significantly from that observed in the natural series, e.g., deoxyvinblastine B (**508**). On the other hand, the chemical shift of C-16′ was almost identical with that exhibited by C-16′-epidecarbomethoxydeacetylvinblastine hydrazide, prepared by isomerizing decarbomethoxydeacetylvinblastine in acid solution.

The unnatural stereochemistry was also the result of the application of the chloroindolenine approach by Kutney and his collaborators who, in addition to reporting an independent preparation of hydrazide **558** and the related vindoline derivative **509**, achieved the first synthesis of a dimeric C-16′ ester (*239*). The route adopted was exactly the same in principle; i.e., chloroindolenine **559** from 16-carbomethoxy-20β-dihydrocleavamine (**560**) was condensed in acid solution with vindoline, 14,15-dihydrovindoline, or deacetylvindoline hydrazide, to give the dimeric bases **511**, **513**, or **561**,

557

562

563

564

	R^1	R^2
558	$NHNH_2$	H
509	OMe	Ac

respectively. The gross structures of these products were unequivocally established by their NMR spectra, and by their reductive acid cleavage to 16-carbomethoxy-20β-dihydrocleavamine and the appropriate vindoline derivative (*239*, *282*). The configuration of the C-16′ asymmetric center was not definitely established, however, until the X-ray crystal structure determination of the methiodide of **511** was carried out, which established beyond doubt that it was 16′-epi-20′-deoxy-20′-epivinblastine (*240*). The structure of the 16′-decarbomethoxy-16′-epi-20′-deoxy-20′-epivinblastine (**509**) was also established by the X-ray method; hence the chloroindolenine approach to the synthesis of dimeric bases would appear to result in the unnatural stereochemistry at C-16′, even where, as in the preparation of **509**, epimerization of C-16′ is possible in the acidic reaction medium.

This coupling reaction proved a very convenient method for the preparation of dimeric bases in high yield; therefore, a detailed investigation was carried out to determine whether the reaction conditions could be modified so that bases having the natural configuration at C-16′ could be obtained. In fact, the yield of dimeric base proved very sensitive to the reaction conditions but, in spite of wide variations in the conditions under which the β-chloroindolenine was prepared, and under which the coupling reaction occurred, only bases of the C-16′-epi series (the unnatural series) were ever obtained. One fortuitous outcome of this study was a greatly improved preparation of the dimeric base **511**; it precipitated in almost quantitative yield when the chloroindolenine **559** was mixed and heated with vindoline in acetyl chloride as solvent (*282*).

Mechanistically, this study proved inconclusive. The intermediacy in the coupling reaction of the isomerized chloroindolenine **562**, which can couple with vindoline by an S_N2' process, or of the species **563** or **564**, would be expected to result in a product whose configuration at C-16′ would be independent of that of the starting material. In fact, 16α-carbomethoxy-20β-dihydrocleavamine (16-epimer of **560**) also coupled with vindoline to give the dimer **511**, and no trace of the stereoisomer with the natural configuration at C-16′ was detected.

The chloroindolenine approach was also used by Atta-ur-Rahman (*283*) in the first synthesis of an epimer of anhydrovinblastine. The mixture of 16α- and 16β-carbomethoxycleavamines (**565**), prepared from catharanthine (**566**), was converted into the corresponding chloroindolenine (**567**) and condensed with vindoline in the presence of methanolic HCl. The product (**515**) exhibited the UV and mass spectra of 15′,20′-anhydrovinblastine, as expected from its mode of formation, but it also belonged to the 16′-epi series; i.e., it was 16′-epi-15′,20′-anhydrovinblastine, because it could be correlated (*240*), by hydrogenation, with 16′-epi-20′-deoxy-20′-vinblastine (**511**) of established stereochemistry.

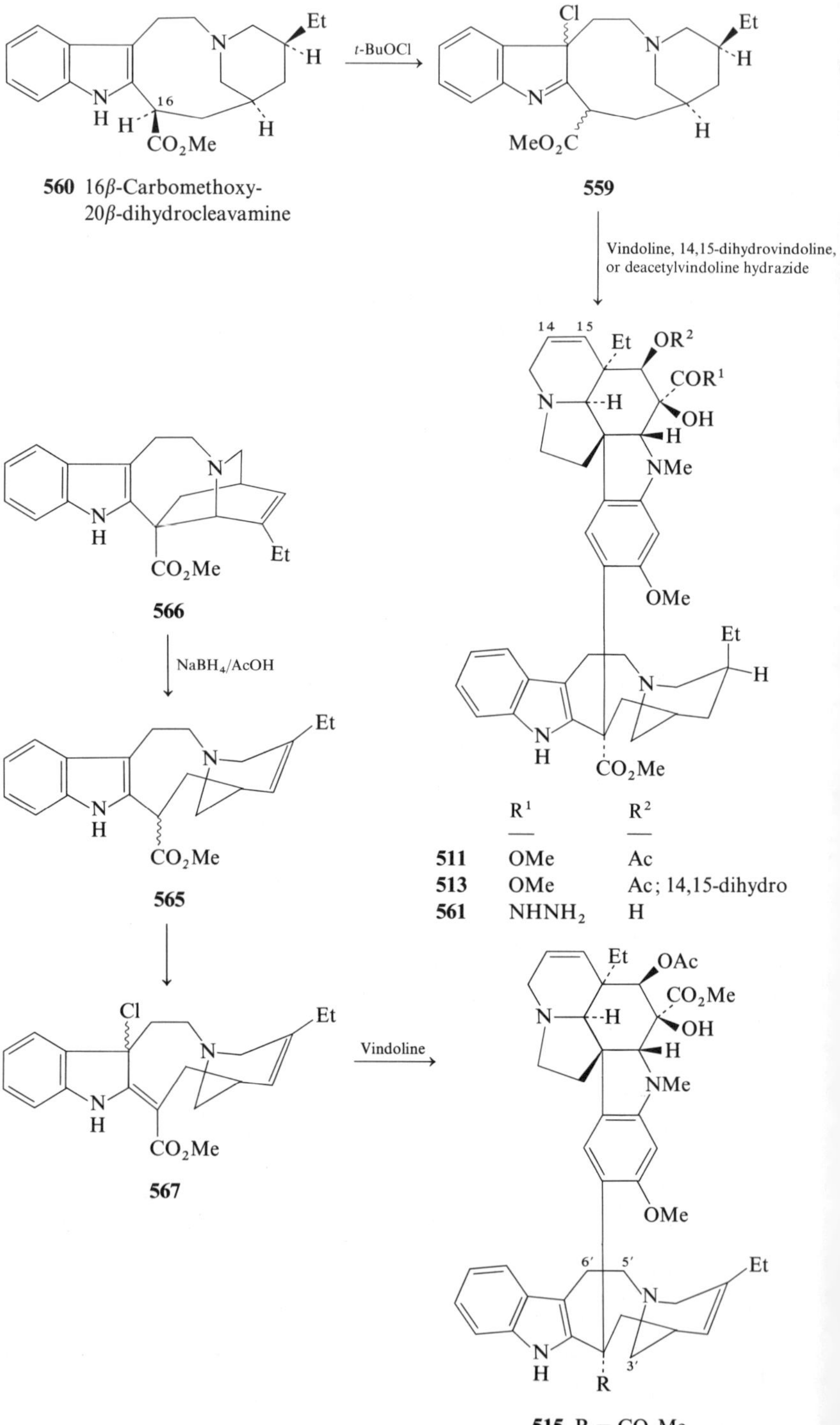

560 16β-Carbomethoxy-20β-dihydrocleavamine

t-BuOCl

559

Vindoline, 14,15-dihydrovindoline, or deacetylvindoline hydrazide

	R^1	R^2
511	OMe	Ac
513	OMe	Ac; 14,15-dihydro
561	$NHNH_2$	H

566

$NaBH_4$/AcOH

565

567

Vindoline

515 R = CO_2Me
568 R = H

The related 16′-decarbomethoxy compound **568** results from the coupling reaction between cleavamine and vindoline and also has the unnatural stereochemistry at C-16′ (*284*). Thus, the ^{13}C chemical shifts of C-3′ and C-6′ were characteristic of 16β-substituted cleavamine derivatives, e.g., 16β-carbomethoxycleavamine, rather than the α-series. Hence the ^{13}C-NMR spectra provide another criterion for the determination of the configuration at C-16′; however, this can obviously not be applied to the alkaloids themselves, where C-16′ is fully substituted.

None of the above attempts resulted in the formation of a compound with the natural stereochemistry at C-16′, and the detailed study by Kutney *et al.* indicated that the chloroindolenine approach was unlikely ever to yield the natural bases; therefore, a different approach was required. The major breakthrough, and the first successful synthesis of an alkaloid of vinblastine type, was achieved by Potier and collaborators (*237*, *285*), who applied the modified Polonovski reaction for the combination of the two monomeric alkaloid units. Almost immediately afterward, the same approach to the synthesis of the vinblastine group of alkaloids was reported by Kutney and his collaborators (*286*, *287*). The principle involved in this approach is that attachment of a good leaving group to N_b in a catharanthine component should result, under appropriate conditions, in the fragmentation of the 16,21-bond, which would leave C-16 susceptible to nucleophilic attack, e.g., by a vindoline molecule.

This proposed utilization of a catharanthine rather than a cleavamine or other 16,21-seco derivative may well provide a laboratory parallel for the biosynthetic process in which the dimeric alkaloids are formed. In the event, treatment of catharanthine N_b-oxide (**569**) with vindoline in the presence of trifluoroacetic anhydride followed by $NaBH_4$ reduction gave a mixture of anhydrovinblastine (**507**) and its 16′*R*-epimer (**515**) (*237*, *285*, *286*); under optimum conditions as much as 50% of anhydrovinblastine (**507**) could be obtained (*237*). Anhydrovinblastine (**507**) was identified by its spectral characteristics, and the 16′*S*-configuration was established by comparison of its CD spectrum with those of the natural alkaloids. Its structure was also confirmed by correlation with vinblastine which, when treated with sulfuric acid, suffered dehydration of the velbanamine component and hydrolysis of the acetate group, to give 15′,20′-anhydro-17-deacetylvinblastine (**570**), identical with the deacetylation product of the synthetic base **507**. This correlation proved rigorously that epimerization at C-14′, which is in principle possible in the postulated intermediate immonium ion **571**, had not occurred (*237*). The isomer **515**, which was shown spectroscopically to have the 16′*R*-configuration, was also shown to be identical with the product obtained earlier by coupling vindoline with the chloroindolenine of carbomethoxycleavamine (*237*).

The viability of this synthetic route was also demonstrated by a number of other coupling reactions, e.g., between the N_b-oxides of dihydrocatharanthine, coronaridine, allocatharanthine, or dihydroallocatharanthine as indole component, and indoline bases such as vindoline, 2,16-dihydro-11-methoxytabersonine, and its N_a-methyl derivative. The best results were obtained with vindoline as nucleophile; certainly, a basic N_a seemed essential, although the presence of oxygen at C-11 was not. Catharanthine and other bases containing a 15,20-double bond gave better results than did the 15,20-dihydro derivatives (*237*, *285*, *287*).

The yields and relative proportions of products **507** and **515** appear to depend critically on the experimental conditions, and indicate that two mechanisms may operate; thus it was suggested that **507** may be the result of a concerted fragmentation–coupling reaction, whereas **515** may be the result of a nonconcerted process. In the preliminary communications (*285*, *286*), no products other than **507** and **515** were reported; however, the reaction follows a more complex course, and in the full papers that followed (*237*, *287*) a more detailed examination of the reaction was described.

Both **507** and **515** are formed by fission of the 16,21-bond, which in catharanthine N_b-oxide is favorably aligned antiparallel to the N_b-oxygen bond. However, the 5,6-bond is also aligned in antiparallel fashion to the N_b-oxygen bond, and can also suffer fission in a competing reaction. Such a fragmentation would expose C-6 to nucleophilic attack by vindoline and the final product, following cyclization of C-5 on to N_a, would be the dimeric 1,3-diazine derivative **572** (*237*). This type of product, also obtained in the Polonovski coupling of vindoline with the N-oxides of other indole bases, is the minor product when the reaction is carried out at $-78°$ using the normal procedure, but becomes increasingly important in the presence of stronger nucleophiles, e.g., if vindoline is replaced by acetate or hydroxide ion (*288*). In all these reactions with vindoline a by-product was the trifluorohydroxyethyl derivative **573**, formed by acylation of vindoline followed by reduction (*237*, *287*).

The results obtained by Kutney *et al.* differed in some respects from those reported by Potier *et al.* At $-50°$ the coupling reaction between catharanthine N-oxide and vindoline was stated to give anhydrovinblastine (**507**) in approximately 50% yield, and no 16′-epi isomer was formed. As the temperature was raised the yield of anhydrovinblastine decreased, and the presence of the 16′-epi isomer was observed in the reaction mixture. At 61° no anhydrovinblastine was formed, but a 34% yield of 16′-epianhydrovinblastine (**515**) was isolated. This experimental evidence is consistent with a concerted α-attack at very low temperatures by vindoline at C-16 in the N-oxide, with simultaneous fission of the 16,21-bond and expulsion of the oxygen attached to N_b. Alternatively, at $-50°$, fragmentation of the 16,21-bond in the

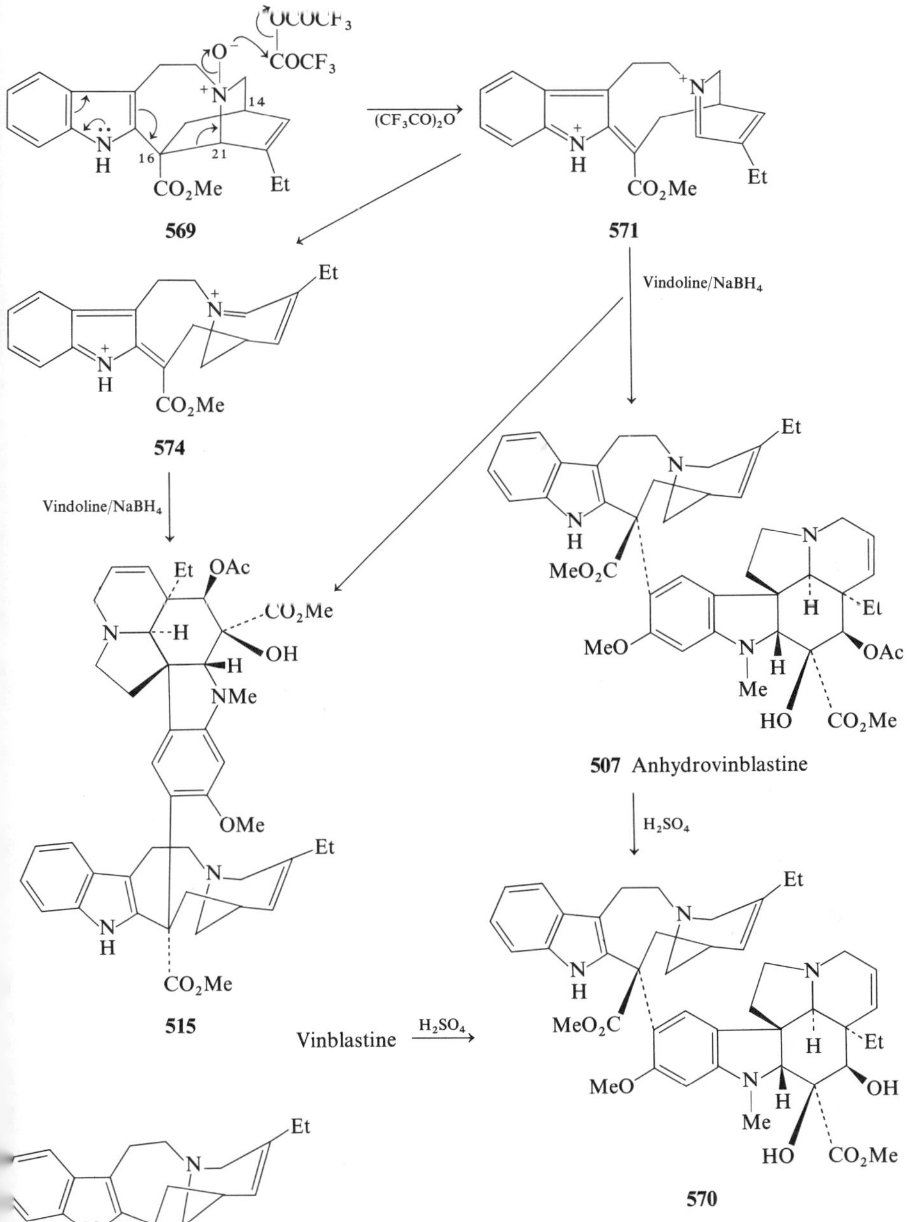

569 $\xrightarrow{(CF_3CO)_2O}$ **571**

574

571 $\xrightarrow{\text{Vindoline}/NaBH_4}$ **507** Anhydrovinblastine

574 $\xrightarrow{\text{Vindoline}/NaBH_4}$ **515**

507 Anhydrovinblastine $\xrightarrow{H_2SO_4}$ **570**

Vinblastine $\xrightarrow{H_2SO_4}$ **570**

515

570

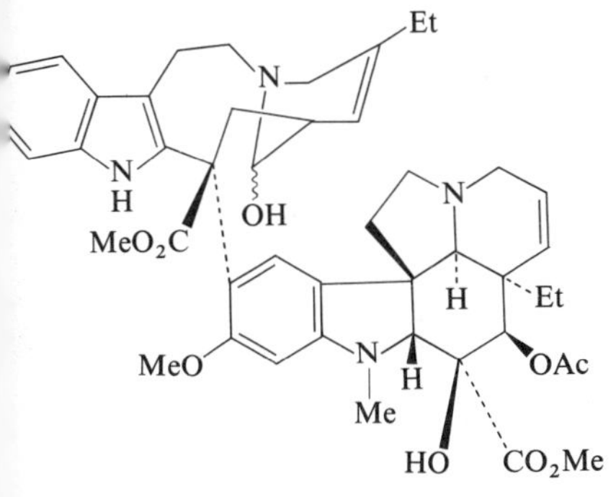

575

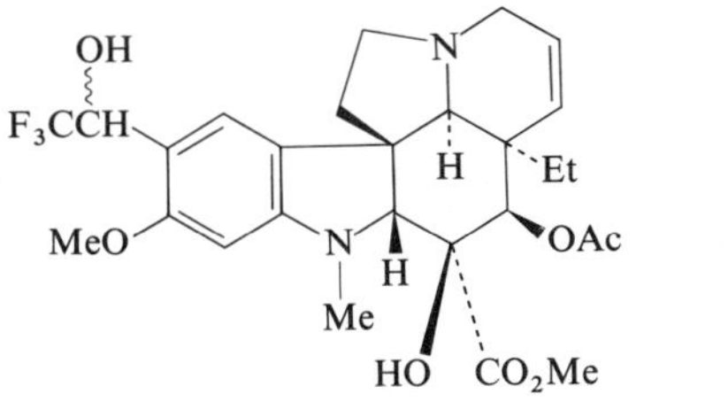

573

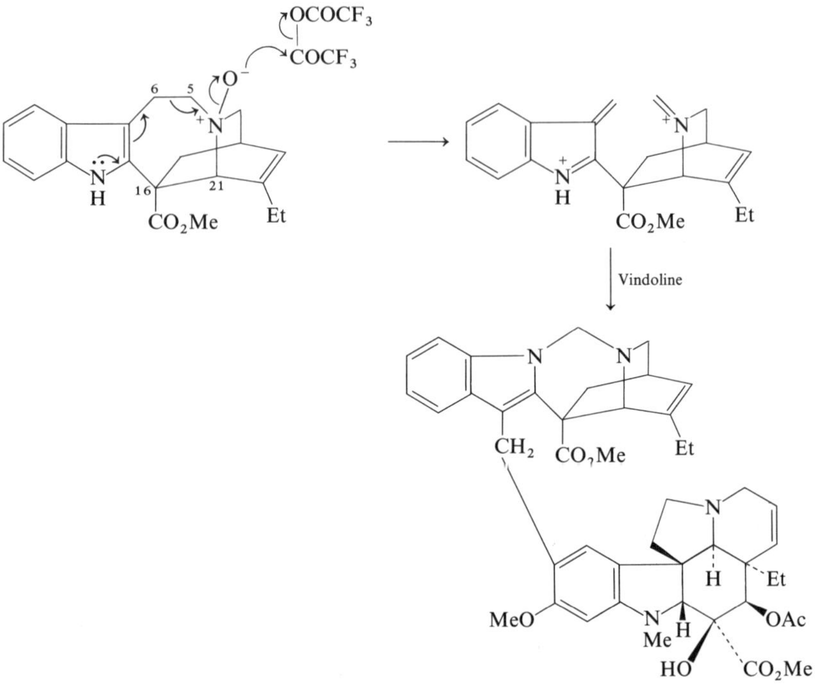

572

catharanthine *N*-oxide could give an immonium ion, which is frozen in its initial conformation (**571**), and suffers preferential attack by vindoline at the α-face to give a product (**507**) having the natural configuration at C-16′. At higher temperatures, ion **571** may undergo a conformational change to the "velbanamine" conformation (**574**), in which the β-face is now more accessible to nucleophilic attack; the preferential product from attack by vindoline is now the 16′-epi product **515** (*287*).

The product (**572**) obtained following fission of the 5,6-bond in catharanthine N_b-oxide appears not to have been obtained by Kutney *et al.*; however, another by-product was obtained if the reaction was not carried out in an inert atmosphere. At −50° in air the 3′-carbinolamine **575** was obtained in 8% yield; its formation could be suppressed if the reaction was conducted in an argon atmosphere, but the yield was augmented to 33% when the reaction was carried out in oxygen. The structure of **575** became clear from the downfield shift of C-3′ to 76.5 ppm, and was confirmed by the reduction (Sn–HCl) of **575** to anhydrovinblastine (**507**).

The coupling reaction of 15,20*S*-dihydrocatharanthine N_b-oxide (**576**) with vindoline under modified Polonovski conditions has also been thoroughly studied, and a number of interesting points have emerged. Potier

+ Vindoline

576

$(CF_3CO)_2O$,
then $NaBH_4$

+

	R^1	R^2
508	Et	H
579	H	Et

	R^1	R^2
577	Et	H
578	H	Et

580

581

et al. (*237*) obtained three deoxyvinblastine isomers (**508**, **577**, and **578**); the fact that the fourth stereoisomer (**579**) was not obtained was interpreted as indicating that the formation of compounds containing the natural (16′*S*) configuration proceeds by a concerted process, thus eliminating the possibility of epimerization at C-20′. The formation of both **577** and **578**, with 16′*R*-configuration, obviously requires the intermediacy of the immonium ion **580**, which can readily epimerize at C-20′ (**580** $\rightleftharpoons$ **581**).

The dimeric base **508** has the structure attributed to deoxyvinblastine B; its stereochemistry was established by conversion to the hydrazide, whose optical properties confirmed that it has the 16′*S*,20′*S*-configuration. It was also shown by comparison with authentic material to be deoxyvinblastine B, and to be obtainable by hydrogenation of anhydrovinblastine in the presence of Adams' catalyst, the double bond in the vindoline component remaining unaffected (*237*).

In the same coupling reaction between dihydrocatharanthine *N* oxide and vindoline, Kutney *et al.* observed the formation of all four stereoisomers (**508** and **577**–**579**) when the reaction was conducted at −10°; at lower temperatures (−50° to −30°), only the two epimers (**508** and **579**) having the natural (16′*S*) stereochemistry were obtained. The elucidation of the configuration at C-16′ in these isomers follows from the CD spectra; however, the stereochemistry at C-20′ is perhaps not so securely established. The premise of Kutney *et al.* is that hydrogenation of 16′-epianhydrovinblastine (**515**) proceeds at that face of the double bond which is opposite the very bulky vindoline component; the isomer **577** is, therefore, the product. Similarly, anhydrovinblastine (**507**) must give deoxyvinblastine (**579**), which was described as the major product of the coupling reaction at low temperatures. This compound, however, was not obtained by Potier *et al.*, who reported that **508** was both the product with the natural 16′*S*-configuration obtained in the coupling reaction, and also the product of hydrogenation of anhydrovinblastine. It should be noted, however, that Kutney's presumed **579**, the hydrogenation product of anhydrovinblastine, was not compared directly with natural deoxyvinblastine (**579**) or with deoxyvinblastine B (**508**) (*287*).

The coupling reaction between decarbomethoxycatharanthine N_b-oxide (**582**) and vindoline gave the expected dimeric products (**583** and **584**), but in low yield (*287*, *289*); the major product (32% yield) was initially (*287*) given structure **585**, but was later (*289*) shown to be analogous to the byproduct (**572**) obtained by Potier *et al.* in the condensation involving catharanthine N_b-oxide, and to have structure **586**.

Finally, coupling reactions were also carried out between decarbomethoxycatharanthine N_b-oxide (**582**) and vindoline *N*-methylamide (**587**), and between catharanthine N_b-oxide (**569**) and vindoline *N*-methylamide. The

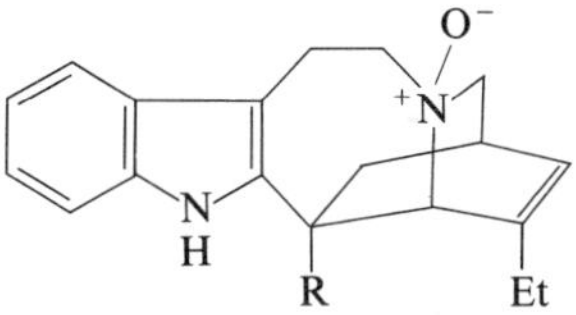

569 R = CO_2Me
582 R = H
+

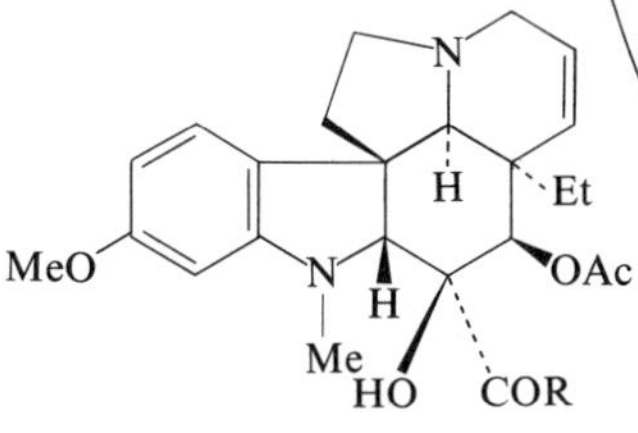

378 R = OMe
587 R = NHMe

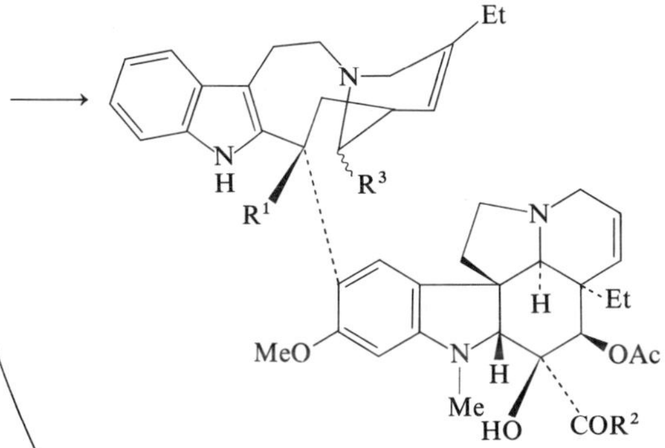

	R^1	R^2	R^3
583	H	OMe	H
588	H	NHMe	H
590	CO_2Me	NHMe	H
591	CO_2Me	NHMe	OH

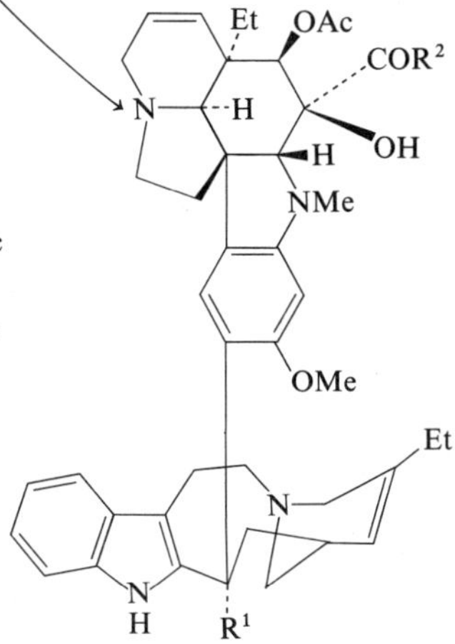

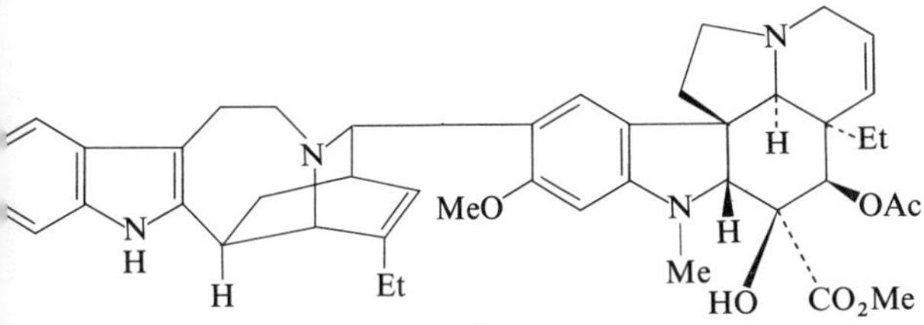

585

586

	R^1	R^2
584	H	OMe
589	H	NHMe
592	CO_2Me	NHMe

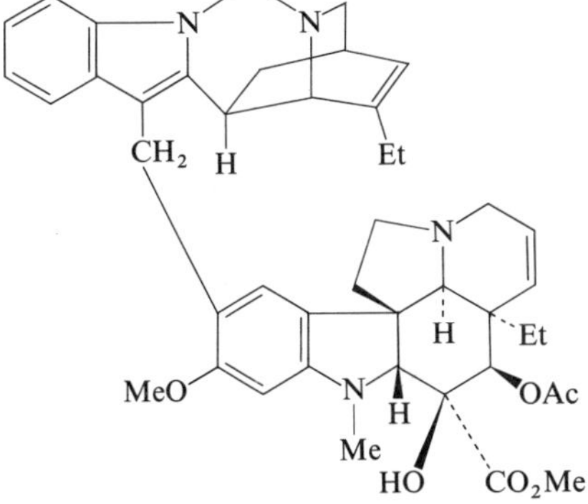

15,20*S*-Dihydrocatharanthine $N_{b'}$-oxide (**576**) + vindoline (**378**) $\xrightarrow{(CF_3CO)_2O}$ **593**

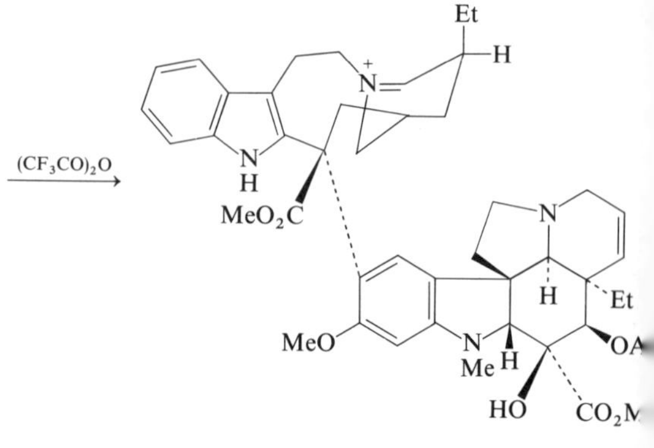

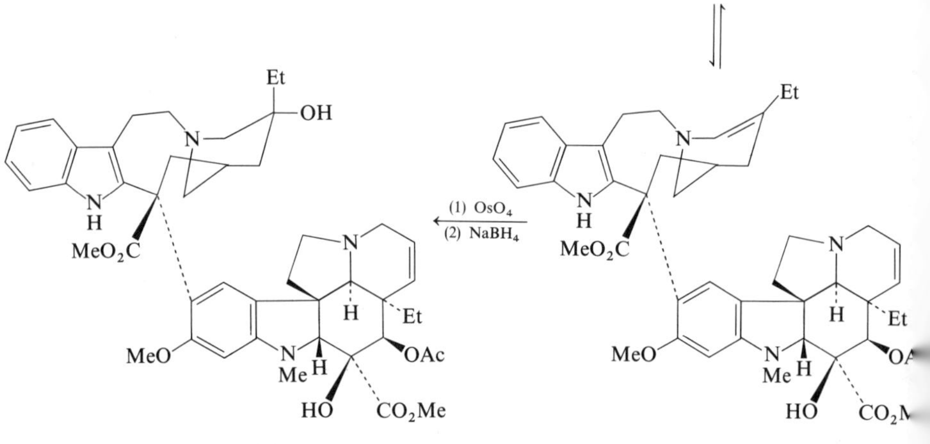

593 ⇌ **594**

594 $\xrightarrow{(1)\ OsO_4,\ (2)\ NaBH_4}$ **506** Leurosidine

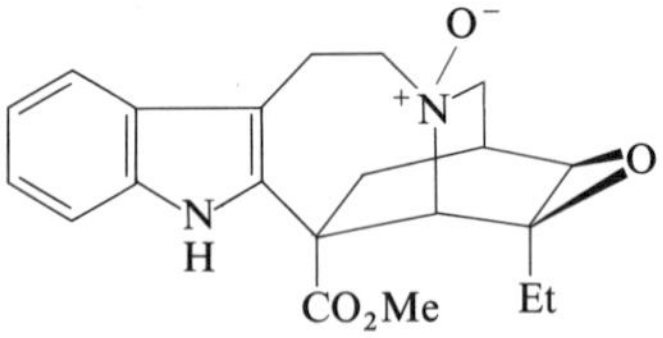

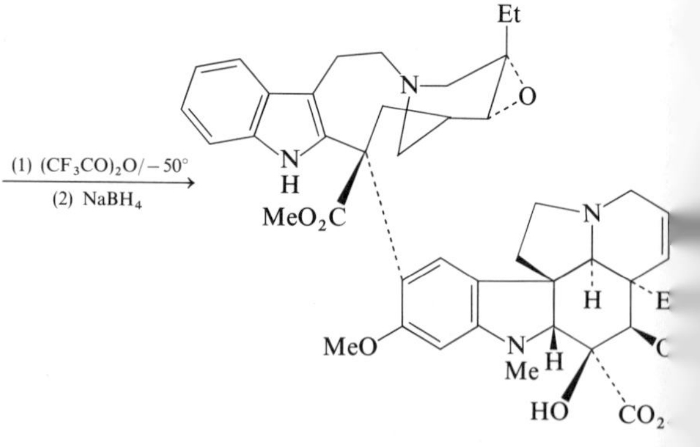

595 + vindoline $\xrightarrow{(1)\ (CF_3CO)_2O/-50^\circ,\ (2)\ NaBH_4}$ **503** Leurosine

former reaction gave the expected dimers **588** and **589**, but the latter reaction gave only isomer **590**, having the natural configuration at C-16′, together with the related 3′-carbinolamine (**591**); apparently, none of the 16′-epi compound **592** was obtained (*287*).

Following these initial studies in which the modified Polonovski reaction was shown to lead to the desired stereochemistry at C-16′ in the dimeric bases, other applications of the same coupling process were reported in rapid succession. One of the first of these involved an ingenious modification of the coupling reaction, and led to the first synthesis of leurosidine (**506**) (*290*). The coupling of 15,20*S*-dihydrocatharanthine N_b-oxide with vindoline in the presence of trifluoroacetic anhydride gives, as initial product, the immonium ion **593** which, in the absence of the reducing agent added in previous couplings, is presumably in equilibrium with the 20′,21′-enamine (**594**).

The vindoline (14,15-) double bond in **594** is considerably more hindered sterically than the enamine (20′,21′-) double bond; accordingly, osmylation of the enamine double bond from the less hindered α-side, followed by reductive removal ($NaBH_4$) of the carbinolamine (C-21′) hydroxyl group, afforded leurosidine (**506**) (*290*).

Kutney's synthesis of leurosine (**503**) provides an independent and unambiguous proof of the configuration of the epoxide function (*291*, *292*).

Condensation of the N_b-oxide epoxide **595**, of known stereochemistry, with vindoline at $-50°$ in the presence of trifluoroacetic anhydride, followed by reduction with sodium borohydride, gave leurosine (**503**) directly. Unlike other condensations with catharanthine derivatives lacking oxygen substituents at positions 15 and 20, the coupling of **595** with vindoline appeared to give no product containing the unnatural configuration at C-16′. In view of this, it was clearly of interest to observe the behavior, in the coupling reaction, of the 15′,20′-epimeric epoxide **596**. Unexpectedly, the only dimeric base that could be isolated was the product (**597**) of fission of the 5,6-bond in **596**, followed by coupling with vindoline and cyclization (*292*).

This reaction in turn prompted the investigation of other coupling reactions with 20α-substituted dihydrocatharanthine derivatives (*292*, *293*). The first such reaction involved the coupling of 20α-hydroxydihydrocatharanthinic acid lactone N_b-oxide (**598**) with vindoline under the usual conditions, a reaction which was investigated simultaneously by Potier (*294*) and Ban (*295*). All groups of workers obtained dimer **599**, the product of fission of the 5,6-bond in **598** followed by condensation with vindoline and cyclization. Kutney's second product proved to be the related lactol **600**, which was simply the product of overreduction of **599** during the reaction; in confirmation, **600** could be obtained by the controlled $NaBH_4$ reduction of **599** in the cold. Both bases **599** and **600** gave a common product, the

596

+ vindoline

$(CF_3CO)_2O$

597

pentahydroxy base **601**, on reduction with LAH (*292*). Potier's major product was the base **599**, which was obtained together with an unidentified dimeric base (*294*). Kutney's third product was shown to be trimeric, and is very probably identical with Ban's major product, which was given structure **602**; this clearly arises by nucleophilic attack by two vindoline molecules on the product of fragmentation of the 5,6-bond in **598** (*295*). All authors were in agreement that the desired product of coupling, i.e., the lactone base **603**, which in principle could be used as a precursor of vinblastine, was not formed.

The first synthesis of vinblastine (**502**) itself has been claimed by Atta-ur-Rahman and collaborators (*296*). Again the modified Polonovski coupling reaction was utilized, the ingredients being the N_b-oxide of 20-acetoxy-15,20-dihydrocatharanthine and vindoline. The catharanthine derivative was represented as **604**, but if the coupling reaction gave 20-acetylvinblastine (**605**), as reported, it must presumably have had the structure and stereochemistry shown in **606**. Mild alkaline hydrolysis of **605** afforded deacetylvinblastine (**607**) which, on acetylation at C-17, gave vinblastine (**502**). This procedure was also stated to give a small amount of 16′-epivinblastine.

This synthesis of vinblastine depends critically on the structure of the catharanthine derivative presumed to be 20α-acetoxy-15,20-dihydrocatharanthine N_b-oxide (**606**). However, following a detailed examination of the reactions of catharanthine, Kutney *et al.* (*297*) concluded that α-approach to the catharanthine molecule is severely hindered, and in fact α-functionalization at C-20 has only been observed through lactone formation by intramolecular attack by the carbomethoxy group. In view of this, the formation of 20α-acetoxydihydrocatharanthine by the modified Prévost reaction on catharanthine, which was the route used (*298*) for the preparation of **606**,

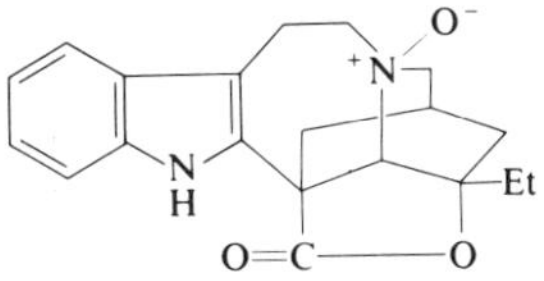

598 + vindoline

(1) $(CF_3CO)_2O$
(2) $NaBH_4$

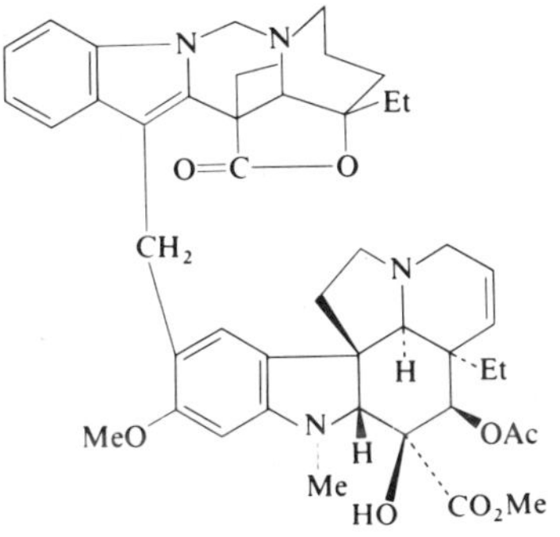

599

$NaBH_4$ MeOH 0

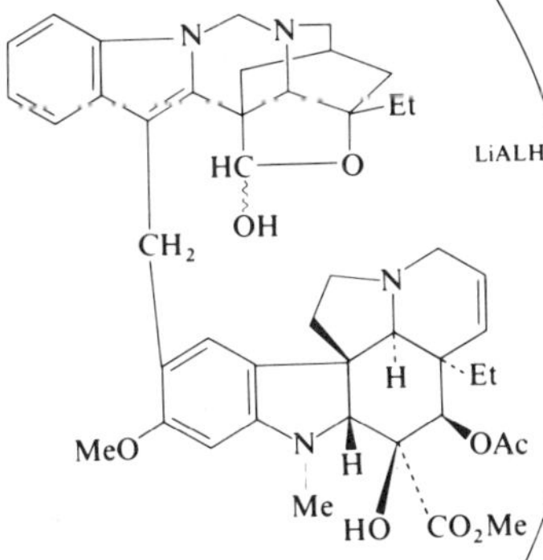

600

$LiAlH_4$

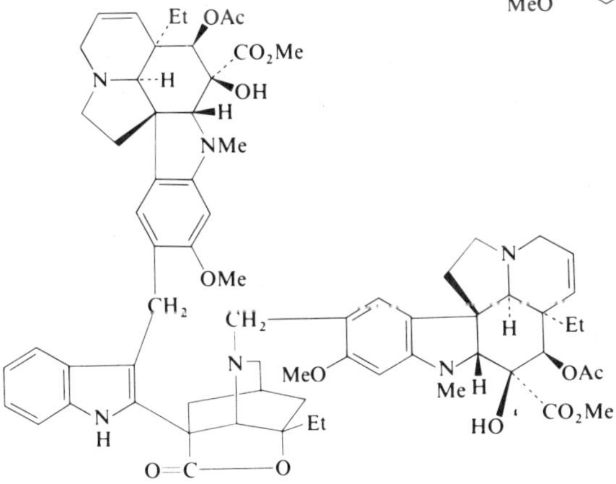

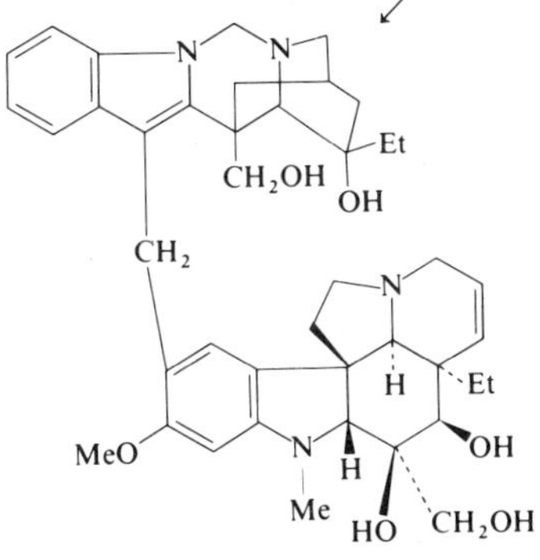

601

$LiALH_4$

602

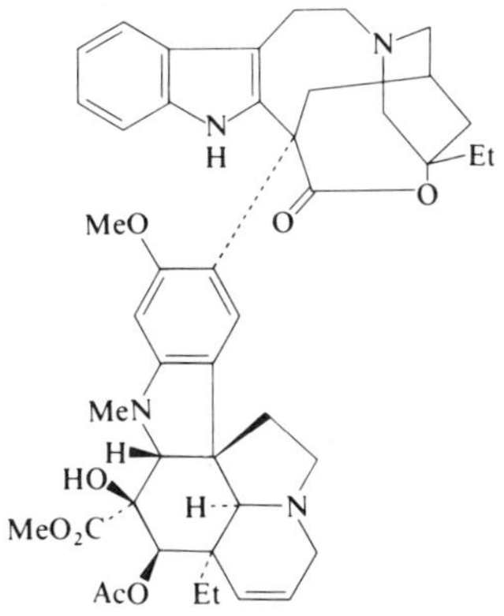

603

is surprising; in fact, Kutney *et al.* have so far been unable to repeat this vitally important preparation.

Recent synthetic work in this area has centered on the synthesis of novel isomers and close relatives of vinblastine, in the continuing search for clinically useful compounds. Thus the Polonovski coupling reaction between 15β-acetoxydihydrocatharanthine *N*-oxide (**608**) and vindoline gave a mixture of 15′α-acetoxy-20′-deoxyleurosidine (**609**) and 15′α-acetoxy-20′-deoxyvinblastine (**610**); no product of type **611**, resulting from fission of the 5,6-bond in **608**, was apparently obtained (*299*).

This result stands in sharp contrast to that reported by Honma and Ban (*300*), who in the same reaction obtained 15′,20′-anhydrovinblastine (**507**) (22%) and the base **611** (32%) produced by fission of the 5,6-bond of **608**. A common product does not appear to have been obtained by the two groups of workers. The natural stereochemistry at C-16′ was also obtained in the product (**612**) of coupling of the epimeric acetates **613** and **614** with vindoline. Because both C-20′ epimeric starting materials gave rise to the same product, epimerization to the preferred epimer must have occurred in

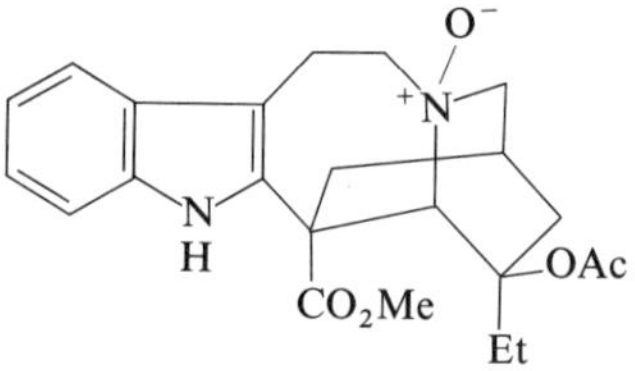

604 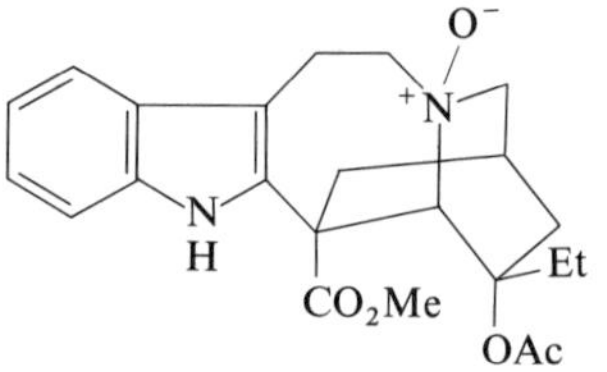 **606**

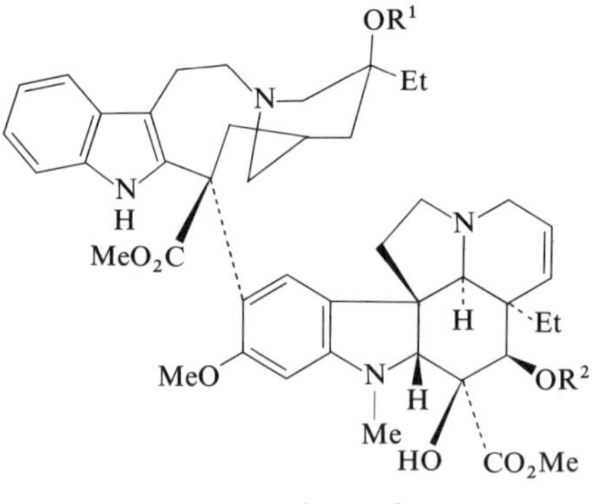

	R^1	R^2
605	Ac	Ac
607	H	H
502	H	Ac

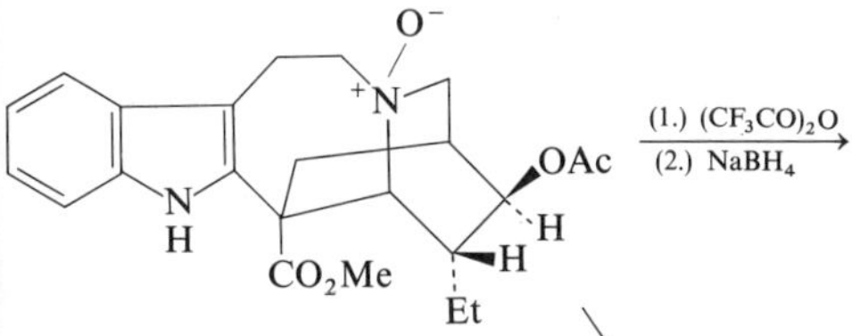

(1.) $(CF_3CO)_2O$
(2.) $NaBH_4$

608 + vindoline

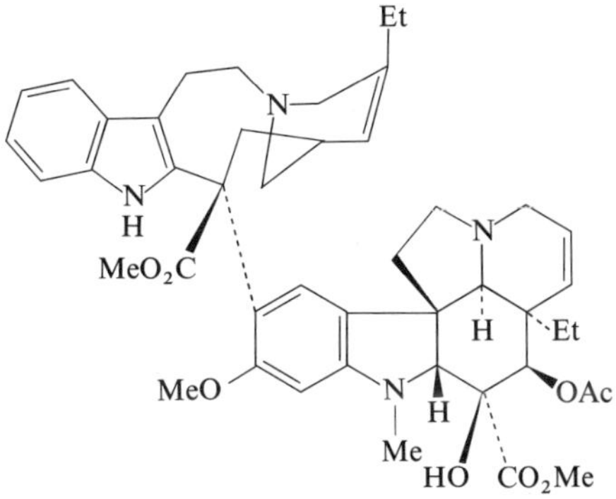

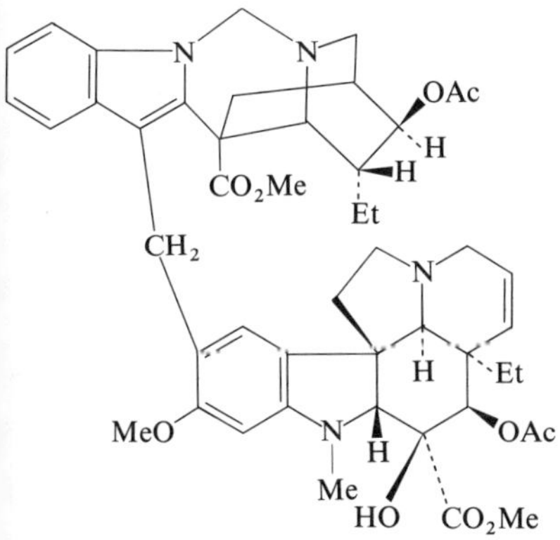

611

507

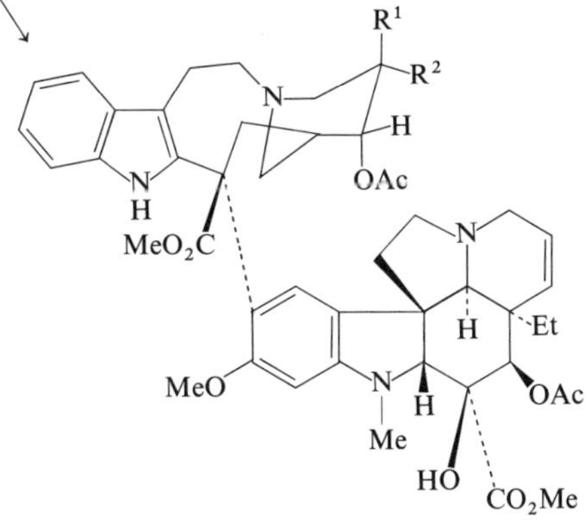

	R^1	R^2
609	Et	H
610	H	Et

the intermediate immonium ion. The other products in these coupling reactions were the bases **615** (from **613**) and **616** (from **614**); the stereoisomerism of the starting materials was preserved in these products, because C-20′ was not involved in their formation (*300*). In neither of these reactions was anhydrovinblastine formed.

It would thus appear from the results of Honma and Ban that the presence of an α-acetoxy group at C-15 in the catharanthine component severely inhibits the approach of the vindoline molecule and the fission of the 16,21-bond in the coupling reaction, as base **612** was obtained in yields of only 6% and 4%, respectively, from **613** and **614**. The effect of a β-acetoxy group is less well-defined. Honma and Ban report the formation of anhydrovinblastine (**507**), but only as the minor product of the reaction, whereas Kutney and Worth (*299*) obtained **609** and **610**, but in unspecified yield. For the synthesis of vinblastine derivatives the absence of a C-15 substituent, as in catharanthine, is preferable; anhydrovinblastine, for example, can be obtained in up to 50% yield by the coupling reaction from catharanthine N_b-oxide (*237*).

The availability of 15′,20′-anhydrovinblastine by synthesis affords another route for the preparation of 15′-oxygenated derivatives, by direct functionalization of the 15′,20′-double bond. Hydroboration–oxidation of **507** gave a mixture of 15′β-hydroxy-20′-deoxyvinblastine (**617**) and the primary alcohol (**618**) formed as a result of concomitant reduction of the vindoline 16-ester group. Hence it would appear that hydroboration–oxidation of catharanthine, used in the preparation of **608**, followed by coupling with

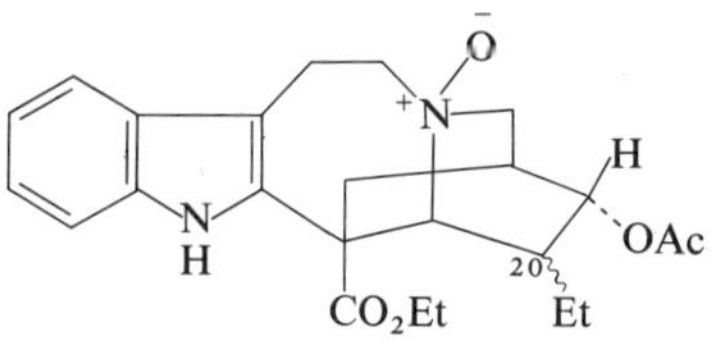

(1) Vindoline, $(CF_3CO)_2O$
(2) $NaBH_4$

613 β-Et at C-20
614 α-Et at C-20

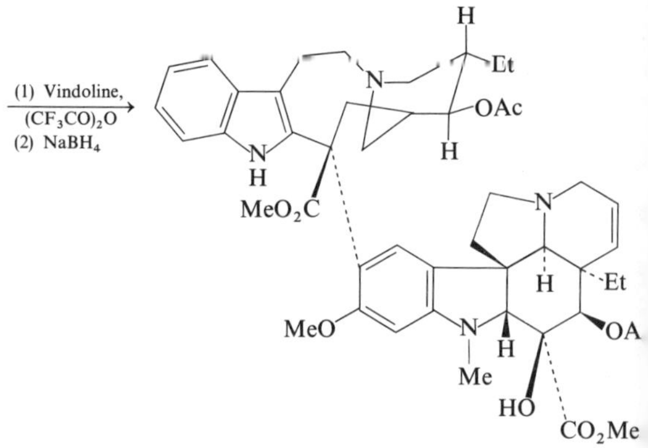

612

+

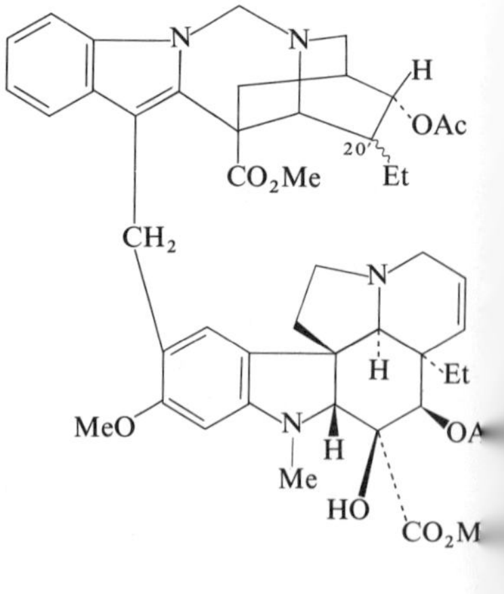

615 β-Et at C-20′
616 α-Et at C-20′

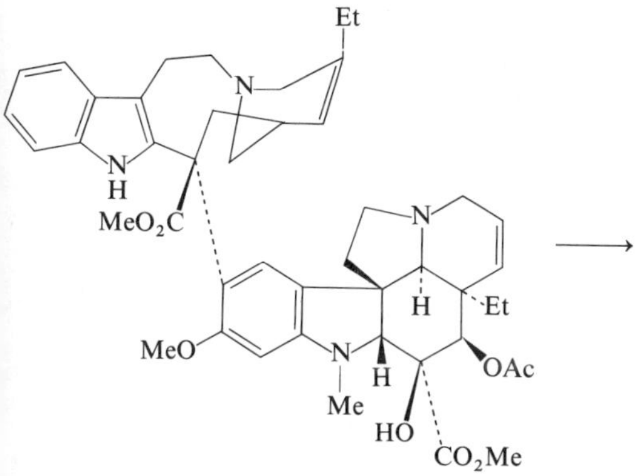

507 → **617** + **618**

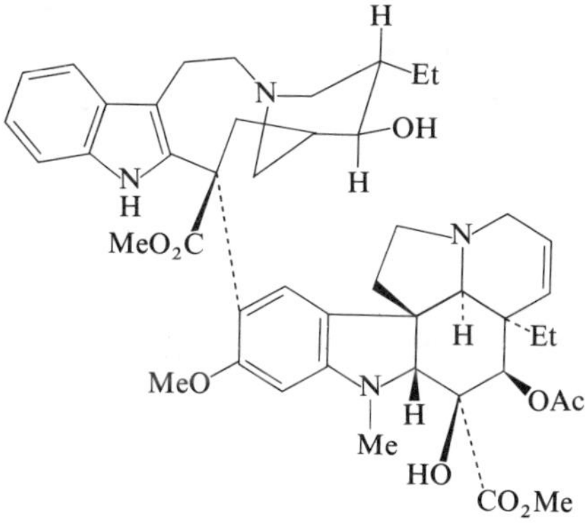

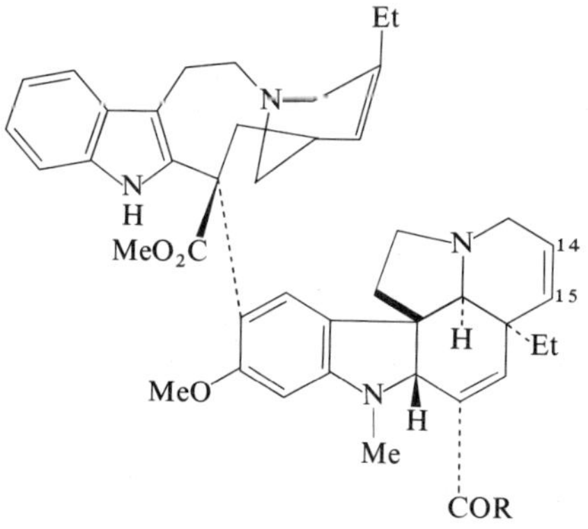

619 R = OMe
621 R = OMe, 14,15-dihydro
620 R = NHMe
622 R = NHMe, 14,15-dihydro

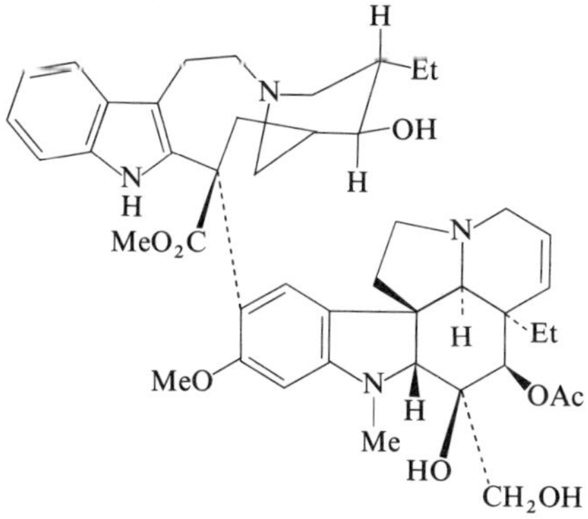

vindoline, affords a synthesis of 15′α-oxygenated dimers (e.g., **609** and **610**), whereas anti-Markovnikov hydration of anhydrovinblastine by the same process leads to 15′β-oxygenated derivatives (e.g., **617** and **618**) (*299*). This latter result is of some interest, because it conflicts with previous observations that electrophilic attack at the 15′,20′-double bond (e.g., epoxidation) normally proceeds preferentially at the α-face (see, for example, the synthesis of leurosine, discussed above). If structures **617** and **618** are confirmed, it must presumably indicate that in the hydroboration reaction initial coordination of borane by $N_{b'}$ occurs, which would result in subsequent attack on the double bond being directed at the β-face.

Other vinblastine derivatives that have been prepared by the Polonovski coupling reaction, primarily for biological evaluation, include the ester **619** and the *N*-methylamide **620**, their 14,15-dihydro derivatives **621** and **622**, and the hydroxyester **623** (*254*, *301*).

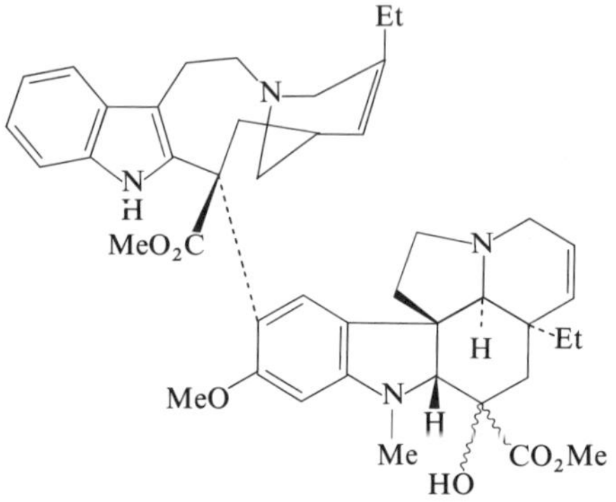

623

Finally, mention may be made of attempts to invert the configuration at C-16′ in compounds of the C-16′ epi series; such a method would rehabilitate the chloroindolenine approach as a viable one to the synthesis of vinblastine derivatives, and would enhance the value of the Polonovski coupling reaction in those cases where the C-16′ epi compound is also formed. One such possibility was found in the hydrazine reaction, but it can only lead to the desired C-16 configurational series in the decarbomethoxy derivatives (*302*). The reaction hinges on the removal of the ester function in Iboga derivatives by reaction with hydrazine which, it has been suggested, proceeds by the mechanism shown in Scheme 27 (*303*). Such a mechanism, which postulates the intermediacy of a species (a) containing a trigonal C-16, would necessarily allow overall stereochemical inversion at this position. The experimental results were of some interest. Vinblastine (**502**), on reaction with hydrazine, suffered loss of the ester group but retained its configuration at C-16′; the product was, therefore, the 16-hydrazide **624**. The dimer **625**, which has the unnatural configuration at C-16′ but is devoid of the C-16′ ester group, was unaffected at this position by hydrazine, and the sole reaction was the replacement of the ester group in the vindoline component by a hydrazide group (→**626**). However, synthetic 16′-epi-20′-deoxy-20′-epivinblastine (**627**) gave a mixture of hydrazide **628** (44%) having the natural configuration at C-16′ and the C-16′-epihydrazide **629** (10%). It would thus appear that in the dimeric series protonation of the intermediate (a) leads preferentially to the compounds having the natural configuration at C-16′.

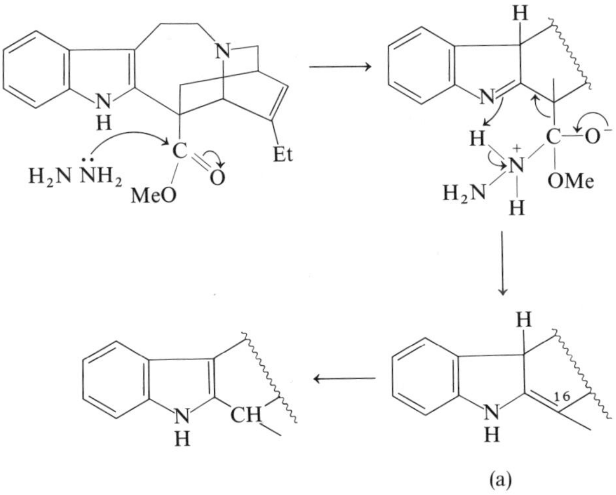

(a)

SCHEME 27.

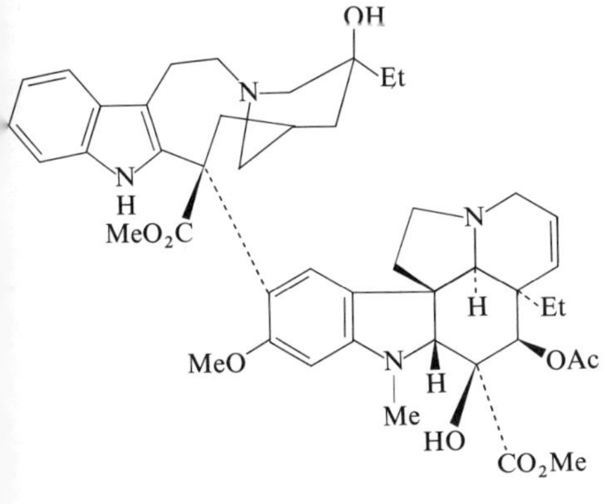

502

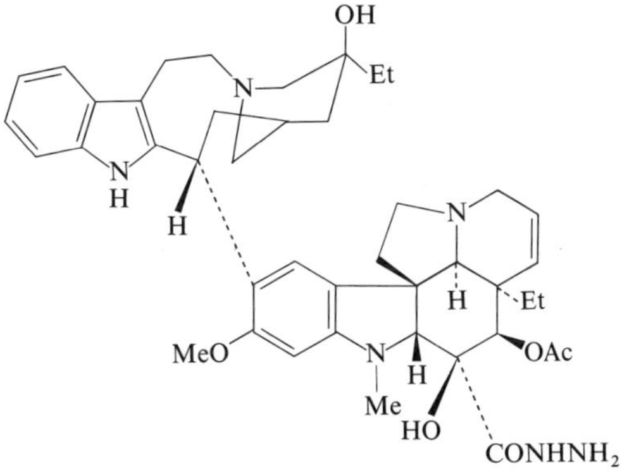

624

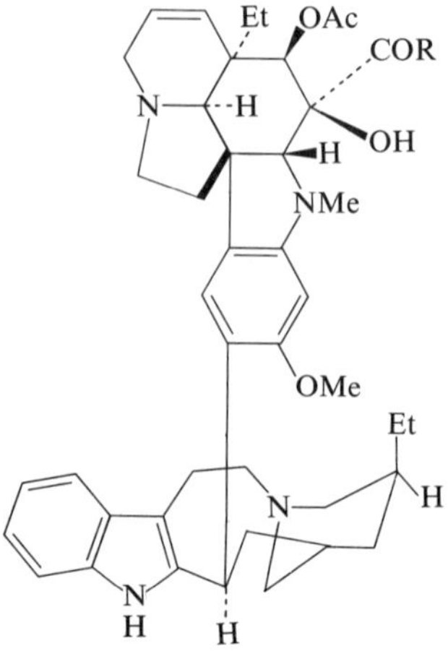

625 R = OMe
626 R = $NHNH_2$

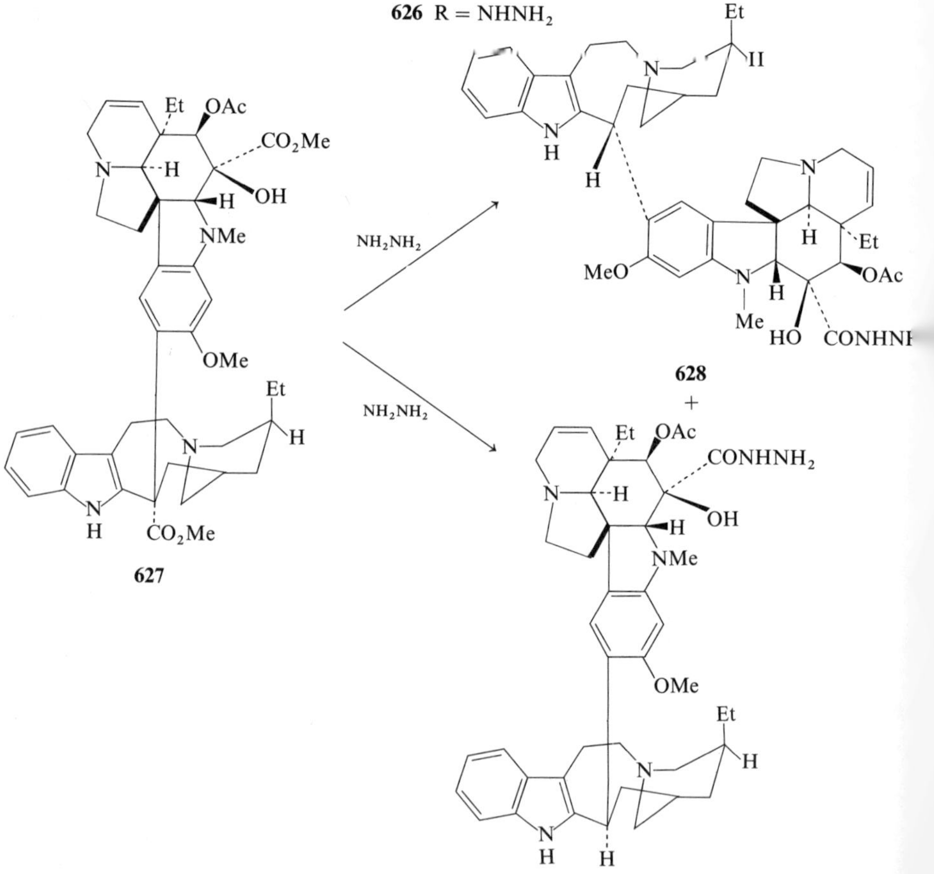

627

NH_2NH_2

NH_2NH_2

628
+
629

XXVI. Summary of the Isolation of the Bisindole Alkaloids

Table XXI summarizes the isolation of structurally identified bisindole alkaloids and Table XXII those alkaloids for which structures remain to be determined.

XXVII. Synthesis of Miscellaneous Bisindole Alkaloids

A. A Dimer in the Camptothecine Series

In their work connected with the biomimetic synthesis of the camptothecine chromophore, Hutchinson and co-workers reported the isolation of a dimer (*455*). Thus, oxidation of **630**, having either the C-3α or C-3β stereochemistry, with DDQ in benzene under reflux gave the dimer **631**, mp 160°, in 73% yield.

The compound showed IR absorptions for acetate (ν_{max} 1761 cm^{-1}) and pyridone (1667 cm^{-1}) functionalities, and a UV spectrum [λ_{max} 245, 253, 290, 335.(sh), 367, and 385 nm] analogous to that of camptothecine. The ^{1}H-NMR spectrum (270 MHz) indicated the presence of an ethyl side chain (triplet at 0.95 ppm, multiplet at 1.90 ppm), eight acetate groups (2.02–2.09 ppm), a multiplet for C-20 at 2.91 ppm, a singlet for the C-5 proton at

OAc H H O O H O glu(OAc)$_4$

630

DDQ
C_6H_6, 20 hr

OAc N O 17 H 20 21 O H 18 O glu(OAc)$_4$ 2

631

5.18 ppm, and a doublet for C-21 at 5.89 ppm. Eight aromatic protons were observed in the region 7.60–8.15 ppm and the H-17 and H-14 protons at 6.45 and 7.24 ppm, respectively.

The dimeric nature of **631** was established by determination of the molecular weight by osmometry and the C-17 carbon appearing as a doublet at 89.7 ppm in the single frequency off-resonance proton decoupled spectrum of **631** (*455*). The *S*-absolute stereochemistry at C-17 was determined from the CD spectrum (*455*).

TABLE XXI
ISOLATION AND PROPERTIES OF THE BISINDOLE ALKALOIDS

Alkaloid	Source (*refs.*)	Molecular formula	Physical data (*refs.*)						
			UV	IR	Mass	^{1}H-NMR	^{13}C-NMR	X-Ray	Synthesis
II. Tryptamine-tryptamine type									
Staurosporine (**1**)	*Streptomyces staurosporeus* (*31*)	$C_{28}H_{26}N_4O_3$	(*31*)	(*31*)	(*31*)	(*31*)		(*32*)	
Trichotomine (**9**)	*Clerodendron trichotomum* Thunb. (*35, 36*)	$C_{30}H_{20}N_4O_6$	(*35, 36*)	(*35, 36*)	(*35, 36*)	(*35, 36*)		(*37*)	(*38–40*)
Trichotomine G_1 (**17**)	*C. trichotomum* (*36*)	$C_{36}H_{30}N_4O_{11}$	(*36*)	(*36*)		(*36*)			
N,N'-Di-(D-glucopyranosyl)trichotomine (**18**)	*C. trichotomum*	$C_{42}H_{40}N_4O_{16}$	(*36*)	(*36*)		(*36*)			
Calycanthine (**19**)	*Bhesa archiboldiana* (Merr. and Perry) Ding Hou (*44*) *Calycanthus floridus* L. (*304, 305*) *C. glaucus* Wild (*306, 307*) *C. occidentalis*, Hook and Arn. (*308*) *Chimonanthus praecox* (L.) Link (*305*) *Palicourea alpina* (Sw.) DC. (*45*)	$C_{22}H_{26}N_4$	(*309*)	(*309*)	(*45*)	(*309*)		(*310, 311*)	(*55*)

Calycanthidine	*Calycanthus glaucus* (312)	$C_{23}H_{28}N_4$	(313)	(314)	(313–315)	(314)			
Chimonanthine (**20**)	*Chimonanthus praecox* (12) *Calycanthus floridus* (55)	$C_{22}H_{26}N_4$	(55)	(55, 316)	(55)	(55)		(48)	(55)
meso-Chimonanthine (**36**)	*C. floridus* (55)	$C_{22}H_{26}N_4$	(55, 69)	(55, 69)	(55, 69)	(55, 69)			(55, 69)
Folicanthine (**22**)	*C. floridus* (317, 318) *C. occidentalis* (319)	$C_{18}H_{23}N_4$	(318, 319)	(318)	(315)				
Chaetocin (**42**)	*Chaetomium minutum* (63)	$C_{30}H_{28}N_6O_6S_4$	(63)	(63)	(63)	(63)	(64)	(63)	
Chaetomin (**41**)	*C. cochlioides* (59–62) *C. globosum* (62)	$C_{31}H_{30}N_6O_6S_4$	(61, 62)	(62)	(62)	(62)			
11α,11′α-dihydroxy-chaetocin (**43**)	*Verticillium tenerum* (64)	$C_{30}H_{28}N_6O_8S_4$	(64)	(64)	(64)	(64)	(64)		
Verticillin A (**44**)	*Verticillium* sp. (65, 66)	$C_{30}H_{28}N_6O_6S_4$	(65, 66)	(65, 66)	(65, 66)	(65, 66)			
Verticillin B (**45**)	*Verticillium* sp. (66)	$C_{30}H_{28}N_6O_7S_4$	(66)	(66)		(66)			
Verticillin C	*Verticillium* sp. (66)	$C_{30}H_{28}N_6O_7S_5$	(66)	(66)		(66)			
Hodgkinsine (**46**)	*Hodgkinsonia frutescens* C. T. White (67)	$C_{33}H_{38}N_6$	(67, 72)	(67, 72)	(69, 71–73)	(73)	(73)	(70, 71)	
Psychotria base	*Psychotria beccarioides* Wernh. (73)	$C_{33}H_{38}N_6$	(73)	(73)	(73)	(73)			
Quadrigemine A (**55**)	*Hodgkinsonia frutescens* (74)	$C_{44}H_{50}N_8$	(74)	(74)	(74)	(74)			

(*Continued*)

TABLE XXI (*Continued*)

Alkaloid	Source (*refs.*)	Molecular formula	Physical data (*refs.*)						
			UV	IR	Mass	^{1}H-NMR	^{13}C-NMR	X-Ray	Synthesis
Quadrigemine B (**56**)	*H. frutescens* (*74*)	$C_{44}H_{50}N_8$	(*74*)	(*74*)	(*74*)	(*74*)			
Psychotridine (**62**)	*Psychotria beccarioides* (*73*)	$C_{55}H_{62}N_{10}$	(*73*)	(*73*)	(*73*)	(*73*)	(*73*)		
III. Tryptamine–tryptamine type with an additional monoterpene unit									
Borreverine (**68**)	*Borreria verticillata* G. F. W. Mey (*76*) *Flindersia fournieri* (*79*)	$C_{32}H_{40}N_4$	(*76*)	(*76*)	(*76*)	(*76*)		(*76*)	(*77*)
Isoborreverine (**70**)	*Borreria verticillata* (*78*) *Flindersia fournieri* (*79*)	$C_{32}H_{40}N_4$	(*78*)	(*78*)	(*78*)	(*78*)			(*77*)
IV. Corynanthe–tryptamine type									
Roxburghine B (**85**)	*Uncaria gambier* Roxb. (*80*)	$C_{31}H_{32}N_4O_2$	(*80*)	(*80*)	(*80*)	(*80, 87, 88*)	(*89*)		
Roxburghine C (**86**)	*U. gambier* (*80*)	$C_{31}H_{32}N_4O_2$	(*80*)	(*80*)	(*80*)	(*80, 87*)	(*89*)		
Roxburghine D (**79**)	*U. gambier* (*80*)	$C_{31}H_{32}N_4O_2$	(*80*)	(*80*)	(*80*)	(*80, 87*)	(*89*)		
Roxburghine E (**87**)	*U. gambier* (*80*)	$C_{31}H_{32}N_4O_2$	(*80*)	(*80*)	(*80*)	(*80, 87*)	(*89*)		
Usambarensine (**133**)	*S. usambarensis* Gilg. (*98, 99*)	$C_{29}H_{28}N_4$	(*98*)	(*98*)	(*98*)	(*98*)			

3′,4′-Dihydro-usambarensine (**120**)	*S. usambarensis* (*98, 99*)	$C_{29}H_{30}N_4$	(*98*)	(*98*)	(*98*)	(*98*)			(*103*)
Tchibangensine (**132**)	*S. tchibangensis* Pellgr. (*106*)	$C_{29}H_{30}N_4$	(*106*)	(*106*)	(*106*)	(*106*)	(*106*)		
N_b′-Methylusambarensine (**123**)	*S. usambarensis* (*98, 99*)	$C_{29}H_{31}\overset{+}{N}_4$	(*98*)	(*98*)	(*98*)				
N_b′-Methyl-3′,4′-dihydrousambarensine (**124**)	*S. usambarensis* (*98, 99*)	$C_{30}H_{33}\overset{+}{N}_4$			(*98*)				
3-Dehydro-ochrolifuanine (**97**)	*Ochrosia lifuana* Guill. (*93*)	$C_{29}H_{32}N_4$	(*93*)	(*93*)	(*93*)				
Ochrolifuanine A (**103**)	*O. lifuana* (*90, 93*) *O. miana* H. Bn ex Guill. (*96*)	$C_{29}H_{32}N_4$	(*90, 93, 96*)	(*90, 93, 96*)	(*90, 93, 96*)	(*90, 93, 96*)	(*97, 106*)		(*94*)
Ochrolifuanine B (**104**)	*O. confusa* Pichon (*95*) *O. lifuana* (*90, 93*) *O. miana* (*96*)	$C_{29}H_{34}N_4$	(*90, 93, 96*)	(*90, 93, 96*)	(*90, 93, 96*)	(*90, 93*)	(*97, 106*)		(*94*)
Usambarine (**127**)	*S. usambarensis* (*92, 104, 108*)	$C_{30}H_{34}N_4$	(*92, 104, 108*)	(*92, 104, 108*)	(*92, 104, 108*)	(*92, 104, 108*)			(*104*)
18,19-Dihydro-usambarine (**134**)	*S. usambarensis* (*102*)	$C_{29}H_{34}N_4O$	(*102*)	(*102*)	(*102*)	(*102*)			
Ochrolifuanine *N*-oxide (**100**)	*Ochrosia lifuana* (*93*)	$C_{29}H_{34}N_4O$	(*93*)	(*93*)	(*93*)				
10-Hydroxy-ochrolifuanine A (**107**)	*O. miana* (*96*)	$C_{29}H_{34}N_4O$	(*96*)	(*96*)	(*96*)				

(*Continued*)

TABLE XXI (*Continued*)

Alkaloid	Source (*refs.*)	Molecular formula	Physical data (*refs.*)						
			UV	IR	Mass	^{1}H-NMR	^{13}C-NMR	X-Ray	Synthesis
10-Hydroxy-ochrolifuanine B (**108**)	*O. miana* (*96*)	$C_{29}H_{34}N_4O$	(*96*)	(*96*)	(*96*)				
Usambaridine	*S. usambarensis* (*108*)	$C_{30}H_{34}N_4O$	(*108*)	(*108*)	(*108*)	(*108*)			
11-Hydroxy usambarine (usambaridine Br) (**136**)	*S. usambarensis* (*102*)	$C_{30}H_{34}N_4O$	(*102*)	(*102*)	(*102*)	(*102*)			
12-Hydroxy usambarine (usambaridine Vi) (**137**)	*S. usambarensis* (*102*)	$C_{30}H_{34}N_4O$	(*102*)	(*102*)	(*102*)	(*102*)			
18,19-Dihydro-11-hydroxy-usambarine (18,-19-dihydro-usambaridine Br) (**139**)	*S. usambarensis* (*102*)	$C_{30}H_{36}N_4O$	(*102*)		(*102*)				
18,19-Dihydro-12-hydroxy-usambarine (18,19-dihydro-usambaridine Vi) (**138**)	*S. usambarensis* (*102*)	$C_{30}H_{36}N_4O$	(*102*)	(*102*)	(*102*)	(*102*)			

Alkaloid B_1 (**148**)	*Rauwolfia obscura* K. Schum. (*109*)	$C_{30}H_{34}N_4O$	(*109*)	(*109*)	(*109*)	(*109*)		
Strychnobaridine (**141/142**)	*S. usambarensis* (*102*)	$C_{30}H_{34}N_4O_2$	(*102*)	(*102*)	(*102*)	(*102*)		
Cinchophyllamine	*C. ledgeriana* Moens (*110*)	$C_{31}H_{36}N_4O_2$	(*110, 111*)	(*110, 111*)	(*110, 111*)	(*111*)		
Isocinchophyllamine (**153**)	*C. ledgeriana* (*110*)	$C_{31}H_{36}N_4O_2$	(*110, 111*)	(*110, 111*)	(*110, 111*)	(*111*)		(*112*)
Strychnopentamine (**154**)	*S. usambarensis* (*102, 107*)	$C_{35}H_{43}N_5O$	(*102*)	(*102*)	(*102*)	(*102*)		(*107*)
Isostrychnopentamine A (**155**)	*S. usambarensis* (*102*)	$C_{35}H_{43}N_5O$	(*102*)	(*102*)	(*102*)	(*102*)		
Isostrychnopentamine B (**156**)	*S. usambarensis* (*102*)	$C_{35}H_{43}N_5O$	(*102*)	(*102*)	(*102*)	(*102*)		
V. Corynanthe–corynanthe type								
Serpentinine (**165**)	*Rauwolfia degeneri* Sherff (*320*) *R. ligustrina* Roem. et Schult. (*321*) *R. mauiensis* Sherff (*320*) *R. sandwicensis* A.D.C. (*320*) *R. serpentina* Benth. ex Kurz (*114, 322*) *R. tetraphylla* L. (*114, 323*) *R. vomitoria* Afz. (*324*)	$C_{42}H_{44}N_4O_6$	(*114–116, 323, 325*)	(*114, 115 325*)	(*116*)	(*116*)		(*117*)

(*Continued*)

TABLE XXI (*Continued*)

Alkaloid	Source (*refs.*)	Molecular formula	Fhysical data (*refs.*)						
			UV	IR	Mass	^{1}H-NMR	^{13}C-NMR	X-Ray	Synthesis
VI. Corynanthe–*Strychnos* type									
Geissospermine (**166**)	*Geissopermum laeve* (Vell.) Baill. (*118, 326, 327*) *G. sericeum* Benth & Hook (*328*)	$C_{40}H_{48}N_4O_3$	(*329, 330*)	(*331, 332*)	(*333*)		(*121*)	(*120*)	
Geissolosimine	*G. vellosii* Allem. (*334, 335*)	$C_{38}H_{44}N_4O$	(*334, 335*)	(*334, 335*)		(*335*)			
VIII. Vobasine–vobasine type									
Accedinisine (**168**)	*Tabernaemontana accedens* Muell.-Arg. (*122*)	$C_{41}H_{48}N_4O_3$	(*122*)	(*122*)	(*122*)	(*122*)			
Accedinine (**169**)	*T. accedens* (*122*)	$C_{41}H_{48}N_4O_4$	(*122*)	(*122*)	(*122*)	(*122*)			
Gardmultine (**180**)	*Gardneria multiflora* Makino (*125*)	$C_{45}H_{54}N_4O_{10}$	(*126*)	(*126*)	(*126*)	(*126*)			
VIII. Bisindole alkaloids of *Alstonia* species									
Macrocarpamine (**202**)	*Alstonia macrophylla* Wall. (*135*)	$C_{41}H_{46}N_4O_3$	(*135*)	(*135*)	(*135*)	(*135*)			
Macralstonidine (**194**)	*A. macrophylla* (*134, 336*) *A. spectabilis* R. Br. (*336, 337*)	$C_{41}H_{48}N_4O_3$	(*134*)	(*134*)	(*134*)	(*134*)			

Villalstonine (**184**)	*A. glabrifolia* Mgf. (*337*) *A. macrophylla* (*131, 133, 336, 338*) *A. muelleriana* Domin. (*128, 130, 339*) *A. spectabilis* (*336, 337*)	$C_{41}H_{50}N_4O_4$	(*131, 338*)	(*131, 132, 338*)	(*131, 132*)	(*131*)	(*142, 188*)	(*136, 139*)
Alstonisidine (**212**)	*A. muelleriana* (*128, 141*)	$C_{42}H_{48}N_4O_4$	(*128, 141*)	(*128, 141*)	(*141*)	(*141*)		(*137, 139*)
Des-N_a'-methyl-anhydroma-cralstonine (**190**)	*A. muelleriana* (*130*)	$C_{42}H_{48}N_4O_4$	(*130*)	(*130*)	(*130*)	(*130*)		
Macralstonine (**186**)	*A. glabrifolia* (*337*) *A. macrophylla* (*133, 134, 336, 340*) *A. muelleriana* (*128–130*)	$C_{43}H_{52}N_4O_5$	(*133*)	(*133*)	(*133*)	(*133*)		
IX. Pleiocarpamine–vincorine type								
Pleiocorine (**214**)	*A. deplanchei* van Heurck et Muell. (*142*)	$C_{41}H_{46}N_4O_5$	(*142*)	(*142*)	(*142*)	(*142*)	(*142*)	
X. Pleiocarpamine–akuammiline type								
Pleiocarpamine (**216**)	*A. deplanchei* (*143*)	$C_{41}H_{46}N_4O_5$	(*143*)	(*143*)	(*143*)	(*143*)	(*143*)	

(*Continued*)

TABLE XXI (*Continued*)

Alkaloid	Source (*refs.*)	Molecular formula	Physical data (*refs.*)						
			UV	IR	Mass	^{1}H-NMR	^{13}C-NMR	X-Ray	Synthesis
XII. Pseudoakuammigine–eburnea type									
Umbellamine (hunterine) (**223**)	*Hunteria eburnea* Pichon (*341*) *H. umbellata* (K. Schum.) Hall. F. (*145*)	$C_{41}H_{48}N_4O_4$	(*145*)	(*145*)	(*145*)	(*145*)			
XIII. *Strychnos–Strychnos* type									
Bisnor-C-curarine	*S. dolichothyrsa* Gilg ex Onochie et Hepper (*146*)	$C_{38}H_{38}N_4O$	(*146*)	(*146*)	(*146*)				(*146*)
Caracurine II	Calebash curare (*342*) *S. toxifera* Schomb. ex Benth. (*343*)	$C_{38}H_{38}N_4O_2$	(*343–345*)			(*345*)		(*346*)	(*347, 348*)
Bisnordihydrotox-iferine (**235**)	*S. dolichothyrsa* (*146*) *S. toxifera* (*349*)	$C_{38}H_{40}N_4$	(*146*)	(*146, 349, 350*)	(*146*)	(*146*)			(*350, 351*)
Bisnordihydrotox-iferine *N*-oxide	*S. dolichothyrsa* (*146*)	$C_{38}H_{40}N_4O$	(*146*)		(*146*)				(*146*)
Bisnordihydrotox-iferine *N,N'*-dioxide	*S. dolichothyrsa* (*146*)	$C_{38}H_{40}N_4O_2$	(*146*)						(*146*)
Caracurine V (**237**)	*S. dolichothyrsa* (*147*) *S. toxifera* (*343*)	$C_{38}H_{40}N_4O_2$	(*147, 343, 350*)	(*147, 350*)	(*147*)	(*345*)	(*147*)		(*147, 148, 352*)
Caracurine V *N*-oxide (**238**)	*S. dolichothyrsa* (*147*)	$C_{38}H_{40}N_4O_3$			(*147*)		(*147*)		(*147*)

Caracurine V *N,N′*-dioxide (**239**)	*S. dolichothyrsa* (*147*)	$C_{38}H_{40}N_4O_4$			(*147*)		(*147*)		(*147*)
Bisnor-C-alkaloid D	*S. dolichothyrsa* (*146*)	$C_{38}H_{44}N_4O_2^{++}$							(*146*)
C-Curarine I	Calebash curare (*342, 353–355*) *S. divaricans* Ducke (*356*) *S. froesii* Ducke (*357*) *S. mittscherlichii* Schomb. (*356, 358*) *S. solimoesana* Kruk. (*359*) *S. trinervis* (Vell.) Mart (*360*)	$C_{40}H_{44}N_4O^{++}$	(*342, 355, 361–364*)	(*153, 361, 364*)		(*352, 364*)			(*352, 365, 366*)
C-Alkaloid G	Calebash curare (*367*) *S. solimoesana* (*359*)	$C_{40}H_{44}N_4O_2^{++}$	(*362, 368*)						(*368*)
Caracurine II methosalt	*S. toxifera* (*369*)	$C_{40}H_{44}N_4O_2^{++}$	(*347*)			(*345*)		(*346*)	(*347*)
C-Alkaloid E	Calebash curare (*367*) *S. froessi* (*357*) *S. solimoesana* (*359*) *S. tomentosa* (*359, 370*)	$C_{40}H_{44}N_4O_3^{++}$	(*362, 371*)			(*352*)			(*352, 371*)

(*Continued*)

TABLE XXI (*Continued*)

Alkaloid	Source (*refs.*)	Molecular formula	Physical data (*refs.*) UV	IR	Mass	^{1}H-NMR	^{13}C-NMR	X-Ray	Synthesis
C-Dihydrotoxiferine	Calebash curare (*342, 367, 372*) *S. froesii* (*357*) *S. trinervis* (*360*)	$C_{40}H_{46}N_4$	(*362*)	(*350*)		(*352, 373*)			(*350, 352, 368, 374*)
C-Alkaloid H	Calebash curare (*367*) *S. trinervis* (*360*)	$C_{40}H_{46}N_4O^{++}$	(*362, 368*)	(*368*)					(*368*)
C-Toxiferine I	Calebash curare *S. froesii* (*357*) *S. toxifera* (*368, 372*)	$C_{40}H_{46}N_4O_2^{++}$	(*362, 369, 375*)	(*350*)		(*373, 376*)			(*140, 350, 352, 377, 378*)
C-Alkaloid	Calebash curare (*366*) *S. mittscherlichii* (*356, 358*) *S. solimoesana* (*359*)	$C_{40}H_{48}N_4O_2^{++}$	(*345, 347, 352*)			(*345*)			(*348*)
C-Calebassine	Calebash curare (*355, 369, 372*) *S. divaricans* (*356*) *S. mittscherlichii* (*356, 358*) *S. solimoesana* (*359*) *S. trinervis* (*360*)	$C_{40}H_{48}N_4O_2$	(*353, 359, 362, 365, 375, 379*)	(*379*)		(*352, 380*)			(*352, 365*)

C-Alkaloid F	Calebash curare (*367*) *S. solimoesana* (*359*)	$C_{40}H_{48}N_4O_2$	(*362, 368*)						(*368*)
C-Alkaloid A	Calebash curare (*367, 381*) *S. toxifera* (*369*)	$C_{40}H_{48}N_4O_4$	(*362, 371*)			(*352, 380*)			(*352, 368, 371*)
XIV. Presecamine and secamine type									
Presecamine	*Rhazya stricta* Decsne (*161, 165*)	$C_{42}H_{52}N_4O_4$	(*161, 165*)	(*161, 165*)	(*161, 165*)	(*161, 165*)			(*161, 165*)
Dihydropre-secamine	*R. stricta* (*161, 165*)	$C_{42}H_{54}N_4O_4$	(*161, 165*)	(*161, 165*)	(*161, 165*)	(*161, 165*)			
Tetrahydropre-secamine (**293**)	*Amsonia tabernae-montana* Walt. (*159*) *Pandaca minuti-flora* Mgf. (*164*) *Rhazya orientalis* A. DC. (*161, 165*) *R. stricta* (*161, 165*)	$C_{42}H_{54}N_4O_4$	(*159, 161, 165*)	(*159, 161, 165*)	(*159, 161, 165*)	(*161, 165*)			(*161, 165*)
Decarbomethoxy-tetrahydro-secamine	*Amsonia tabernae-montana* (*159*)	$C_{40}H_{54}N_4O_2$	(*159*)	(*159*)	(*159*)				
Secamine	*Rhazya stricta* (*155, 165*)	$C_{42}H_{52}N_4O_4$	(*155, 165*)	(*155, 165*)	(*155, 165*)	(*155, 165*)			(*161, 165*)
Dihydrosecamine	*R. stricta* (*155, 165*)	$C_{42}H_{54}N_4O_4$	(*155, 165*)	(*155, 165*)	(*155, 165*)	(*155, 165*)			

(*Continued*)

TABLE XXI (*Continued*)

Alkaloid	Source (*refs.*)	Molecular formula	Physical data (*refs.*)						
			UV	IR	Mass	^{1}H-NMR	^{13}C-NMR	X-Ray	Synthesis
Tetrahydro-secamine (**288**)	*Amsonia elliptica* Roem. et Schult. (*157*, *158*) *A. tabernaemontana* (*159*) *Rhazya orientalis* (*156*, *165*) *R. stricta* (*155*, *165*)	$C_{42}H_{56}N_4O_4$	(*155–159*, *165*)	(*155–159*, *165*)	(*155–159*, *165*)	(*155–159*, *165*)			(*161*, *165*)
XV. Iboga–canthinone type									
Bonafousine (**296**)	*Bonafuosia tetrastachya* (Humb., Bon. & Kunth.) Mgf. (*167*)	$C_{35}H_{40}N_4O_3$	(*167*)	(*167*)	(*167*)	(*167*)		(*167*)	
XVI. Iboga–vobasine type									
Tabernamine (**303**)	*Tabernaemontana johnstonii* Pichon (*170*, *172*)	$C_{40}H_{48}N_4O_2$	(*170*)	(*170*)	(*170*)	(*170*, *172*)			(*170*)

Decarbomethoxy-voacamine	*Conopharyngia longiflora* Stapf. (*382*) *Ervatamia orientalis* (R. Br.) Turtill (*174*) *Pandaca speciosa* Mgf. (*384*) *Voacanga africana* Stapf. (*385*) *V. thouarsii* Roehm & Schult. (*382, 386*)	$C_{41}H_{50}N_4O_3$	(*174, 386*)	(*385, 386*)	(*385, 386*)	(*174, 386*)	(*385*)
20′-*βH*-16-Decarbomethoxydihydrovoacamine (**319**)	*Ervatamia orientalis* (*174*)	$C_{41}H_{52}N_4O_3$	(*174*)	(*174*)	(*174*)	(*174*)	(*174*)
20′-*αH*-16-Decarbomethoxydihydrovoacamine (**322**)	*E. orientalis* (*172*)	$C_{41}H_{52}N_4O_3$	(*174*)	(*174*)	(*174*)	(*174*)	(*174*)
N-Demethylvoacamine (**315**)	*Tabernaemontana accedens* (*122*)	$C_{42}H_{50}N_4O_5$	(*122*)	(*122*)	(*122*)	(*122*)	
Gabunine (**310**)	*Gabunia odoratissima* Stapf. (*387*) *Tabernaemontana holstii* K. Schum. (*180*) *T. johnstonii* (*172*)	$C_{42}H_{50}N_4O_5$	(*387*)	(*387*)	(*180*)	(*180, 387*)	(*172*)

(*Continued*)

TABLE XXI (*Continued*)

Alkaloid	Source (*refs.*)	Molecular formula	Physical data (*refs.*)						
			UV	IR	Mass	^{1}H-NMR	^{13}C-NMR	X-Ray	Synthesis
Gabunamine (**308**)	*T. johnstonii* (*172*)	$C_{42}H_{50}N_4O_5$	(*172*)	(*172*)	(*172*)	(*172*)			(*172*)
Conodurine (**311**)	*Conopharyngia durissima* Stapf. (*303*) *Gabunia odoratissima* (*387, 388*) *Tabernaemontana holstii* (*180*) *T. johnstonii* (*172*)	$C_{43}H_{52}N_4O_5$	(*303, 387*)	(*303, 387*)	(*303*)	(*178*)			(*178*)
Conoduramine (**306**)	*Conopharyngia durissima* (*303*) *Gabunia odoratissima* (*387, 388*) *Tabernaemontana elegans* Stapf. (*176*) *T. holstii* (*180*) *T. johnstonii* (*172*)	$C_{43}H_{52}N_4O_5$	(*176, 180, 303*)	(*176, 303*)	(*176*)	(*176, 178*)			(*178*)
3′-Oxoconodurine (**335**)	*T. holstii* (*180*)	$C_{43}H_{50}N_4O_6$	(*180*)	(*180*)	(*180*)	(*180*)			
3′-(2-Oxopropyl)-conodurine (**334**)	*T. holstii* (*180*)	$C_{45}H_{56}N_4O_6$	(*180*)	(*180*)	(*180*)	(*180*)			
Voacamidine (**314**)	*T. accedens* (*122*) *Voacanga africana* (*178, 389*)	$C_{43}H_{52}N_4O_5$	(*389*)	(*389*)		(*171, 178*)			(*171*)

Voacamine (**299**)	*Conophayngia longiflora* (*382*) *Gabunia eglandulosa* Stapf. (*390*) *Hedranthera barteri* (Hook. f.) Pichon (*207*) *Stemmadenia donnell-smithii* (Rose) Woodson (*391*) *Tabernaemontana accedens* (*392*) *T. arborea* J. N. Rose ex. J. D. Smith (*393*) *T. australis* Muell. Arg. (*394*) *T. oppositifolia* Urb. (*394*) *T. psychotrifolia* H. B. K. (*394*) *Voacanga africana* (*199, 385, 394–396*) *V. bracteata* Stapf. var. *bracteata* Pichon (*397*)	$C_{43}H_{52}N_4O_5$	(*171, 207, 391, 402, 405, 406*)	(*171, 207, 391, 395, 402, 405, 406*)	(*171, 207, 385, 396, 403, 407, 408*)	(*171, 396, 409*)		(*122, 171, 385, 409*)

(*Continued*)

TABLE XXI (*Continued*)

Alkaloid	Source (*refs.*)	Molecular formula	Physical data (*refs.*)						
			UV	IR	Mass	^{1}H-NMR	^{13}C-NMR	X-Ray	Synthesis
	V. globosa Merrill (*397, 398*) *V. grandifolia* (Miq.) Rolfe (*399*) *V. megacarpa* Merrill (*400*) *V. papuana* (F. Muell.) K. Schum. (*401*) *V. schweinfurthii* Stapf. (*402, 403*) *V. thouarsii* (*382, 386, 392, 404*)								
19,20-Epoxycono-duramine (**305**)	*Tabernaemontana johnstonii* (*172*)	$C_{43}H_{52}N_4O_6$	(*172*)	(*172*)	(*172*)	(*172*)			
Voacamine *N*-oxide	*T. accedens* (*122*) *Voacanga africana* (*410*)	$C_{43}H_{52}N_4O_6$	(*122, 410*)	(*122, 410*)	(*122, 410*)	(*122, 410*)			(*122, 410*)
Voacorine (**300**)	*Conopharyngia longiflora* (*382, 411*) *Tabernaemontana brachyantha* Stapf. (*168*)	$C_{43}H_{52}N_4O_6$	(*405*)	(*395, 405*)	(*385, 395*)	(*395, 406*)			(*171*)

	Voacanga africana (*385, 395, 396, 412–414*) *V. bracteata* var. *bracteata* (*397*) *V. schweinfurthii* (*397, 415*) *V. thouarsii* var. *obtusa* (*382*)						
19-Epivoacorine (**302**)	*Tabernaemontana arborea* (*393*) *Voacanga bracteata* var. *bracteata* (*169, 397*)	$C_{43}H_{52}N_4O_6$	(*169, 397*)	(*169, 397*)	(*169*)	(*169, 397*)	
19′-Epivoacorine (**301**)	*Tabernaemontana brachyantha* (*168*)	$C_{43}H_{52}N_4O_6$	(*168*)	(*168*)	(*168*)	(*168*)	
Tabernaelegantine A (**328**)	*T. elegans*(*176, 177*)	$C_{43}H_{54}N_4O_5$	(*177*)	(*177*)	(*177*)	(*177*)	(*177*)
Tabernaelegantine B (**329**)	*T. elegans* (*176, 177*)	$C_{43}H_{54}N_4O_5$	(*177*)	(*177*)	(*177*)	(*177*)	(*177*)
Tabernaelegantine C (**330**)	*T. elegans* (*176, 177*)	$C_{43}H_{54}N_4O_5$	(*177*)	(*177*)	(*177*)	(*177*)	(*177*)
Tabernaelegantine D (**331**)	*T. elegans* (*176, 177*)	$C_{43}H_{54}N_4O_5$	(*177*)	(*177*)	(*177*)	(*177*)	(*177*)
Tabernaelegantinine A (**332**)	*T. elegans* (*176, 177*)	$C_{46}H_{58}N_4O_6$	(*177*)	(*177*)	(*177*)	(*177*)	(*177*)
Tabernaelegantinine B (**333**)	*T. elegans* (*176, 177*)	$C_{46}H_{58}N_4O_6$	(*177*)	(*177*)	(*177*)	(*177*)	(*177*)

(*Continued*)

TABLE XXI (*Continued*)

Alkaloid	Source (*refs.*)	Molecular formula	Physical data (*refs.*)						
			UV	IR	Mass	^{1}H-NMR	^{13}C-NMR	X-Ray	Synthesis
XVIII. Cleavamine–vobasine type									
Capuvosine (**343**)	*Capuronetta elegans* (*180*)	$C_{40}H_{50}N_4O_3$	(*180*)	(*180*)	(*180*)	(*180*)			(*180*)
Dehydroxy-capuvosine (**344**)	*C. elegans* (*182*)	$C_{40}H_{50}N_4O_2$	(*182*)	(*182*)	(*182*)	(*182*)	(*417*)		
	Pandaca boiteaui (*416*)								
Demethyl-capuvosine (**345**)	*Capuronetta elegans* (*182*)	$C_{39}H_{48}N_4O_3$	(*182*)	(*182*)	(*182*)	(*182*)			
XVIII. Pseudo-*Aspidosperma*–vobasine type									
Capuvosidine (**347**)	*C. elegans* (*182*)	$C_{40}H_{48}N_2O_2$	(*182*)	(*182*)	(*182*)	(*181, 182*)			
XIX. Iboga–iboga type									
12,12′-Bis(11-hydroxy-coronaridine) (**348**)	*Bonafousia tetrastachya* (Humb., Bon. and Kunth) Mgf. (*183*)	$C_{42}H_{50}N_4O_6$	(*183*)	(*183*)	(*183*)	(*183*)	(*183*)		
Gabonine	*Tabernanthe iboga* Baill. (*418*)	$C_{42}H_{56}N_4O_8$	(*418, 419*)	(*418, 419*)	(*419*)	(*419*)			
XX. *Aspidosperma–pleiocarpa* alkaloids									
Pycnanthine (**369**)	*Pleiocarpa pycnantha* (K. Schum.) Stapf. var *pycnantha* Pichon (*184*)	$C_{40}H_{44}N_4O_2$	(*184*)	(*184*)	(*184*)	(*184*)			

14′,15′-Dihydropycnanthine (pleiomutinine) (**368**)	*Gonioma malagasy* Mgf. et P. Bt. (*188*)	$C_{40}H_{46}N_4O_2$	(*184, 186, 188*)	(*184, 188*)	(*184, 188*)	(*184*)	(*188*)
	Pleiocarpa mutica Benth. (*186*)						
Pycnanthinine (**372**)	*P. pycnantha* var. *pycnantha* (*185*)	$C_{40}H_{46}N_4O_2$	(*185*)	(*185*)	(*185*)	(*185*)	
XXI. *Aspidosperma–Aspidosperma* alkaloids							
Folicangine (**468**, **478**)	*Voacanga africana* (*205, 208, 210, 214*)	$C_{42}H_{46}N_4O_5$	(*205*)	(*205*)	(*205*)	(*205*)	(*210*)
Bistabersonine (**373**)	*Crioceras dipladeniiflorus* (Stapf.) K. Schum. (*190*)	$C_{42}H_{48}N_4O_4$	(*190*)		(*190*)	(*190*)	
Isovoafoline (**465**)	*Voacanga africana* (*205, 210*)	$C_{42}H_{48}N_4O_4$	(*205*)	(*205*)	(*205*)	(*205*)	(*210*)
Voafoline (**464**)	*V. africana* (*205, 208, 210*)	$C_{42}H_{48}N_4O_4$	(*205*)	(*205*)	(*205*)	(*205*)	(*210*)
Callichiline (**442**)	*Callichilia subsessilis* Stapf. (*215*) *Hedranthera barteri* (Hook f.) Pichon (*207, 216–218*)	$C_{42}H_{48}N_4O_5$	(*215, 217, 218*)	(*215, 217, 218*)	(*217, 218*)	(*217, 218*)	
Voafolidine (**455**)	*Voacanga africana* (*205, 208, 210*)	$C_{42}H_{48}N_4O_5$	(*205*)	(*205*)	(*205*)	(*205*)	(*210*)

(*Continued*)

TABLE XXI (*Continued*)

Alkaloid	Source (*refs.*)	Molecular formula	Physical data (*refs.*)						
			UV	IR	Mass	^{1}H-NMR	^{13}C-NMR	X-Ray	Synthesis
12-Demethyl-vobtusine (**399**)	*Hedranthera barteri* (*202*) *Voacanga thouarsii* (*203, 204*)	$C_{42}H_{48}N_4O_6$	(*202–204*)	(*202–204*)	(*202–204*)	(*202–204*)			(*202*)
18-Oxoamataine (**431**, **480**)	*V. thouarsii* (*203, 210, 214*)	$C_{43}H_{46}N_4O_7$	(*203*)	(*203*)	(*203*)	(*203*)	(*210*)		
Owerreine (anhydrodihy-droamataine) (**420**)	*Hedranthera barteri* (*207*)	$C_{43}H_{48}N_4O_5$	(*207*)	(*207*)	(*207*)	(*207*)			
Amataine (grandifoline, subsessiline) (**426**, **479**)	*Callichilia subsessilis* (*205*) *Hedranthera barteri* (*207, 210, 214*) *Voacanga chalotiana* Pierre ex Stapf. (*197*) *V. grandifolia* (Miq.) Rolfe (*212*) *V. thouarsii* (*203*)	$C_{43}H_{48}N_4O_6$	(*197, 203, 207, 213*)	(*197, 203, 207, 213*)	(*197, 203 207, 213*)	(*197, 203, 207, 213*)	(*210*)		
2-Deoxy-3′-oxo-vobtusine (**418**)	*V. thouarsii* (*203, 204, 206*)	$C_{43}H_{48}N_4O_6$	(*203, 204*)	(*203, 204*)	(*203, 204*)	(*203, 204*)			

2-Deoxy-18-oxo-vobtusine (**409**)	*V. africana* (*205, 208, 209*)	$C_{43}H_{48}N_4O_6$	(*209*)	(*209*)	(*209*)	(*209*)	(*210*)		
Quimbeline (**421**)	*V. chalotiana* (*197, 211*)	$C_{43}H_{48}N_4O_6$	(*211*)	(*211*)	(*211*)	(*211*)			
3′-Oxovobtusine (**416**)	*V. thouarsii* (*203, 204, 206, 210*)	$C_{43}H_{48}N_4O_7$	(*203*)	(*203*)	(*203*)	(*203*)	(*210*)		
18-Oxovobtusine (**408**)	*V. africana* (*205, 208, 209*) *V. grandifolia* (*206, 399*) *V. thouarsii* (*203*)	$C_{43}H_{48}N_4O_7$	(*203, 209*)	(*203, 209*)	(*203, 209*)	(*203, 209*)	(*210*)		(*204*)
3′-Oxovobtusine *N*-oxide (**417**)	*V. thouarsii* (*203, 204*)	$C_{43}H_{50}N_4O_5$	(*203, 204*)	(*203, 204*)	(*203, 204*)	(*203, 204*)	(*210*)		
2-Deoxyvobtusine (Goziline) (**400**)	*Hedranthera barteri* (*202, 207*) *Voacanga africana* (*205, 207*) *V. grandifolia* (*206*)	$C_{43}H_{50}N_4O_5$	(*202, 205, 206, 207*)	(*202, 205, 206, 207*)	(*202, 205, 206, 207*)	(*202, 205, 206, 207*)			
Vobtusine (**390**)	*Callichilia subsessilis* (*215*) *C. stenosepala* Stapf. (*420*) *Conopharyngia longiflora* (*382*) *Crioceras dipladeniiflorus* (Stapf.) K. Schum. (*190*)	$C_{43}H_{50}N_4O_6$	(*197–199, 202, 392, 402*)	(*197–199, 202, 392, 402, 422*)	(*197–199, 202*	(*197–199, 202*)	(*210*)	(*200*)	(*213*)

(*Continued*)

TABLE XXI (*Continued*)

Alkaloid	Source (*refs.*)	Molecular formula	Physical data (*refs.*)						
			UV	IR	Mass	^{1}H-NMR	^{13}C-NMR	X-Ray	Synthesis
	Hedranthera barteri (*199, 207, 216, 386, 420*)								
	Rejoua aurantiaca Gaudich (*421*)								
	Voacanga africana (*198, 199, 208, 385, 392*)								
	V. chalotiana (*197, 198*)								
	V. dregei E. Mey (*397, 422*)								
	V. globosa (*364*)								
	V. grandifolia (*206, 399*)								
	V. megacarpa (*400*)								
	V. papuana (*397, 421*)								
	V. schweinfurthii (*402, 403*)								
	V. thouarsii (*203, 382, 386*)								

10,10′-Bis-*N*-acetyl-11,12-dihydroxyaspidospermidine (**377**)	*Aspidosperma melanocalyx* Muell.-Arg. (*191*)	$C_{43}H_{54}N_4O_6$	(*191*)	(*191*)	(*191*)	(*191*)	
Vindolicine (**382**)	*Catharanthus longifolius* (Pichon) Pich. (*194*) *C. roseus* (L.) G. Don (*192, 193*)	$C_{51}H_{66}N_4O_{12}$	(*193, 194*)	(*193, 194*)	(*193, 194*)	(*193*)	(*193*)
XXII. Bisindole alkaloids from *Melodinus* species							
19-Episcandomelonine (**484**)	*Melodinus scandens* Forst. (*219*)	$C_{41}H_{42}N_4O_5$	(*219*)	(*219*)	(*219*)	(*219*)	(*219*)
Scandomelonine (**483**)	*M. scandens* (*219*)	$C_{41}H_{42}N_4O_5$	(*219*)	(*219*)	(*219*)	(*219*)	(*219*)
19-Episcandomeline (**487**)	*M. scandens* (*219*)	$C_{42}N_{46}N_4O_6$	(*219*)	(*219*)	(*219*)	(*219*)	(*219*)
Scandomeline (**486**)	*M. scandens* (*219*)	$C_{42}H_{46}N_4O_6$	(*219*)	(*219*)	(*219*)	(*219*)	(*219*)
Methylene bis-1,1′-melonine	*M. celastroides* (*223*)	$C_{39}H_{52}N_4$	(*223*)		(*223*)	(*223*)	
XXIII. *Aspidosperma*–eburnea type							
Criophylline (**492**)	*Crioceras dipladeniiflorus* (*224, 225*)	$C_{40}H_{46}N_4O_4$	(*224, 225*)	(*224, 225*)	(*224, 225*)	(*224, 225*)	(*224*)
Paucivenine (**493**)	*Melodinus balansae* Baillon var. *paucivenosus* (S. Moore) Boiteau (*226*)	$C_{40}H_{48}N_4O_2$	(*226*)	(*226*)	(*226*)		

(*Continued*)

TABLE XXI (*Continued*)

Alkaloid	Source (*refs.*)	Molecular formula	Physical data (*refs.*)						
			UV	IR	Mass	^{1}H-NMR	^{13}C-NMR	X-Ray	Synthesis
Pleiomutine (**499**)	*Pleiocarpa mutica* (*186, 219, 423*) *P. tubicana* Stapf. (*423*)	$C_{41}H_{50}N_4O_2$	(*186, 219*)	(*219, 424*)	(*219, 424*)	(*219, 424*)			(*219, 424*)
XXIV. Haplophytine									
Haplophytine (**500**)	*Haplophyton cimicidum* A. DC. (*228, 229, 425, 426*)	$C_{37}H_{40}N_4O_7$	(*228, 229*)	(*228, 229*)	(*228, 229*)	(*228, 229*)	(*229*)	(*230*)	
XXV. *Aspidosperma*–Cleavamine Type									
Vinblastine (**502**)	*Catharanthus roseus* (*427–429*) *C. ovalis* Mgf. (*430*) *C. longifolius* Pichon (*431*)	$C_{46}H_{58}N_4O_9$	(*427, 432*)	(*432*)	(*432, 433*)	(*432, 434*)	(*233–235*)		
Vincristine (**501**) (leurocristine)	*C. roseus* (*428, 435*)	$C_{46}H_{56}N_4O_{10}$	(*432, 435*)	(*432*)	(*435*)	(*434*)		(*232*)	
Leurosidine (**506**)	*C. roseus* (*435*)	$C_{46}H_{58}N_4O_9$	(*432, 435*)	(*432*)	(*435*)		(*234*)		
Deacetoxy-vinblastine (**529**)	*C. roseus* (*262, 436*)	$C_{44}H_{56}N_4O_7$			(*262*)	(*262*)	(*235*)		

Leurosine (**503**)	*C. roseus* (*427, 428, 437*) *C. ovalis* (*430*) *C. longifolius* (*431*) *C. lanceus* (Boj. ex A.DC) Pichon (*438*) *C. pusillus* (Murr.) G. Don (*439*)	$C_{46}H_{56}N_4O_9$	(*427, 432*)		(*263*)		(*233, 234*)		(*259, 265, 266*)
Isoleurosine (**504**)	*C. roseus* (*263*)	$C_{46}H_{58}N_4O_8$	(*192, 432*)	(*192*)		(*263*)			
Pleurosine (**505**)	*C. roseus* (*263*)	$C_{46}H_{56}N_4O_{10}$	(*432, 440*)	(*440*)					
De-*N*-methyl-vinblastine **516**)	*C. roseus* (*243, 244, 246, 247*)	$C_{45}H_{56}N_4O_9$					(*235*)		
Catharine (**542**)	*C. roseus* (*192, 269*) *C. ovalis* (*340*) *C. longifolius* (*431*)	$C_{46}H_{54}N_4O_{10}$	(*192, 432*)	(*192, 270*)		(*270*)		(*271, 272*)	(*260, 265, 269*)
Vincathicine (**543**)	*C. roseus* (*273*)	$C_{46}H_{56}N_4O_9$	(*273*)	(*273*)		(*273, 274*)	(*274*)		(*275*)
Vinamidine (catharinine) (**545**)	*C. roseus* (*276*) *C. ovalis* (*277*) *C. longifolius* (*277*)	$C_{46}H_{56}N_4O_{10}$	(*277*)	(*276, 277*)	(*277*)	(*276, 277*)	(*276*)		
Leurocolombine (**549**)	*C. roseus* (*276*)	$C_{46}H_{58}N_4O_{10}$	(*276*)	(*276*)		(*276*)	(*276*)		
Pseudovinblastine diol (**550**)	*C. roseus* (*276*)	$C_{44}H_{56}N_4O_8$			(*276*)	(*276*)			
Vincadioline (**553**)	*C. roseus* (*279*)	$C_{46}H_{58}N_4O_{10}$					(*235*)		
Vincovaline (**554**)	*C. ovalis* (*280*)	$C_{46}H_{58}N_4O_9$	(*280*)		(*280*)	(*280*)			

TABLE XXII
SUMMARY OF THE ISOLATION OF BISINDOLE ALKALOIDS OF UNKNOWN STRUCTURE

Alkaloid	Source	Molecular formula	Reference
I. Corynanthe–tryptamine type			
Alkaloid C_2	*Rauwolfia obscura*	$C_{30}H_{34}N_4O_2$	(*109*)
Alkaloid D_1	*R. obscura*	$C_{30}H_{34}N_4O$	(*109*)
Alkaloid D_2	*R. obscura*	$C_{30}H_{36}N_4O$	(*109*)
Roxburghine A	*Uncaria gambier*	$C_{31}H_{32}N_4O_2$	(*80*)
II. *Alstonia* alkaloids			
Macrophylline	*Alstonia macrophylla*	$C_{45}H_{54}N_4O_5$	(*441*)
Unknown	*A. constricta*		(*442*)
Unknown	*A. macrophylla*		(*369*)
III. *Strychnos–Strychnos* type			
Caracurine VI	*Strychnos toxifera*	$C_{38}H_{40}N_4O$	(*343*)
Toxiferine XII	*S. toxifera*	$C_{39}H_{46}N_4O^{++}$	(*343, 443*)
C-Isodihydro-toxiferine I	Calebash curare	$C_{40}H_{46}N_4^{++}$	(*372*)
C-Alkaloid B	Calebash curare	$C_{40}H_{46}N_4^{++}O_2$	(*367, 444*)
	S. mittscherlichii		(*358*)
Toxiferine II	Calebash curare	$C_{40}H_{48}N_4^{++}O_4$	(*372*)
	S. toxifera		(*372*)
C-Alkaloid BL	Calebash curare	$C_{40}H_{50}N_4^{++}O_2$	(*444*)
IV. Iboga–vobasine type			
Tabernaelegantidine	*Tabernaemontana elegans*		(*176, 177*)
V. *Aspidosperma–Aspidosperma* type			
Ervafoline	*Ervatamia pandacaqui*	$C_{40}H_{44}N_4O_4$	(*445*)
Ervafolidine	*E. pandacaqui*	$C_{40}H_{46}N_4O_5$	(*445*)
Isoervafolidine	*E. pandacaqui*	$C_{40}H_{46}N_4O_5$	(*445*)
Alkaloid VI	*Voacanga africana*	$C_{42}H_{48}N_4O_5$	(*208*)
Alkaloid G	*V. thouarsii*	$C_{42}G_{50}N_4O_5$	(*203*)
Alkaloid C	*V. thouarsii*	$C_{42}H_{54}N_4O_{10}$	(*203*)
Alkaloid I	*V. thouarsii*	$C_{43}H_{48}N_4O_6$	(*203*)
Alkaloid E	*V. thouarsii*	$C_{43}H_{50}N_4O_7$	(*203*)
Alkaloid S_2	*V. schweinfurthii*		(*403*)
Alkaloid S_3	*V. schweinfurthii*		(*403*)
VI. *Aspidosperma*–cleavamine type			
Carosine	*Catharanthus roseus*	$C_{46}H_{56}N_4O_{10}$	(*440*)
Catharicine	*C. roseus*	$C_{46}H_{52}N_4O_{10}$	(*440*)
Leurosivine	*C. roseus*	$C_{41}H_{54}N_3O_9$	(*446, 447*)
Neoleurocristine	*C. roseus*	$C_{46}H_{56}N_4O_{12}$	(*440*)
Neoleurosidine	*C. roseus*	$C_{48}H_{62}N_4O_{11}$	(*440*)
Rovidine	*C. roseus*	—	(*273*)
Vinaphamine	*C. roseus*	—	(*273*)
Vincovalinine	*C. ovalis*	$C_{44}H_{54}N_4O_7$	(*280*)
Vincovalicine	*C. ovalis*	$C_{46}H_{54}N_4O_{10}$	(*280*)

(*Continued*)

TABLE XXII (*Continued*)

Alkaloid	Source	Molecular formula	Reference
VII. Series unknown			
Alkaloid D	*Pandaca caducifolia* Mgf.	$C_{40}H_{44}N_4O_3$ $C_{40}H_{46}N_4O_4$	(*448*)
Alkaloid G	*P. caducifolia*	$C_{40}H_{44}N_4O_4$	(*448*)
Alkaloid E	*P. caducifolia*	$C_{40}H_{46}N_4O_4$	(*448*)
Alkaloid F	*P. caducifolia*	$C_{40}H_{46}N_4O_4$	(*448*)
Voacaline	*Voacanga africana*	$C_{41}H_{50}N_4O_6$	(*448*)
Aspidexcin	*Aspidosperma excelsum*	$C_{42}H_{56}N_4O_4$	(*449*)
Vinosidine	*Catharanthus roseus*	$C_{44}H_{52}N_4O_5$	(*446, 447*)
Vindolidine	*C. roseus*	$C_{48}H_{64}N_4O_{10}$	(*440*)
Aspidexicin	*Aspidosperma excelsum*		(*449*)
Alkaloid	*Melodinus aenus* Baill		(*450*)
Alkaloid E	*Muntafaria sessilifolia* (Bak.) Pichon		(*451*)
Alkaloid F	*M. sessilifolia*		(*451*)
Alkaloid I	*M. sessilifolia*		(*451*)
Alkaloid J	*M. sessilifolia*		(*451*)
Alkaloid N	*M. sessilifolia*		(*451*)
Alkaloid O	*M. sessilifolia*		(*451*)
Alkaloid P	*M. sessilifolia*		(*451*)
Alkaloid Q	*M. sessilifolia*		(*451*)
Alkaloid R	*M. sessilifolia*		(*451*)
Alkaloid B	*Strychnos camptoneura* Gilg et. Busse		(*452*)
Carosidine	*Catharanthus roseus*		(*440*)
Vincamicine	*C. roseus*		(*192*)
Vinsedicine	*C. roseus*		(*453*)
Vinsedine	*C. roseus*		(*453*)
Voacafrine	*Voacanga africana*		(*454*)
Voacafricine	*V. africana*		(*454*)

B. Oxidative Coupling of Indolines

As part of a general study of the oxidative coupling of indolines (*456*), it was observed that oxidation of (+)-*N*-deacetylaspidospermine (**632**) with potassium permanganate in acetone gave a 16% yield of the corresponding indolenine **633** and a 60% yield of a dimer (M^+ *m/e* 620) having the structural formula **634**. An analogous oxidation of (+)-*O*-methyl-*N*-depropionyl-aspidolimine (**635**) gave the indolenine **636** and the dimer **637** (*456*).

In the aromatic region of the NMR spectrum of **634**, a three-proton multiplet was observed in the region 6.78–7.16 ppm and two doublets

632 R = H
635 R = OCH_3

$KMnO_4$ acetone

633 R = H (16%)
636 R = OCH_3 (22%)

634 R = H (60%)
637 R = OCH_3 (21.5%)

($J = 2$ Hz) for the meta-related C-9′ and C-11′ protons. Substitution is therefore at C-10′ in one of the units. The IR spectrum of **634** showed no NH absorption but did show an indolenine absorption at 1590^{-1}. The UV spectrum (λ_{max} 260, 310, and 345 nm) was in agreement with the proposed structure.

References

1. J. E. Saxton, *Alkaloids* (*N.Y.*) **7**, 1 (1960).
2. J. E. Saxton, *Alkaloids* (*N.Y.*) **8**, 159 (1965).
3. W. I. Taylor, *Alkaloids* (*N.Y.*) **8**, 203 (1965).
4. W. I. Taylor, *Alkaloids* (*N.Y.*) **8**, 250 (1965).
5. A. R. Battersby and H. F. Hodson, *Alkaloids* (*N.Y.*) **8**, 515 (1965).
6. R. H. F. Manske, *Alkaloids* (*N.Y.*) **8**, 581 (1965).
7. R. H. F. Manske and W. Ashley Harrison, *Alkaloids* (*N.Y.*) **8**, 679 (1965).
8. W. I. Taylor, *Alkaloids* (*N.Y.*) **11**, 73 (1968).
9. W. I. Taylor, *Alkaloids* (*N.Y.*) **11**, 79 (1968).
10. W. I. Taylor, *Alkaloids* (*N.Y.*) **11**, 125 (1968).
11. A. R. Battersby and H. F. Hodson, *Alkaloids* (*N.Y.*) **11**, 189 (1968).
12. B. Gilbert, *Alkaloids* (*N.Y.*) **11**, 205 (1968).
13. A. A. Gorman, M. Hesse, H. Schmid, P. G. Waser, and W. H. Hopff, *Alkaloids* (*London*) **1**, 200 (1971).
14. M. Gorman, N. Neuss, and K. Biemann, *J. Am. Chem. Soc.* **84**, 1058 (1962).
15. G. H. Büchi, *Pure Appl. Chem.* **9**, 21 (1964).
16. J. A. Joule, *Alkaloids* (*London*) **2**, 209 (1972).
17. J. A. Joule, *Alkaloids* (*London*) **3**, 187 (1973).
18. J. A. Joule, *Alkaloids* (*London*) **4**, 280 (1974).
19. J. A. Joule, *Alkaloids* (*London*) **5**, 183 (1975).
20. J. E. Saxton, *Alkaloids* (*London*) **6**, 236 (1976).

21. J. E. Saxton, *Alkaloids* (*London*) **7**, 237 (1977).
22. K. Bernauer, *Fortschr. Chem. Org. Naturst.* **17**, 183 (1959).
23. A. R. Battersby and H. F. Hodson, *Q. Rev. Chem. Soc.* **14**, 27 (1960).
24. H. Schmid, *Bull. Schweiz. Akad. Med. Wiss.* **22**, 415 (1967).
25. G. B. Marini-Bettolo, *Farmaco., Ed. Sci.* **25**, 150 (1970).
26. S. Ghosal, *Pharmastudent* **16**, 49 (1972); *CA* **81**, 96355r (1974).
27. A. Chatterjee, J. Banerji, and A. Banerji, *J. Indian Chem. Soc.* **51**, 156 (1974).
28. S. Kohlmuenzer, *Postepy Biochem.* **14**, 209 (1968).
29. A. M. Aliev and N. A. Babaev, *Farmatsiya* (*Moscow*) **22**, 77 (1973); *CA* **79**, 96871q (1973).
30. "The *Catharanthus* Alkaloids" (W.I. Taylor and N. R. Farnsworth, eds.). Dekker, New York, 1975.
31. S. Omura, Y. Iwai, A. Hirano, A. Nakagawa, J. Awaya, H. Tsuchiya, Y. Takahashi, and R. Masuma, *J. Antibiot.* **30**, 275 (1977).
32. A. Furusaki, N. Hashiba, T. Matsumoto, A. Hirano, Y. Iwai, and S. Dmura, *Chem. Commun.* 800 (1978).
33. G. Bionda, *Ann. Chim. Appl.* **36**, 210 (1946).
34. V. Plouvier, *C. R. Acad. Sci.* **231**, 1546 (1950).
35. S. Iwadare, Y. Shizuri, K. Sasaki, and Y. Hirata, *Tetrahedron Lett.* 1051 (1974).
36. S. Iwadare, Y. Shizuri, K. Sasaki, and Y. Hirata, *Tetrahedron* **30**, 4105 (1974).
37. K. Sasaki, S. Iwadare, and Y. Hirata, *Tetrahedron Lett.* 1055 (1974).
38. S. Iwadare, Y. Shizuri, K. Yamada, and Y. Hirata, *Tetrahedron Lett.* 1177 (1974).
39. S. Iwadare, Y. Shizuri, K. Yamada, and Y. Hirata, *Jpn. Kokai Tokkyo Koho* **76**, 39,696 (1976); *CA* **85**, 177760d (1976).
40. G. J. Kapadia and R. E. Rao, *Tetrahedron Lett.* 975 (1977).
41. G. Hahn and H. Werner, *Justus Liebigs Ann. Chem.* **520**, 123 (1935).
42. M. L. Wilson and C. J. Coscia, *J. Am. Chem. Soc.* **97**, 431 (1975).
43. G. J. Kapadia, G. S. Rao, E. Leete, M. B. E. Fayez, Y. N. Vaishnav, and H. M. Fales, *J. Am. Chem. Soc.* **92**, 6943 (1970).
44. C. C. J. Culvenor, S. R. Johns, J. A. Lamberton, and L. W. Smith, *Aust. J. Chem.* **23**, 1279 (1970).
45. R. B. Woo-Ming and K. L. Stuart, *Phytochemistry* **14**, 2529 (1975).
46. K. L. Stuart and R. B. Woo-Ming, *Tetrahedron Lett.* 3853 (1974).
47. K. L. Stuart and R. B. Woo-Ming, *Phytochemistry* **14**, 594 (1975).
48. I. J. Grant, T. A. Hamor, J. M. Robertson, and G. A. Sim, *J. Chem. Soc.* 5678 (1965).
49. M. Takamizawa, H. Takahashi, H. Sakakibara, and T. Kobayashi, *Tetrahedron* **23**, 2959 (1967).
50. R. H. F. Manske, *Can. J. Res., Sect. B* **4**, 275 (1939).
51. H. M. Gordin, *J. Am. Chem. Soc.* **27**, 1426 (1905).
52. S. F. Mason and G. W. Vane, *J. Chem. Soc. B* 370 (1966).
53. W. S. Brickell, S. F. Mason, and D. R. Roberts, *J. Chem. Soc. B* 691 (1971).
54. J. B. Hendrickson, R. Rees, and G. Goschke, *Proc. Chem. Soc., London* 383 (1962).
55. E. S. Hall, F. McCapra, and A. I. Scott, *Tetrahedron* **23**, 4131 (1967).
56. H. R. Schütte and B. Maier, *Arch. Pharm. Ber. Dtsch. Pharm. Ges.* **298**, 459 (1965).
57. D. G. O'Donovan and M. F. Keogh, *J. Chem. Soc. C* 1570 (1966).
58. R. Robinson and H. T. Teuber, *Chem. Ind.* (*London*) 783 (1954).
59. S. A. Waksman and E. Bugie, *J. Bacteriol.* **48**, 527 (1944).
60. W. B. Geiger, J. E. Conn, and S. A. Waksman, *J. Bacteriol.* **48**, 531 (1944).
61. W. B. Geiger, *Arch. Biochem.* **21**, 125 (1949).
62. S. Safe and A. Taylor, *J. Chem. Soc. Perkin Trans. 1* 472 (1972).
63. D. Hauser, H. P. Weber, and H. P. Sigg, *Helv. Chim. Acta* **53**, 1061 (1970).

64. D. Hauser, H. R. Loosli, and P. Niklaus, *Helv. Chim. Acta* **55**, 2182 (1972).
65. H. Minato, M. Matsumoto, and T. Katayama, *Chem. Commun.* 44 (1971).
66. H. Minato, M. Matsumoto, and T. Katayama, *J. Chem. Soc. Perkin Trans. 1* 1819 (1973).
67. E. F. L. J. Anet, G. K. Hughes, and E. Ritchie, *Aust. J. Chem.* **14**, 173 (1961).
68. B. Robinson, *Chem. Ind.* (*London*) 218 (1963).
69. J. B. Hendrickson, R. Goschke, and R. Rees, *Tetrahedron* **20**, 565 (1964).
70. J. P. Fridrichsons, M. F. Mackay, and A. McL. Mathieson, *Tetrahedron Lett.* 3521 (1967).
71. J. Fridrichson, M. F. Mackay, and A. McL. Mathieson, *Tetrahedron* **30**, 85 (1974).
72. R. Attitulah, W. G. Bardsley, G. F. Smith, and N. Lahey, *Int. Symp. Chem. Nat. Prod. 4th, Stockholm* Abstr., 2B-7, p. 84.
73. N. K. Hart, S. R. Johns, J. A. Lamberton, and R. E. Summons, *Aust. J. Chem.* **27**, 639 (1974).
74. K. P. Parry and G. F. Smith, unpublished data.
75. J. L. Pousset, J. Kerharo, G. Maynart, X. Monseur, A. Cavé, and R. Goutarel, *Phytochemistry* **12**, 2308 (1973).
76. J.-L. Pousset, A. Cavé, A. Chiaroni, and C. Riche, *Chem. Commun.* 261 (1977).
77. F. Tillequin, M. Koch, J.-L. Pousset, and A. Cavé, *Chem. Commun.* 826 (1978).
78. J. L. Pousset, unpublished data, cited in ref. 77.
79. F. Tillequin, M. Koch, M. Bert, and T. Sevenet, *J. Nat. Prod.* **42**, 92 (1979).
80. L. Merlini, R. Mondelli, and G. Nasini, *Tetrahedron* **26**, 2259 (1970).
81. H. J. Monteiro, *Alkaloids* (*N.Y.*) **11**, 145 (1968).
82. A. J. Gaskell and J. A. Joule, *Tetrahedron* **24**, 5115 (1968).
83. G. A. Cordell, *Lloydia* **36**, 219 (1974).
84. H. Riesner and E. Winterfeldt, *Chem. Commun.* 786 (1972).
85. E. Winterfeldt, H. Radunz, and T. Korth, *Chem. Ber.* **101**, 3172 (1968).
86. C. Cistaro, L. Merlini, R. Mondelli, and G. Nasini, *Chem. Commun.* 785 (1972).
87. C. Cistaro, L. Merlini, R. Mondelli, and G. Nasini, *Gazz. Chim. Ital.* **103**, 153 (1973).
88. C. Cistaro, R. Mondelli, and M. Anteunis, *Helv. Chim. Acta* **59**, 2249 (1976).
89. L. Merlini, R. Mondelli, G. Nasini, F. Wehrli, E. W. Hagaman, and E. Wenkert, *Helv. Chim. Acta* **59**, 2254 (1976).
90. N. Peube-Locou, M. Koch, M. Plat, and P. Potier, *C. R. Acad. Sci., Ser. C* **273**, 905 (1971).
91. E. Wenkert, C.-j. Chang, H. P. S. Chawla, D. W. Cochran, E. W. Hagaman, J. D. King, and K. Orito, *J. Am. Chem. Soc.* **98**, 3645 (1976).
92. M. Koch and M. Plat, *C. R. Acad. Sci., Ser. C* **273**, 753 (1971).
93. N. Peube-Locou, M. Koch, M. Plat, and P. Potier, *Ann. Pharm. Fr.* **30**, 775 (1972).
94. M. Koch, M. Plat, and N. Préaux, *Bull. Soc. Chim. Fr.* 2868 (1973).
95. J. Bruneton and A. Cavé, *Phytochemistry* **11**, 2618 (1972).
96. N. Préaux, M. Koch, M. Plat, and T. Sévenet, *Plant. Med. Phytother.* **8**, 250 (1974).
97. M. C. Koch, M. M. Plat, N. Préaux, H. E. Gottlieb, E. W. Hagaman, F. M. Schell, and E. Wenkert, *J. Org. Chem.* **40**, 2836 (1975).
98. L. Angenot and N. G. Bisset, *J. Pharm. Belg.* **26**, 585 (1971).
99. L. Angenot and A. Denoël, *Planta Med.* **21**, 96 (1972).
100. M. Dubois, C. H. Ginion, L. Angenot, W. van Dorsser, and A. Dresse, *Arch. Int. Physiol. Biochim.* **82**, 823 (1974).
101. L. Angenot, M. Dubois, C. H. Ginion, W. van Dorsser, and A. Dresse, *Arch. Int. Pharmacodyn. Ther.* **215**, 246 (1975).
102. L. Angenot, C. Coune, and M. Tits, *J. Pharm. Belg.* **33**, 11 (1978).
103. K. Yamada, K. Aoki, and D. Uemura, *J. Org. Chem.* **40**, 2572 (1975).
104. L. J. G. Angenot, C. A. Coune, M. J. G. Tits, and K. Yamada, *Phytochemistry* **17**, 1687 (1978).

105. M. Rueffer, N. Nagakura, and M. H. Zenk, *Tetrahedron Lett.* 1593 (1978).
106. C. Richard, C. Delaude, L. Le Men-Olivier, and J. Le Men, *Phytochemistry* **17**, 539 (1978).
107. L. Dupont, J. Lamotte-Brasseur, O. Dideberg, H. Campsteyn, M. Vermeire, and L. Angenot, *Acta Crystallogr., Sect. B* **33**, 1801 (1977).
108. M. Koch, E. Fellion, and M. Plat, *Ann. Pharm. Fr.* **31**, 45 (1973).
109. P. Timmins and W. E. Court, *Planta Med.* **27**, 105 (1975).
110. J. Le Men, C. Kan, P. Potier, and M.-M. Janot, *Ann. Pharm. Fr.* **23**, 691 (1965).
111. P. Potier, C. Kan, J. Le Men, M.-M. Janot, H. Budzikiewicz, and C. Djerassi, *Bull. Soc. Chim. Fr.* 2309 (1966).
112. J. Guilhem, *Acta Crystallogr., Sect. B* **30**, 742 (1974).
113. P. Potier, personal communication (1977).
114. C. Djerassi, J. Fishman, M. Gorman, J. P. Kutney, and S. C. Pakrashi, *J. Am. Chem. Soc.* **79**, 1217 (1957).
115. H. Kaneko, *Yakugaku Zasshi* **80**, 1357, 1362, 1365, 1370, 1374, 1378, 1382 (1960).
116. M. Hanaoka, G. Englert, A. A. Gorman, M. Hesse, and H. Schmid, unpublished data, cited in reference 13.
117. H. Irie, K. Ishizuka, S. Kawashima, N. Masaki, K. Osaki, T. Shingu, S. Uyeo, H. Kaneko, and S. Naruto, *Chem. Commun.* 871 (1972).
118. O. Hesse, *Ber. Dtsch. Chem. Ges.* **10**, 2162 (1877).
119. F. Puisieux, A. Le Hir, R. Goutarel, M.-M. Janot, and J. Le Men, *Ann. Pharm. Fr.* **17**, 626 (1959).
120. A. Chiaroni, C. Riche, M. Pais, and R. Goutarel, *Tetrahedron Lett.* 4729 (1976).
121. R. Goutarel, M. Pais, H. E. Gottlieb, and E. Wenkert, *Tetrahedron Lett.* 1235 (1978).
122. H. Achenbach and E. Schaller, *Chem. Ber.* **109**, 3527 (1976).
123. S. Sakai, N. Aimi, A. Kubo, M. Kitagawa, M. Shiratori, and J. Haginiwa, *Tetrahedron Lett.* 2057 (1971).
124. S. Sakai, N. Aimi, K. Yamaguchi, H. Ohhira, K. Hori, and J. Haginiwa, *Tetrahedron Lett.* 715 (1975).
125. J. Haginiwa, S. Sakai, A. Kubo, K. Takahashi, and M. Taguchi, *Yakugaku Zasshi* **90**, 219 (1970).
126. S. Sakai, N. Aimi, K. Yamaguchi, E. Yamanaka, and J. Haginiwa, *Tetrahedron Lett.* 719 (1975).
127. J. E. Saxton, *Alkaloids* (*N.Y.*) **14**, 157 (1973).
128. R. C. Elderfield and R. E. Gilman, *Phytochemistry* **11**, 339 (1972).
129. J. M. Cook and P. W. Le Quesne, *Phytochemistry* **10**, 437 (1971).
130. D. E. Burke, G. A. Cook, J. M. Cook, K. G. Haller, H. A. Lazar, and P. W. Le Quesne, *Phytochemistry* **12**, 1467 (1973).
131. M. Hesse, H. Hürzeler, C. W. Gemenden, B. S. Joshi, W. I. Taylor, and H. Schmid, *Helv. Chim. Acta* **48**, 689 (1965).
132. M. Hesse, F. Bodmer, C. W. Gemenden, B. S. Joshi, W. I. Taylor, and H. Schmid, *Helv. Chim. Acta* **49**, 1173 (1966).
133. T. Kishi, M. Hesse, W. Vetter, C. W. Gemenden, W. I. Taylor, and H. Schmid, *Helv. Chim. Acta* **49**, 946 (1966).
134. E. E. Waldner, M. Hesse, W. I. Taylor, and H. Schmid, *Helv. Chim. Acta* **50**, 1926 (1967).
135. F. Mayerl and M. Hesse, *Helv. Chim. Acta* **61**, 337 (1978).
136. D. E. Burke and P. W. Le Quesne, *Chem. Commun.* 678 (1972).
137. D. E. Burke, J. M. Cook, and P. W. Le Quesne, *Chem. Commun.* 697 (1972).
138. D. E. Burke, C. A. Demarkey, P. W. Le Quesne, and J. M. Cook, *Chem. Commun.* 1346 (1972).
139. D. E. Burke, J. M. Cook, and P. W. Le Quesne, *J. Am. Chem. Soc.* **95**, 546 (1973).

140. R. C. Elderfield and G. D. Manalo, *J. Philipp. Pharm. Assoc.* **50**, 91 (1964).
141. J. W. Cook and P. W. Le Quesne, *J. Org. Chem.* **36**, 582 (1971).
142. B. C. Das, J.-P. Cosson, G. Lukacs, and P. Potier, *Tetrahedron Lett.* 4299 (1974).
143. B. C. Das, J.-P. Cosson, and G. Lukacs, *J. Org. Chem.* **42**, 2785 (1977).
144. S. Sakai and N. Shinma, *Heterocycles* **4**, 985 (1976).
145. Y. Morita, M. Hesse, and H. Schmid, *Helv. Chim. Acta* **52**, 89 (1969).
146. R. Verpoorte and A. Baerheim Svendsen, *Lloydia* **39**, 357 (1976).
147. R. Verpoorte and A. Baerheim Svendsen, *J. Pharm. Sci.* **67**, 171 (1978).
148. A. R. Battersby and H. F. Hodson, *J. Chem. Soc.* 736 (1960).
149. H. Fritz and G. Rubach, *Justus Liebigs Ann. Chem.* **715**, 135 (1968).
150. H. Fritz and S. H. Eggers, *Justus Liebigs Chem. Ann. Chem.* **736**, 33 (1970).
151. H. Fritz and R. Oehl, *Justus Liebigs Ann. Chem.* 1628 (1973).
152. H. Fritz and R. Oehl, *Angew. Chem., Int. Ed. Engl.* **5**, 957 (1966).
153. W. von Philipsborn, W. Arnold, J. Nagyyari, K. Bernauer, H. Schmid, and P. Karrer, *Helv. Chim. Acta* **43**, 141 (1960).
154. A. Chatterjee, J. Banerji, and A. Banerji, *J. Indian Chem. Soc.* **51**, 156 (1974).
155. D. A. Evans, G. F. Smith, G. N. Smith, and K. S. J. Stapleford, *Chem. Commun.* 859 (1968).
156. D. A. Evans, J. A. Joule, and G. F. Smith, *Phytochemistry* **7**, 1429 (1968).
157. S. Sakai, N. Aimi, K. Kato, H. Ido, and J. Haginiwa, *Chem. Pharm. Bull.* **19**, 1503 (1971).
158. S.-I. Sakai, N. Aimi, K. Kato, H. Ido, K. Masuda, Y. Watanabe, and J. Haginiwa, *Yakugaku Zasshi* **95**, 1152 (1975).
159. B. Zsadon, M. Szilasi, J. Tamas, and P. Kaposi, *Phytochemistry* **14**, 1438 (1975).
160. G. A. Cordell, G. F. Smith, and G. N. Smith, *Chem, Commun.* 189 (1970).
161. G. A. Cordell, G. F. Smith, and G. N. Smith, *Chem. Commun.* 191 (1970).
162. G. A. Cordell, *Alkaloids (N.Y.)* **17**, 199 (1979).
163. G. A. Cordell, G. N. Smith, and G. F. Smith, unpublished data (1970).
164. N. Petitfrére, A. M. Morfaux, M. M. Debray, L. Le Men-Oliver, and J. Le Men, *Phytochemistry* **14**, 1648 (1975).
165. G. A. Cordell, Ph.D. Thesis, Univ. of Manchester, 1970.
166. G. A. Cordell, G. N. Smith, and G. F. Smith, *J. Indian Chem. Soc.* **55** 1083 (1978).
167. M. Damak, A. Ahond, H. Doucerain, and C. Riche, *Chem. Commun.* 510 (1976).
168. M. B. Patel, L. Thompson, C. Miet, and J. Poisson, *Phytochemistry* **12**, 451 (1973).
169. J. Poisson, F. Puisieux, C. Miet, and M. B. Patel, *Bull. Soc. Chim. Fr.* 3549 (1965).
170. D. G. I. Kingston, B. B. Gerhart, and F. Ionescu, *Tetrahedron Lett.* 649 (1976).
171. G. Büchi, R. E. Manning, and S. A. Monti, *J. Am. Chem. Soc.* **86**, 4631 (1964).
172. D. G. I. Kingston, B. B. Gerhart, F. Ionescu, M. M. Mangino, and S. M. Sami, *J. Pharm. Sci.* **67**, 249 (1978).
173. D. G. I. Kingston, *J. Pharm. Sci.* **67**, 272 (1978).
174. J. R. Knox and J. Slobbe, *Aust. J. Chem.* **28**, 1813 (1975).
175. D. W. Thomas and K. Biemann, *J. Am. Chem. Soc.* **87**, 5447 (1965).
176. B. Gabetta, E. M. Martinelli, and G. Mustich, *Planta Med.* **27**, 195 (1975).
177. E. Bombardelli, A. Bonati, B. Gabetta, E. M. Martinelli, G. Mustich, and B. Danieli, *J. Chem. Soc. Perkin Trans. 1* 1432 (1976).
178. U. Renner and H. Fritz, *Tetrahedron Lett.* 283 (1964).
179. A. Husson, Y. Langlois, C. Riche, H.-P. Husson, and P. Potier, *Tetrahedron* **29**, 3095 (1973).
180. D. G. I. Kingston, B. T. Li, and F. Ionescu, *J. Pharm. Sci.* **66**, 1135 (1977).
181. I. Chardon-Loriaux and H.-P. Husson, *Tetrahedron Lett.* 1845 (1975).

182. I. Chardon-Loriaux, M. M. Debray, and H.-P. Husson, *Phytochemistry* **17**, 1605 (1978).
183. M. Damak, C. Poupat, and A. Ahond, *Tetrahedron Lett.* 3531 (1976).
184. A. A. Gorman, N. J. Dastoor, M. Hesse, W. von Philipsborn, U. Renner, and H. Schmid, *Helv. Chim. Acta* **52**, 33 (1969).
185. A. A. Gorman and H. Schmid, *Monatsh. Chem.* **98**, 1554 (1967).
186. W. G. Kump and H. Schmid, *Helv. Chim. Acta.* **44**, 1503 (1961).
187. M. Hesse, W. von Philipsborn, D. Schumann, G. Spiteller, M. Spiteller-Friedmann, W. I. Taylor, H. Schmid, and P. Karrer, *Helv. Chim. Acta* **47**, 878 (1964).
188. P. Rasoanaivo and G. Lukacs, *J. Org. Chem.* **41**, 376 (1976).
189. A. Ahond, M.-M. Janot, N. Langlois, G. Lukacs, P. Potier, P. Rasoanaivo, M. Sangaré, N. Neuss, M. Plat, J. Le Men, E. W. Hagaman, and E. Wenkert, *J. Am. Chem. Soc.* **96**, 633 (1974).
190. J. Bruneton, A. Bouquet, and A. Cavé, *Phytochemistry* **12**, 1475 (1973).
191. E. C. Miranda and B. Gilbert, *Experientia* **25**, 575 (1969).
192. G. H. Svoboda, M. Gorman, N. Neuss, and A. J. Barnes, *J. Pharm. Sci.* **50**, 409 (1961).
193. A. El-Sayed, G. A. Handy, and G. A. Cordell, unpublished data, (1978).
194. A. Rabaron, M. Plat, and P. Potier, *Plant. Med. Phytother.* **7**, 53 (1973).
195. G. E. Mallet, D. Fukuda, and M. Gorman, *Lloydia* **27**, 334 (1964).
196. T. Nabih, L. Youel, and J. P. Rosazza, *J.C.S. Perkin I* 757 (1978).
197. B. Gabetta, E. M. Martinelli, and G. Mustich, *Fitoterapia* **45**, 32 (1974).
198. J. Poisson, M. Plat, H. Budzikiewicz, L. J. Durham, and C. Djerassi, *Tetrahedron* **22**, 1075 (1966).
199. A. A. Gorman, V. Agwada, M. Hesse, U. Renner, and H. Schmid, *Helv. Chim. Acta* **49**, 2072 (1966).
200. O. Lefebvre-Soubeyran, *Acta Crystallogr., Sect. B* **29**, 2855 (1973).
201. G. H. Aynilian, S. G. Weiss, G. A. Cordell, D. J. Abraham, F. A. Crane, and N. R. Farnsworth, *J. Pharm. Sci.* **63**, 536 (1974).
202. J. Naranjo, M. Hesse, and H. Schmid, *Helv. Chim. Acta* **55**, 1849 (1972).
203. Y. Rolland, G. Croquelois, N. Kunesch. P. Boiteau, M. Debray, J. Pecher, and J. Poisson, *Phytochemistry* **12**, 2039 (1973).
204. Y. Rolland, N. Kunesch, F. Libot, J. Poisson, and H. Budzikiewicz, *Bull. Soc. Chim. Fr.* 2503 (1975).
205. N. Kunesch, B. C. Das, and J. Poisson, *Bull. Soc. Chim. Fr.* 4370 (1970).
206. P. L. Majumder, T. K. Chanda, and B. N. Dinda, *Phytochemistry* **13**, 1261 (1974).
207. V. Agwada, M. B. Patel, M. Hesse, and H. Schmid, *Helv. Chim. Acta* **53**, 1567 (1970).
208. N. Kunesch, C. Miet, M. Troly, and J. Poisson, *Ann. Pharm. Fr.* **26**, 79 (1968).
209. N. Kunesch, J. Poisson, and B. C. Das, *Tetrahedron Lett.* 1745 (1968).
210. Y. Rolland, N. Kunesch, J. Poisson, E. W. Hagaman, F. M. Schell, and E. Wenkert, *J. Org. Chem.* **41**, 3270 (1976).
211. E. Bombardelli, A. Bonati, B. Danieli, B. Gabetta, E. M. Martinelli, and G. Mustich, *Experientia* **31**, 139 (1975).
212. P. L. Majumder, R. Raychaudhuri, and A. Chatterjee, personal communication, cited in ref. 213.
213. V. C. Agwada, J. Naranjo, M. Hesse, H. Schmid, Y. Rolland, N. Kunesch, J. Poisson, and A. Chatterjee, *Helv. Chim. Acta* **60**, 2830 (1977).
214. N. Kunesch, Y. Rolland, J. Poisson, P. L. Majumder, R. Majumder, A. Chatterjee, V. C. Agwada, J. Naranjo, M. Hesse, and H. Schmid, *Helv. Chim. Acta* **60**, 2854 (1977).
215. R. Goutarel, A. Rassat, M. Plat, and J. Poisson, *Bull. Soc. Chim. Fr.* 893 (1959).
216. M. B. Patel and J. A. Rowson, *Planta Med.* **13**, 206, 270 (1965).

217. V. Agwada, A. A. Gorman, M. Hesse, and H. Schmid, *Helv. Chim. Acta* **50**, 1939 (1967).
218. M. Plat, N. Kunesch. J. Poisson, C. Djerassi, and H. Budzikiewicz, *Bull. Soc. Chim. Fr.* 2669 (1967).
219. M. Daudon, M. H. Mehri, M. M. Plat, E. Hagaman, and E. Wenkert, *J. Org. Chem.* **41**, 3275 (1976).
220. M. Plat, H. Mehri, M. Koch, V. Scheidegger, and P. Potier, *Tetrahedron Lett.* 3395 (1970).
221. K. Bernauer, cited in reference 222.
222. D. W. Thomas, H. Achenbach, and K. Biemann, *J. Am. Chem. Soc.* **88**, 1537 (1966).
223. A. Rabaron, M. H. Mehri, T. Sevenet, and M. M. Plat, *Phytochemistry* **17**, 1452 (1978).
224. A. Cavé, J. Bruneton, A. Ahond, A. M. Bui, H. P. Husson, C. Kan, G. Lukacs, and P. Potier, *Tetrahedron Lett.* 5081 (1973).
225. J. Bruneton, A. Bouquet, and A. Cavé, *Phytochemistry* **13**, 1963 (1974).
226. M. H. Mehri, A. Rabaron, T. Sevenet, and M. M. Plat, *Phytochemistry* **17**, 1451 (1978).
227. S. Takano, S. Hatakeyama, and K. Ogasawara, *Heterocycles* **6**, 1311 (1977).
228. I. D. Rae, M. Rosenberger, A. G. Szabo, C. R. Willis, P. Yates, D. E. Zacharias, G. A. Jeffrey, B. Douglas, J. L. Kirkpatrick, and J. A. Weisbach, *J. Am. Chem. Soc.* **89**, 3061 (1967).
229. P. Yates, F. N. MacLachlan, I. D. Rae, M. Rosenberger, A. G. Szabo, C. R. Willis, M. P. Cava, M. Behforonz, M. V. Lakshmikantham, and W. Zeiger, *J. Am. Chem. Soc.* **95**, 7842 (1973).
230. P.-T. Cheng, S. C. Nyburg, F. N. MacLachlan, and P. Yates, *Can. J. Chem.* **54**, 726 (1976).
231. J. E. Saxton, *Alkaloids (N.Y.)* **8**, 673 (1965).
232. J. W. Moncrief and W. N. Lipscomb, *J. Am. Chem. Soc.* **87**, 4963 (1965); *Acta Crystallogr.* **21**, 322 (1966).
233. E. Wenkert, D. W. Cochran, E. W. Hagaman, F. M. Schell, N. Neuss, A. S. Katner, P. Potier, C. Kan, M. Plat, M. Koch, H. Mehri, J. Poisson, N. Kunesch, and Y. Rolland, *J. Am. Chem. Soc.* **95**, 4990 (1973).
234. E. Wenkert, E. W. Hagaman, B. Lal, G. E. Gutowski, A. S. Katner, J. C. Miller, and N. Neuss, *Helv. Chim. Acta* **58**, 1560 (1975).
235. D. E. Dorman and J. W. Paschal, *Org. Magn. Res.* **8**, 413 (1976).
236. N. Neuss, L. L. Huckstep, and N. J. Cone, *Tetrahedron Lett.* 811 (1967).
237. N. Langlois, F. Guéritte, Y. Langlois, and P. Potier, *J. Am. Chem. Soc.* **98**, 7017 (1976).
238. J. P. Kutney, D. E. Gregonis, R. Imhof, I. Itoh, E. Jahngen, A. I. Scott, and W. K. Chan, *J. Am. Chem. Soc.* **97**, 5013 (1975).
239. J. P. Kutney, J. Beck, F. Bylsma and W. J. Cretney, *J. Am. Chem. Soc.* **90**, 4504 (1968).
240. J. P. Kutney, J. Cook, K. Fuji, A. M. Treasurywala, J. Clardy, J. Fayos, and H. Wright, *Heterocycles* **3**, 205 (1975).
241. J. P. Bloxsidge, J. A. Elvidge, J. R. Jones, R. B. Mane, V. M. A. Chambers, E. A. Evans, and D. Greenslade, *J. Chem. Res., Synop.* 42 (1977).
242. K. Jovánovics E. Bittner, E. Dezseri, J. Eles, and K. Szasz, Can. Patent 948, 625; *CA* **82**, 103133 (1975).
243. G. Richter, Fr. Patent 2,210,392 (1974); *CA* **82**, 116076 (1975).
244. K. Jovánovics, K. Szasz, G. Fekete, E. Bittner, E. Dezseri, and J. Eles, Ger. Patent 2,259,388; *CA* **81**, 82369 (1974).
245. K. Jovánovics, K. Szasz, G. Fekete, E. Bittner, E. Dezseri, and J. Eles, Br. Patent 1,382,460; *CA* **83**, 79459 (1975).
246. G. Richter, Neth. Patent 72, 17069; *CA* **83**, 84848 (1975).
247. W. E. Jones, Ger. Patent 2,442,245 *CA* **83**, 120821 (1975).
248. S. Görög, B. Herenyi and K. Jovánovics, *J. Chromatogr.* **139**, 203 (1977).
249. G. Richter, Fr. Patent 2,210,393; *CA* **82**, 73294 (1975).

250. K. Jovánovics, K. Szasz, G. Fekete, E. Bittner, E. Dezseri, and J. Eles, Ger. Patent 2,259,447; *CA* **81**, 91780 (1974); U.S. Patent 3,899,493; *CA* **83**, 179360 (1975); Can. Patent 989,829; *CA* **85**, 177762 (1976).
251. K. Szasz, K. Jovánovics, G. Fekete, E. Bittner, E. Dezseri, and J. Eles, Hung. Patent 8,058; *CA* **81**, 37695 (1974).
252. G. Richter, Belg. Patent 823,560; *CA* **84**, 59835 (1976).
253. J. P. Kutney, J. Balsevich, T. Honda, P. H. Liao, H. P. M. Thiellier, and B. R. Worth, *Heterocycles* **9**, 201 (1978).
254. J. P. Kutney, J. Balsevich, T. Honda, P. H. Liao, H. P. M. Thiellier, and B. R. Worth, *Can. J. Chem.* **56**, 2560 (1978).
255. D. R. Brannon and N. Neuss, Ger. Patent 2,440,931 (1975); *CA* **83**, 7184 (1975).
256. N. Neuss, G. E. Mallett, D. R. Brannon, J. A. Mabe, H. R. Horton, and L. L. Huckstep, *Helv. Chim. Acta* **57**, 1886 (1974).
257. G. J. Cullinan and K. Gerzon, Ger. Patent 2,415,980; *CA* **82**, 57996 (1975).
258. J. C. Miller, G. E. Gutowski, G. A. Poore, and G. B. Boder, *J. Med. Chem.* **20**, 409 (1977); G. E. Gutowski and J. C. Miller, Ger. Patent 2,630,392 (1977); *CA* **87**, 6243 (1977).
259. J. P. Kutney, J. Balsevich, G. H. Bokelman, T. Hibino, I. Itoh, and A. H. Ratcliffe, *Heterocycles* **4**, 997 (1976).
260. J. P. Kutney, J. Balsevich, and G. H. Bokelman, *Heterocycles* **4**, 1377. (1976).
261. G. E. Gutowski, A. S. Katner, and J. C. Miller, unpublished observations, cited in ref. 234.
262. N. Neuss, A. J. Barnes and L. L. Huckstep, *Experientia* **31**, 18 (1975).
263. N. Neuss, M. Gorman, N. J. Cone, and L. L. Huckstep, *Tetrahedron Lett.* 783 (1968).
264. D. J. Abraham and N. R. Farnsworth, *J. Pharm. Sci.* **58**, 694 (1969).
265. J. P. Kutney, J. Balsevich, G. H. Bokelman, T. Hibino, T. Honda, I. Itoh, A. H. Ratcliffe, and B. R. Worth, *Can. J. Chem.* **56**, 62 (1978).
266. Y. Langlois, N. Langlois, P. Mangeney, and P. Potier, *Tetrahedron Lett.* 3945 (1976).
267. N. Langlois and P. Potier, *J. Chem. Soc., Chem Commun.* 102 (1978).
268. K. L. Stuart, J. P. Kutney, and B. R. Worth, *Heterocycles* **9**, 1015 (1978).
269. J. P. Kutney, J. Balsevich, and B. R. Worth, *Heterocycles* **9**, 493 (1978).
270. D. J. Abraham, N. R. Farnsworth, R. N. Blomster, and R. E. Rhodes, *J. Pharm. Sci.* **56**, 401 (1967).
271. P. Rasoanaivo, A. Ahond, J. P. Cosson, N. Langlois, P. Potier, J. Guilhem, A. Ducruix, C. Riche, and C. Pascard, *C. R. Acad. Sci., Ser. C* **279**, 75 (1974); J. Guilhem, A. Ducruix, C. Riche, and C. Pascard, *Acta Crystallogr., Sect. B* **32**, 936 (1976).
272. P. Rasoanaivo, N. Langlois, A. Chiaroni, and C. Riche, *Tetrahedron* **35**, 641 (1978).
273. G. H. Svoboda and A. J. Barnes, *J. Pharm. Sci.* **53**, 1227 (1964).
274. S. S. Tafur, J. L. Occolowitz, T. K. Elzey, J. W. Paschal, and D. E. Dorman, *J. Org. Chem.* **41**, 1001 (1976).
275. N. Neuss and S. S. Tafur, U.S. Patent 3,968,113 (1976); *CA* **85**, 143352 (1976).
276. S. S. Tafur, W. E. Jones, D. E. Dorman, E. E. Logsdon, and G. H. Svoboda, *J. Pharm. Sci.* **64**, 1953 (1975).
277. R. Z. Andriamialisoa, N. Langlois, P. Potier, A. Chiaroni, and C. Riche, *Tetrahedron* **34**, 677 (1978).
278. S. Tafur, personal communication to authors of ref. 277.
279. W. E. Jones and G. J. Cullinan, U.S. Patent 3,887,565; *CA* **84**, 5230 (1976).
280. R. Z. Andriamialisoa, N. Langlois, and P. Potier, *Tetrahedron Lett.* 2849 (1976).
281. J. Harley-Mason and Atta-ur-Rahman, *Chem. Commun.* 1048 (1967); *Tetrahedron* **36**, 1057 (1980).
282. J. P. Kutney, J. Beck, F. Bylsma, J. Cook, W. J. Cretney, K. Fuji, R. Imhof, and A. M. Treasurywala, *Helv. Chim. Acta* **58**, 1690 (1975).
283. Atta-ur-Rahman, *Pak. J. Sci. Ind. Res.* **14**, 487 (1971).

284. E. Wenkert, E. W. Hagaman, N. Kunesch, N. Wang, and B. Zsadon, *Helv. Chim. Acta* **59**, 2711 (1976).
285. P. Potier, N. Langlois, Y. Langlois, and F. Guéritte, *J. Chem. Soc., Chem. Commun.* 670 (1975).
286. J. P. Kutney, A. H. Ratcliffe, A. M. Treasurywala, and S. Wunderly, *Heterocycles* **3**, 639 (1975).
287. J. P. Kutney, T. Hibino, E. Jahngen, T. Okutani, A. H. Ratcliffe, A. M. Treasurywala, and S. Wunderly, *Helv. Chim. Acta* **59**, 2858 (1976).
288. N. Langlois, F. Guéritte, Y. Langlois, and P. Potier, *Tetrahedron Lett.* 1487 (1976).
289. R. Z. Andriamialisoa, Y. Langlois, N. Langlois, and P. Potier, *C. R. Acad. Sci., Ser. C* **284**, 751 (1977).
290. N. Langlois and P. Potier, *Tetrahedron Lett.* 1099 (1976).
291. J. P. Kutney and B. R. Worth, *Heterocycles* **4**, 1777 (1976).
292. J. P. Kutney, A. V. Joshua, P. H. Liao, and B. R. Worth, *Can J. Chem.* **55**, 3235 (1977).
293. J. P. Kutney, A. V. Joshua, and P. H. Liao, *Heterocycles* **6**, 297 (1977).
294. P. Mangeney, R. Costa, Y. Langlois, and P. Potier, *C. R. Acad. Sci., Ser. C* **284**, 701 (1977).
295. Y. Honma and Y. Ban, *Heterocycles* **6**, 291 (1977).
296. Atta-ur-Rahman, A. Basha, and M. Ghazala, *Tetrahedron Lett.* 2351 (1976).
297. J. P. Kutney, T. Honda, A. V. Joshua, N. G. Lewis, and B. R. Worth, *Helv. Chim. Acta* **61**, 690 (1978).
298. Atta-ur-Rahman, N. Waheed, and M. Ghazala, *Z. Naturforsch., Teil B* **31**, 265 (1976).
299. J. P. Kutney and B. R. Worth, *Heterocycles* **6**, 905 (1977).
300. Y. Honma and Y. Ban, *Tetrahedron Lett.* 155 (1978).
301. J. P. Kutney, W. B. Evans, and T. Honda, *Heterocycles* **6**, 443 (1977).
302. J. P. Kutney, E. Jahngen, and T. Okutani, *Heterocycles* **5**, 59 (1976).
303. U. Renner, D. A. Prins, and W. G. Stoll, *Helv. Chim. Acta* **42**, 1572 (1959).
304. E. Späth and W. Stroh, *Ber. Dtsch. Chem. Ges.* **58**, 2131 (1925).
305. R. H. F. Manske, *J. Am. Chem. Soc.* **51**, 1836 (1929).
306. G. R. Eccles, *Proc. Am. Pharm. Assoc.* **84**, 382 (1888).
307. H. M. Gordin, *J. Am. Chem. Soc.* **31**, 1305 (1909).
308. R. H. F. Manske, and L. Marion, *Can. J. Res., Sect. B* **17**, 293 (1939).
309. R. B. Woodward, N. C. Yang, T. J. Katz, V. M. Clark, J. Harley-Mason, R. F. Ingleby, and N. Sheppard, *Proc. Chem. Soc., London* 76 (1960).
310. T. A. Hamor, J. M. Robertson, H. N. Shrivastava, and J. V. Silverton, *Proc. Chem. Soc., London* 78 (1960).
311. T. A. Hamor and J. M. Robertson, *J. Chem. Soc.* 194 (1962).
312. G. Barger, A. Jacob, and J. Madinaveitia, *Recl. Trav. Chim. Pays-Bas* **57**, 548 (1938).
313. J. E. Saxton, W. G. Bardsley, and G. F. Smith, *Proc. Chem. Soc., London* 148 (1962).
314. T. R. Govindachari, K. Nagarajan, and S. Rajappa, *J. Sci. Ind. Res., Sect. B* **21**, 455 (1962).
315. E. Clayton, R. I. Reed, and J. M. Wilson, *Tetrahedron* **18**, 1495 (1962).
316. H. F. Hodson, B. Robinson, and G. F. Smith, *Proc. Chem. Soc., London* 465 (1961).
317. K. Eiter and O. Svierak, *Monatsh. Chem.* **82**, 186 (1951).
318. K. Eiter and O. Svierak, *Monatsh. Chem.* **83**, 1453 (1952).
319. H. F. Hodson and G. F. Smith, *J. Chem. Soc.* 1877 (1957).
320. M. Gorman, N. Neuss, C. Djerassi, J. P. Kutney, and P. J. Scheuer, *Tetrahedron* **1**, 328 (1957).
321. J. Müller, *Experientia* **13**, 479 (1957).
322. S. Siddiqui and R. H. Siddiqui, *J. Indian Chem. Soc.* **8**, 667 (1931).
323. C. Djerassi and J. Fishman, *Chem. Ind. (London)* 627 (1955).
324. R. Paris, *Ann. Pharm. Fr.* **1**, 138 (1943).
325. E. Schlittler, H. U. Huber, F. E. Bader, and H. Zahnd, *Helv. Chim. Acta* **37**, 1912 (1954).

326. A. Bertho and G. V. Schuckmann, *Ber. Dtsch. Chem. Ges.* **64**, 2278 (1931).
327. M. Aurousseau, *Ann. Pharm. Fr.* **19**, 104 (1961).
328. R. A. Paris and M. Pointet, *Ann. Pharm. Fr.* **12**, 547 (1954).
329. B. Bertho and H. F. Sarx, *Justus Liebigs Ann. Chem.* **556**, 22 (1944).
330. H. Rapoport, R. J. Windgassen, N. A. Hughes, and T. P. Onak, *J. Am. Chem. Soc.* **82**, 4404 (1960).
331. A. Bertho, M. Koll, and M. I. Ferosie, *Chem. Ber.* **91**, 2581 (1958).
332. M.-M. Janot, R. Goutarel, A. Le Hir, and F. Puisieux, *C. R. Acad. Sci.* **248**, 108 (1959).
333. M. Hesse, unpublished data, cited in ref. 13.
334. H. Rapoport, T. P. Onak, N. A. Hughes, and M. E. Reinecke, *J. Am. Chem. Soc.* **80**, 1601 (1958).
335. H. Rapoport and R. E. Moore, *J. Org. Chem.* **27**, 2981 (1962).
336. T. M. Sharp, *J. Chem. Soc.* 1227 (1934).
337. N. K. Hart, S. R. Johns, and J. A. Lamberton, *Aust. J. Chem.* **25**, 2739 (1972).
338. A. Chatterjee, S. K. Talapatra, and N. Adityachaudhury, *Chem. Ind.* (*London*) 667 (1961).
339. C. E. Nordman and S. K. Kumra, *J. Am. Chem. Soc.* **87**, 2059 (1965).
340. S. K. Talapatra and N. Adityachaudhury, *Sci. Cult.* **24**, 243 (1958).
341. N. Neuss and N. J. Cone, *Experientia* **16**, 302 (1960).
342. A. Zurcher, O. Ceder, and V. Boekelheide, *J. Am. Chem. Soc.* **80**, 1500 (1958).
343. H. Asmis, H. Schmid, and P. Karrer, *Helv. Chim. Acta* **37**, 1983 (1954).
344. A. R. Battersby, H. F. Hodson, G. V. Rao, and D. A. Yeowell, *Proc. Chem. Soc., London* 412 (1961).
345. A. R. Battersby, D. A. Yeowell, L. M. Jackman, H.-D. Schroeder, M. Hesse, H. H. Hiltebrand, W. von Philipsborn, H. Schmid, and P. Karrer, *Proc. Chem. Soc., London* 413 (1961).
346. A. T. McPhail and G. A. Sim, *Proc. Chem. Soc., London* 416 (1961).
347. H.-D. Schroeder, H. Hiltebrand, H. Schmid, and P. Karrer, *Helv. Chim. Acta* **44**, 34 (1961).
348. H. Asmis, E. Bachli, H. Schmid, and P. Karrer, *Helv. Chim. Acta* **37**, 1993 (1954).
349. H. Asmis, P. Waser, H. Schmid, and P. Karrer, *Helv. Chim. Acta* **38**, 1661 (1955).
350. K. Bernauer, F. Berlage, W. von Philipsborn, H. Schmid, and P. Karrer, *Helv. Chim. Acta* **41**, 2293 (1958).
351. K. Bernauer and H. Schmid, U.S. Patent 3,073,832; *CA* **58**, 9160 (1963).
352. M. Grdinic, D. A. Nelson, and V. Boekelheide, *J. Am. Chem. Soc.* **86**, 3357 (1964).
353. H. Wieland, W. Konz, and R. Sanderhoff, *Justus Liebigs Ann. Chem.* **527**, 160 (1937).
354. H. Wieland and H. J. Pistor, *Justus Liebigs Ann. Chem.* **536**, 68 (1938).
355. P. Karrer and H. Schmid, *Helv. Chim. Acta* **29**, 1853 (1946).
356. G. B. Marini-Bettòlo, M. A. Jorio, A. Pimenta, A. Ducke, and D. Bovet, *Gazz. Chim. Ital.* **84**, 1161 (1954).
357. A. Pimenta, M. A. Jorio, K. Adank, and G. B. Marini-Bettòlo, *Gazz. Chim. Ital.* **84**, 1147 (1954).
358. J. Kebrle, H. Schmid, P. Waser, and P. Karrer, *Helv. Chim. Acta* **36**, 345 (1953).
359. G. B. Marini-Bettòlo, P. de Berredo Carneiro, and G. C. Casinovi, *Gazz. Chim. Ital.* **86**, 1148 (1956).
360. K. Adank, D. Bovet, A. Ducke, and G. B. Marini-Bettòlo, *Gazz. Chim. Ital.* **83**, 966 (1953).
361. H. Schmid, A. Ebnöther, and P. Karrer, *Helv. Chim. Acta* **33**, 1486 (1950).
362. J. Kebrle, H. Schmid, P. Waser, and P. Karrer, *Helv. Chim. Acta* **36**, 102 (1953).
363. W. von Philipsborn, H. Schmid, and P. Karrer, *Helv. Chim. Acta* **39**, 913 (1956).
364. J. Nagyvary, W. Arnold, W. von Philipsborn, H. Schmid, and P. Karrer, *Tetrahedron* **14**, 138 (1961).

365. K. Bernauer, H. Schmid, and P. Karrer, *Helv. Chim. Acta* **40**, 1999 (1957).
366. H. Fritz and R. Oehl, *Angew. Chem.* **78**, 978 (1966).
367. H. Schmid, J. Kebrle, and P. Karrer, *Helv. Chim. Acta* **35**, 1864 (1952).
368. F. Berlage, K. Bernauer, H. Schmid, and P. Karrer, *Helv. Chim. Acta* **42**, 2650 (1959).
369. H. King, *J. Chem. Soc.* 3263 (1949).
370. G. B. Marini-Bettòlo, M. Lederer, M. A. Jorio, and A. Pimenta, *Gazz. Chim. Ital.* **84**, 1155 (1954).
371. K. Bernauer, F. Berlage, H. Schmid, and P. Karrer, *Helv. Chim. Acta* **41**, 1202 (1958).
372. H. Wieland, K. Bähr, and B. Witkop, *Justus Liebigs Ann. Chem.* **547**, 156 (1941).
373. V. Boekelheide, O. Ceder, T. Crabb, Y. Kawazoe, and R. N. Knowles, *Tetrahedron Lett.* **26**, 1 (1960).
374. W. von Philipsborn, K. Bernauer, H. Schmid, and P. Karrer, *Helv. Chim. Acta* **42**, 461 (1959).
375. H. Schmid and P. Karrer, *Helv. Chim. Acta* **30**, 1162 (1947).
376. W. Arnold, M. Hesse, H. Hiltebrand, A. Melera, W. von Philipsborn, H. Schmid, and P. Karrer, *Helv. Chim. Acta* **44**, 620 (1961).
377. F. Berlage, K. Bernauer, W. von Philipsborn, P. Waser, H. Schmid, and P. Karrer, *Helv. Chim. Acta* **42**, 394 (1959).
378. J. C. Dacre, *N.Z. Med. J.* **65**, 549 (1966).
379. K. Bernauer, H. Schmid, and P. Karrer, *Helv. Chim. Acta* **41**, 673 (1958).
380. M. Hesse, H. Hiltebrand, C. Weissmann, W. von Philipsborn, K. Bernauer, H. Schmid, and P. Karrer, *Helv. Chim. Acta* **44**, 2211 (1961).
381. H. Schmid and P. Karrer, *Helv. Chim. Acta* **33**, 512 (1950).
382. R. H. Martin, unpublished data, cited in ref. 383.
383. M. Hesse, "Indolalkaloide in Tabellen." Springer-Verlag, Berlin and New York, 1964; *ibid.* "Ergängsungwerk."
384. M. C. Lévy, M.-M. Debray, L. Le Men-Olivier, and J. Le Men, *Phytochemistry* **14**, 579 (1975).
385. D. W. Thomas and K. Biemann, *Lloydia* **31**, 1 (1968).
386. A. Goldblatt, C. Hootele, and J. Pecher, *Phytochemistry* **9**, 1293 (1970).
387. M. P. Cava, S. K. Talapatra, J. A. Weisbach, B. Douglas, R. F. Raffauf, and J. L. Beal, *Tetrahedron Lett.* 931 (1965).
388. J. A. Weisbach and B. Douglas, *Chem. Ind.* (*London*) 623 (1965).
389. U. Renner, *Experientia* **13**, 468 (1957).
390. V. C. Agwada, Y. Morita, U. Renner, M. Hesse, and H. Schmid, *Helv. Chim. Acta* **58**, 1001 (1975).
391. F. Walls, O. Collera, and A. Sandoval, *Tetrahedron* **2**, 173 (1958).
392. M.-M. Janot and R. Goutarel, *C. R. Acad. Sci.* **240**, 1719 (1955).
393. D. G. I. Kingston, *J. Pharm. Sci.* **67**, 271 (1978).
394. M. Gorman, N. Neuss, N. J. Cone, and J. A. Deyrup, *J. Am. Chem. Soc.* **82**, 1142 (1960).
395. H. Budzikiewicz, C. Djerassi, F. Puisieux, F. Percheron, and J. Poisson, *Bull. Soc. Chem. Fr.* 1899 (1963).
396. G. H. Büchi, R. E. Manning, and S. A. Monti, *J. Am. Chem. Soc.* **86**, 4631 (1963).
397. F. Puisieux, M. B. Patel, J. M. Rowson, and J. Poisson, *Ann. Pharm. Fr.* **23**, 33 (1965).
398. M. Quirin, F. Quirin, and J. Le Men, *Ann. Pharm. Fr.* **22**, 361 (1964).
399. P. L. Majumder, B. N. Dinda, and T. K. Chanda, *Indian J. Chem.* **11**, 1208 (1973).
400. F. S. Magno, G. Aguilar-Santos, and A. C. Santos, *Philipp. J. Sci.* **93**, 597 (1964); *CA* **63**, 12003 (1965).
401. G. B. Guise, M. Rasmussen, E. Ritchie, and W. C. Taylor, *Aust. J. Chem.* **18**, 1279 (1965).
402. F. Fish, F. Newcombe, and J. Poisson, *J. Pharm. Pharmacol.* **12**, 41T (1960).

403. F. Newcombe and B. A. Patel, *Planta Med.* **17**, 276 (1969).
404. M.-M. Janot and R. Goutarel, *C. R. Acad. Sci.* **240**, 1800 (1955).
405. M.-M. Janot, F. Percheron, M. Chaigneau, and R. Goutarel, *C. R. Acad. Sci.* **244**, 1955 (1957).
406. R. Goutarel, F. Percheron, and M.-M. Janot, *C. R. Acad. Sci.* **243**, 1670 (1956).
407. H. Budzikiewicz, *Chem. Ztg.* **67**, 342 (1966).
408. M. Hesse, *Helv. Chim. Acta* **50**, 42 (1967).
409. S. A. Monti, W. O. Johnson, and D. H. White, *Tetrahedron Lett.* 4459 (1966).
410. F. Puisieux, J.-P. Devissaguet, C. Miet, and J. Poisson, *Bull. Soc. Chim. Fr.* 251 (1967).
411. J. J. Dugan, M. Hesse, U. Renner, and H. Schmid, *Helv. Chim. Acta* **50**, 60 (1967).
412. R. Goutarel and M.-M. Janot, *C. R. Acad. Sci.* **242**, 2981 (1956).
413. J. LaBarre and L. Gillo, *C. R. Soc. Biol.* **150**, 1628 (1956).
414. U. Renner, D. A. Prins, A. L. Burlingame, and K. Biemann, *Helv. Chim. Acta* **46**, 2186 (1963).
415. F. Fish and F. Newcombe, *J. Pharm. Pharmacol.* **16**, 832 (1964).
416. F. Picot, M. Andriantsiferana, and H.-P. Husson, unpublished data, cited in ref. 182.
417. I. Chardon Loriaux, A. Ahond, C. Riche, and H.-P. Husson, unpublished data, cited in ref. 182.
418. D. F. Dickel, C. L. Holden, R. C. Maxfield, L. E. Paszek, and W. I. Taylor, *J. Am. Chem. Soc.* **80**, 123 (1958).
419. W. I. Taylor, *J. Org. Chem.* **30**, 309 (1965).
420. M. D. Patel, J. M. Rowson, and D. A. H. Taylor, *J. Chem. Soc.* 2587 (1961).
421. G. B. Guise, M. Rasmussen, E. Ritchie, and W. C. Taylor, *Aust. J. Chem.* **18**, 927 (1965).
422. B. O. G. Schuler, A. A. Verbeek, and F. L. Warren, *J. Chem. Soc.* 4776 (1958).
423. C. Kump, Ph.D. Thesis, Univ. of Zurich, 1964.
424. M. Hesse, F. Bodmer, and H. Schmid, *Helv. Chim. Acta* **49**, 964 (1966).
425. E. F. Rogers, H. R. Snyder, and R. F. Fischer, *J. Am. Chem. Soc.* **74**, 1987 (1952).
426. H. R. Snyder, R. F. Fischer, J. F. Walker, H. E. Els, and G. A. Nussberger, *J. Am. Chem. Soc.* **76**, 2819 (1954).
427. N. Neuss, M. Gorman, G. H. Svoboda, G. Maciak, and C. T. Beer, *J. Am. Chem. Soc.* **81**, 4754 (1959); M. Gorman, N. Neuss, and G. H. Svoboda, *J. Am. Chem. Soc.* **81**, 4745 (1959).
428. S. Kohlmunzer and H. Tomczyk, *Diss. Pharm. Pharmacol.* **19**, 403 (1967).
429. R. L. Noble, C. T. Beer, and J. H. Cutts, *Ann. N.Y. Acad. Sci.* **76**, 882 (1958); *Biochem. Pharmacol.* **1**, 347 (1958).
430. N. Langlois and P. Potier, *Phytochemistry* **11**, 2617 (1972).
431. P. Rasoanaivo, N. Langlois, and P. Potier, *Phytochemistry* **11**, 2616 (1972).
432. G. H. Svoboda, I. S. Johnson, M. Gorman, and N. Neuss, *J. Pharm. Sci.* **51**, 707 (1962).
433. P. Bommer, W. McMurray, and K. Biemann, *J. Am. Chem. Soc.* **86**, 1439 (1964).
434. N. Neuss, M. Gorman, H. E. Boaz, and N. J. Cone, *J. Am. Chem. Soc.* **84**, 1509 (1962).
435. G. H. Svoboda, *Lloydia* **24**, 173 (1961).
436. N. Neuss and A. J. Barnes, U.S. Patent 3,954,773; *CA* **85**, 108887 (1976).
437. G. H. Svoboda, *J. Am. Pharm. Assoc., Sci. Ed.* **47**, 834 (1959).
438. N. R. Farnsworth, *J. Pharm. Sci.* **61**, 1840 (1972).
439. M. Tin-Wa, N. R. Farnsworth, H. H. S. Fong, and J. Trojánek, *Lloydia* **33**, 261 (1970).
440. G. H. Svoboda, M. Gorman, A. J. Barnes, and A. T. Oliver, *J. Pharm. Sci.* **51**, 518 (1962).
441. F. Manas-Santos and A. C. Santos, *Nat. Appl. Sci. Bull.* **5**, 136 (1936); *CA* **31**, 6243 (1937).
442. W. D. Crow, N. C. Hancox, S. R. Johns, and J. A. Lamberton, *Aust. J. Chem.* **23**, 2489 (1970).
443. A. R. Battersby, R. Binks, H. F. Hodson, and D. A. Yeowell, *J. Chem. Soc.* 1848 (1960).

444. I. Schmidt, P. Waser, H. Schmid, and P. Karrer, *Helv. Chim. Acta* **43**, 1218 (1960).
445. P. Lathuillière, L. Olivier, J. Lévy, and J. Le Men, *Ann. Pharm. Fr.* **28**, 57 (1970).
446. G. H. Svoboda, *J. Pharm. Sci.* **52**, 407 (1963).
447. G. H. Svoboda, A. T. Oliver, and D. R. Bedwell, *Lloydia* **26**, 141 (1963).
448. J. M. Panas, B. Richard, C. Potron, R. S. Razafrindrambao, M.-M. Debray, L. Le Men-Olivier, J. Le Men, A. Husson and H.-P. Husson, *Phytochemistry* **14**, 1120 (1975).
449. P. Relyveld, *Pharm. Weekbl.* **100**, 614 (1965).
450. S. Baassou, H. Mehri, and M. Plat, *Phytochemistry* **17**, 1449 (1978).
451. M. Zeches, M.-M. Debray, G. Ledouble, L. Le Men-Olivier and J. Le Men, *Phytochemistry* **14**, 1122 (1975).
452. R. Verpoorte and A. B. Svendsen, *Acta Pharm. Suec.* **12**, 455 (1975).
453. M. Gorman, G. H. Svoboda, and N. Neuss, *Lloydia* **28**, 259 (1965).
454. K. V. Rao, *J. Org. Chem.* **23**, 1455 (1958).
455. C. R. Hutchinson, M.-T. S. Hsia, A. H. Heckendorf, and G. J. O'Loughlin, *J. Org. Chem.* **41**, 21 (1976).
456. H. J. Rosenkranz, B. Winkler-Lardelli, H.-J. Hansen, and H. Schmid, *Helv. Chim. Acta* **57**, 887 (1974).

Appendix

Since this manuscript was first submitted, a number of significant papers have been published in this area. Space permits only limited discussion of selected portions of this recent work. A summary of the bisindole alkaloids isolated during this period appears in Table XXIII. The discussion is organized along the same lines as in the main body of the text; therefore, section numbers here correspond to those in the text.

An early summary of the isolation and structure determination of bisindole alkaloids was inadvertently omitted from the main body of the text (*457*).

II. Tryptamine–Tryptamine Type

Hino *et al.* (*458*) have described an oxidative dimerization of N_b-methoxycarbonyltryptamine (**638**) in oxygen-saturated formic acid by irradiation at 10°–15°. The reaction products (**639** and **640**) could be reduced with LAH

NH CO_2CH_3 N H

$Pt/h\nu/O_2$, HCO_2H

N N CHO CO_2CH_3 2

638

639 racemic
640 meso

TABLE XXIII
RECENT ISOLATIONS OF BISINDOLE ALKALOIDS

Alkaloid	Source	Reference
II. Tryptamine–tryptamine type		
Quadrigemine A	*Hodgkinsonia frutescens*	(*459*)
Quadrigemine B	*H. frutescens*	(*459*)
meso-Chimonanthine	*H. frutescens*	(*459*)
III. Tryptamine–tryptamine type with an additional monoterpene unit		
Borreverine (**68**)	*Flindersia fournieri*	(*461, 462*)
Isoborreverine (**70**)	*F. fournieri*	(*461, 462*)
4,4-Dimethylisoborreverine (**642**)	*F. fournieri*	(*461, 462*)
4-Methylborreverine (**643**)	*F. fournieri*	(*461, 462*)
4-Methylisoborreverine (**644**)	*F. fournieri*	(*461, 462*)
15′-Hydroxy-14′,15′-dihydroborreverine	*F. fournieri*	(*462*)
15′-Hydroxy-14′,15′-dihydroisoborreverine	*F. fournieri*	(*462*)
IV. Corynanthe–tryptamine type		
Roxburghine D (**79**)	*Uncaria elliptica*	(*463*)
Roxburghine X	*U. elliptica*	(*463*)
12-Hydroxyusambarine (**137**)	*Strychnos usambarensis*	(*464*)
X. Pleiocarpamine–akuammiline type		
Pleiocorine (**214**)	*Alstonia odontophora* Boiteau	(*466*)
Pleiocraline (**216**)	*A. odontophora*	(*466*)
1′-Demethylpleiocorine	*A. odontophora*	(*466*)
XIII. *Strychnos–Strychnos* type		
12′-Hydroxyisostrychnobiline (**647**)	*Strychnos variabilis* Wild.	(*468*)
Isostrychnobiline (**646**)	*S. variabilis*	(*467*)
Strychnobiline (**645**)	*S. variabilis*	(*467*)
Sungucine (**649**)	*S. icaja* Baill.	(*469*)
XVI. Iboga–vobasine type		
Voacamine (**299**)	*Tabernaemontana* sp.	(*470*)
Tabernaelegantinine C (**650**)	*T. elegans*	(*471*)
Tabernaelegantinine D (**651**)	*T. elegans*	(*471*)
XVII. Cleavamine–vobasine type		
Capuvosidine (**652**)	*Capuronetta elegans*	(*472*)
	Pandaca boiteaui	(*472, 473*)
15-Dehydroxycapuvosine (**653**)	*Capuronetta elegans*	(*472*)
	Pandaca boiteaui	(*472, 473*)
Dihydrocapuvosidine (**652**)	*Capuronetta elegans*	(*472*)
	Pandaca boiteaui	(*472, 473*)
15-Dehydroxyisocapuvosine (**654**)	*P. boiteaui*	(*473*)
XXI. *Aspidosperma–Aspidosperma* type		
3-Hydroxyvobtusine (**655**)	*Voacanga chalotiana*	(*475*)
XXIV. *Aspidosperma*-Canthinone Type		
Cimiciphytine (**656**)	*Haplophyton cimicidum*	(*476*)
Norcimiciphytine (**657**)	*H. cimicidum*	(*476*)

(*Continued*)

TABLE XXIII (*Continued*)

Alkaloid	Source	Reference
XXV. *Aspidosperma*–cleavamine type		
Catharine (**542**)	*Catharanthus ovalis*	(*493*)
Catharinine (**545**)	*C. ovalis*	(*493*)
20′-Deoxyleurosidine (**508**)	*C. ovalis*	(*493*)
20′-Deoxyvinblastine	*C. ovalis*	(*493*)
21′-Hydroxyleurosine	*C. ovalis*	(*493*)
Leurocristine (**501**)	*C. ovalis*	(*493*)
Leurosine (**503**)	*C. ovalis*	(*493*)
Leurosine N'_b-Oxide (**505**)	*C. ovalis*	(*493*)
21′-Oxoleurosine (**541**)	*C. ovalis*	(*493*)
	C. roseus	(*492*)
Vinblastine (**502**)	*C. ovalis*	(*493*)
Vincathicine (**543**)	*C. ovalis*	(*493*)
Vincovalicine (**665**)	*C. ovalis*	(*493*)
Vincovaline (**554**)	*C. ovalis*	(*493*)
Vincovalinine	*C. ovalis*	(*493*)
Vindolicine (**382**)	*C. ovalis*	(*493*)
XXVII. *Aspidosperma*–Pseudoaspidosperma type		
Ervafoline (**666**)	*Stenosolen heterophyllus* (Vahl) Mgf.	(*495*)
XXIX. Bisindoline alkaloids from *Petchia ceylanica*		
Peceyline (**667**)	*Petchia ceylanica* Wight.	(*497*)
Peceylanine (**668**)	*P. ceylanica*	(*497*)
Pelankine (**669**)	*P. ceylanica*	(*497*)

to give (±)-folicanthine (**22**) in 29% yield. Alkaline hydrolysis of **639** and LAH reduction afforded (±)-chimonanthine (**35**). N_a-Methyl-N_b-methoxycarbonyltryptamine and N_b-benzyloxycarbonyltryptamine reacted similarly.

A full text on the work by Parry and Smith concerned with the tetrameric tryptamine derivatives, quadrigemines A and B, from *Hodgkinsonia*

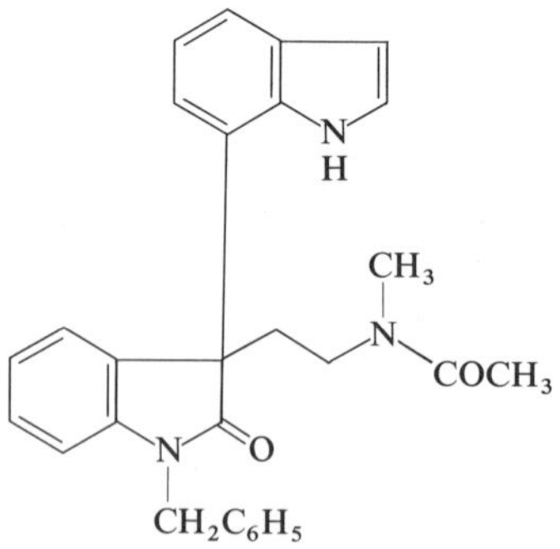

641

frutescens has appeared (*459*). Preliminary results relating to the synthesis of **641**, a possible intermediate in the synthesis of quadrigemine A, have been described (*460*).

III. Tryptamine–Tryptamine Type with an Additional Monoterpene Unit

Three new alkaloids, 4,4′-dimethylisoborreverine (**642**), 4-methylborreverine (**643**), and 4-methylisoborreverine (**644**) were isolated from *Flindersia fournieri* Panch. et Seb. (*461*, *462*), together with the major alkaloids borreverine (**68**) and isoborreverine (**70**). Two other new compounds in this

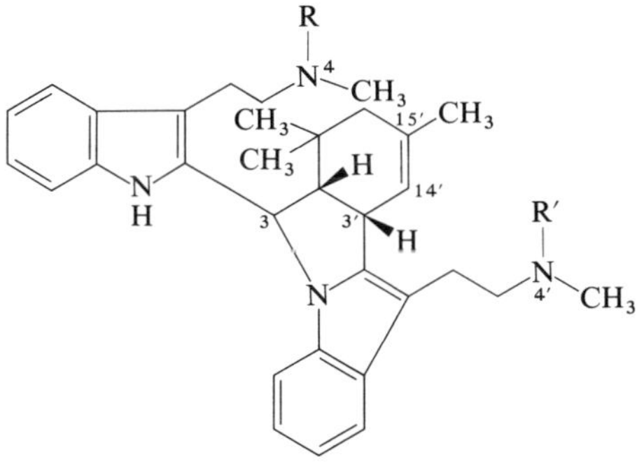

70 R = R′ = H
642 R = R′ = CH_3
644 R = CH_3, R′ = H

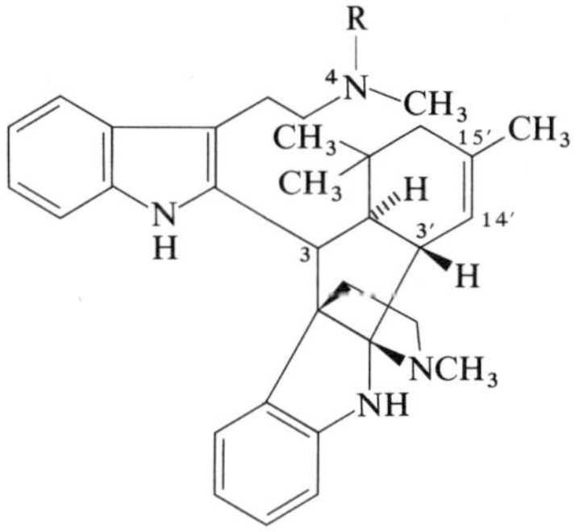

68 R = H
643 R = CH_3

series, 15′-hydroxy-14′,15′-dihydroisoborreverine and 15′-hydroxy-14′,15′-dihydroborreverine were also isolated (*462*).

IV. Corynanthe–Tryptamine Type

Roxburghine X, a stereoisomer of roxburghine D, has been isolated from *Uncaria elliptica* (R. B. ex G. Don.) (*463*).

Spectral properties of 12-hydroxyusambarine (usambaridine Vi) (**137**) have been discussed (*464*), and Seguin and Koch (*465*) have reported on the semisynthesis of usambarine (**127**) from corynantheal and *N*-methyltryptamine.

X. Pleiocarpamine–Akuammiline Type

Alstonia odontophora Boiteau yielded pleiocorine (**214**) and pleiocraline (**216**), together with the new alkaloid 1′-demethylpleiocorine (*466*).

XIII. *Strychnos–Strychnos* Type

A new series of *Strychnos–Strychnos* alkaloids has been isolated from the roots of *Strychnos variabilis* Wild (*467*, *468*). Two of the isolates, strychnobiline (**645**) and isostrychnobiline (**646**) were isomers, whereas the third

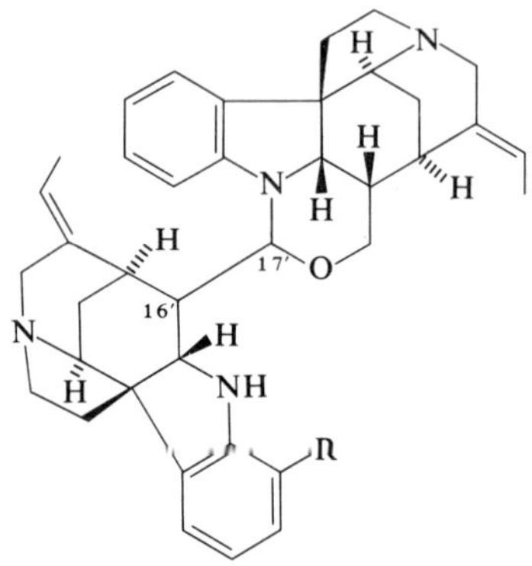

645 R = H Strychnobiline,
646 R = H Isostrychnobiline, 16′α-H, 17′α-H
647 R = OH 12′-Hydroxyisostrychnobiline, 16′α-H, 17′α-H

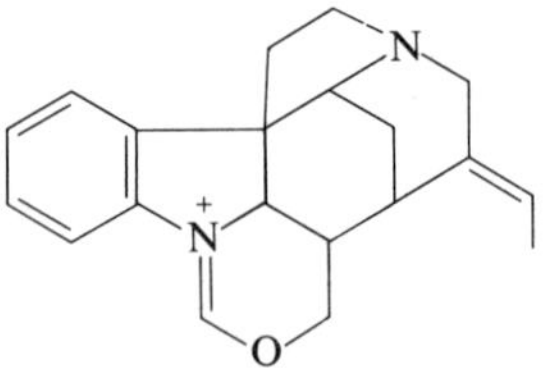

648 *m*/*e* 307

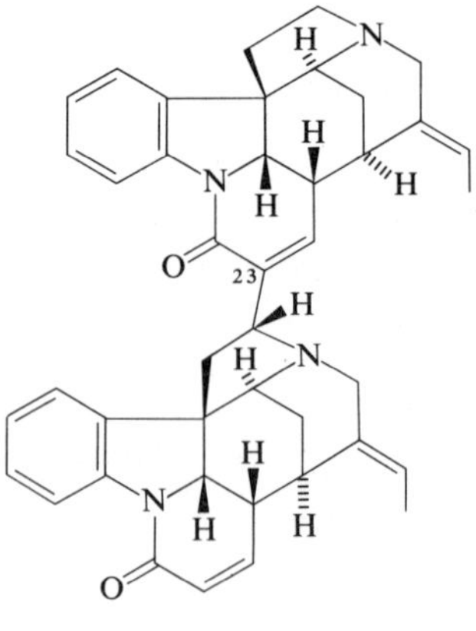

649 Sungucine

compound was the 12′-hydroxy derivative **647**, having a strongly hydrogen-bonded phenolic group. All the isolates displayed a base peak at *m/e* 307, assigned the structure **648**.

Sunguicine (**649**) from *S. icaja* Baill. (*469*) contains a novel C_{23}—C_5 link, and its structure was solved by X-ray crystallography.

XVI. Iboga–Vobasine Type

Voacamine (**299**) has been isolated from a *Tabernaemontana* sp. (*470*). *T. elegans* had previously yielded a number of bisindole alkaloids (*176*, *177*),

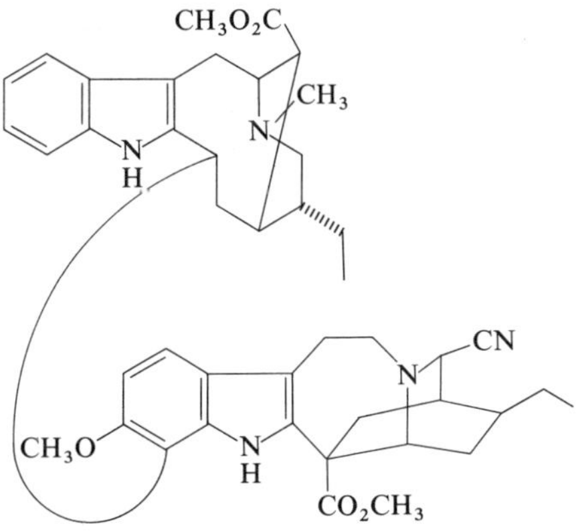

650 Tabernaelegantinine C

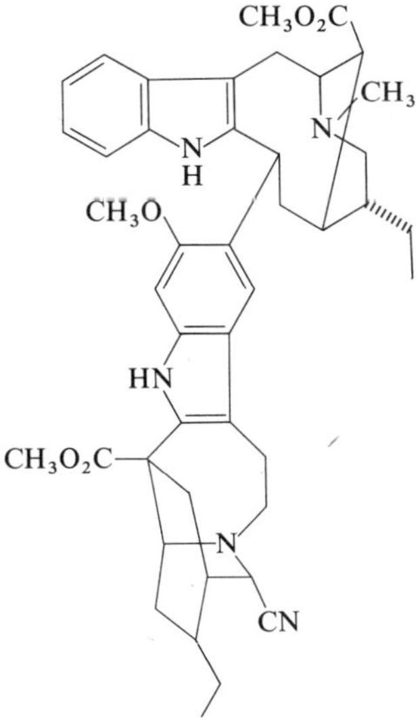

651 Tabernaelegantinine D

and further work (*471*) has yielded two interesting alkaloids, tabernaelegantinines C (**650**) and D (**651**), each possessing a cyano group at C-3′. These groups appeared at 120 ppm in the ^{13}C-NMR spectra of the isolates.

XVII. Cleavamine–Vobasine Type

A number of new members (**652–654**) in the series (see Table XXIII) have been isolated from *Capuronetta elegans* (*472*) and *Pandaca boiteaui* (*472*, *473*), and a more complete discussion of the chemistry of capuvosidine (**347**) has appeared (*472*). Sodium borohydride reduction of **347** afforded dihydrocapuvosidine (**652**) and 15-dehydroxycapuvosine (**653**). 15-Dehydroxyisocapuvosine (**654**) represents a new skeletal variation in which the vobasinyl unit is linked to C-10 rather than C-11 (*473*).

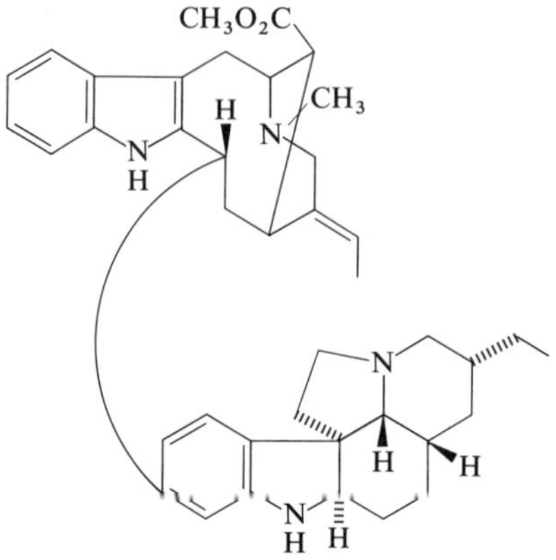

652 Dihydrocapuvosidine

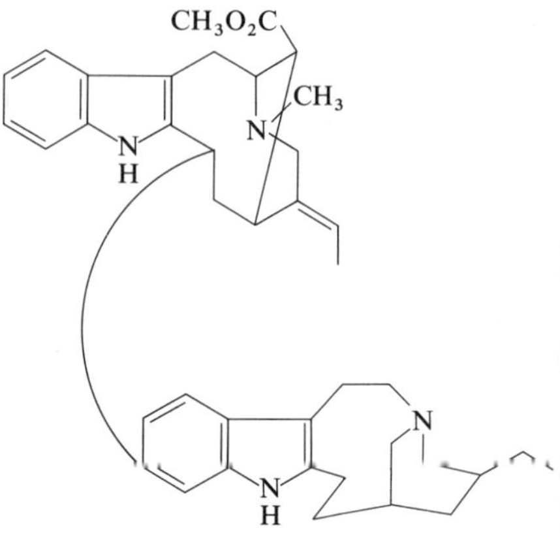

653

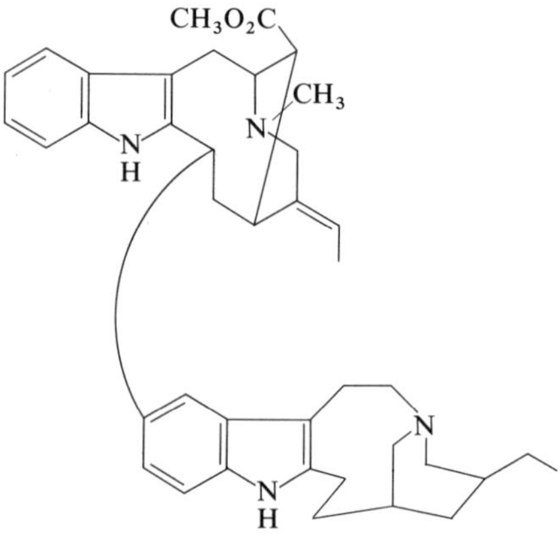

654 15-Dehydroxyisocapuvosine

XXI. *Aspidosperma–Aspidosperma* Type

The chemistry of amataine (grandifoline) (**426**) has been reviewed (*474*). *Voacanga chalotiana* has yielded 3′-hydroxyvobtusine (**655**) (*475*), which is of possible interest as a link between vobtusine (**390**) and amataine (**426**). The 14*S*-configuration was established via the CD spectrum, and the location of the hydroxy group was assigned from the observation of a singlet for H-3′ at 4.53 ppm.

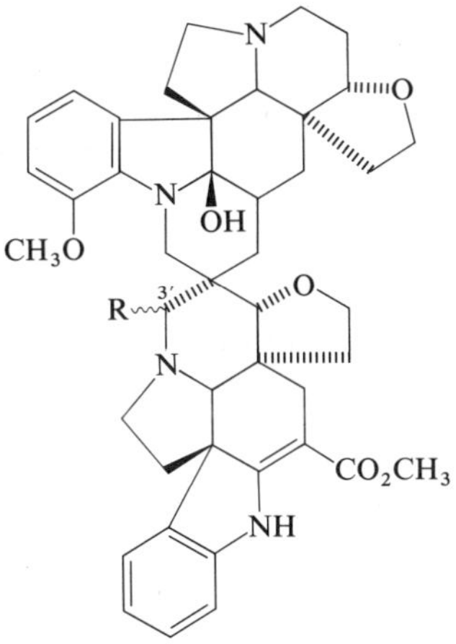

390 R = H Vobtusine
655 R = OH 3′-Hydroxyvobtusine

XXIV. *Aspidosperma*–Canthinone Type

Two new alkaloids, cimiciphytine (**656**) and its nor derivative **657**, have been obtained from *Haplophyton cimicidum* (*476*). Their structures were

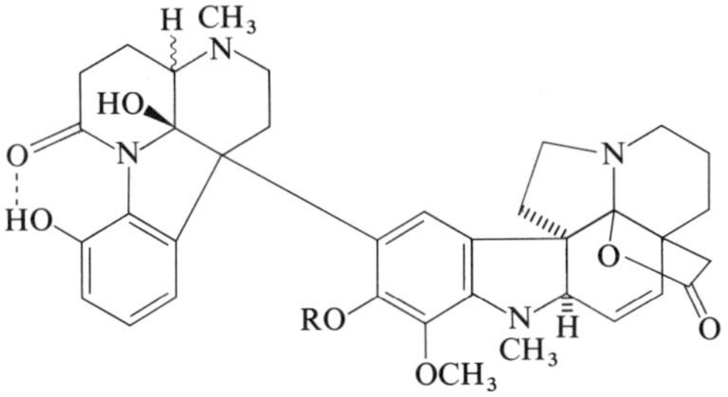

656 R = CH_3 Cimiciphytine
657 R = H Norcimiciphytine

determined on the basis of spectral evidence and reduction of haplophytine with Zn–HOAc to **656** (5% yield), and reaction of **657** with diazomethane to give **656**.

XXV. *Aspidosperma*–Cleavamine Type

15′,20′-Anhydrovinblastine (**507**), as mentioned previously, has for some time been regarded as the biosynthetic precursor of the vinblastine group of alkaloids, although its presence in the plant has never been established, probably owing to the facility with which it can be transformed into other alkaloids, e.g., leurosine (**503**) and catharine (**542**). Evidence has now been obtained, however, that anhydrovinblastine is a natural product and can actually be isolated from *Catharanthus roseus*, provided that the extraction procedure is carried out quickly (*477*). Evidence has also been provided which demonstrates that catharine is a *bona fide* alkaloid, and not an artifact (*478*).

Whether leurosine, catharine, and the other alkaloids of the vinblastine group are true alkaloids or artifacts derived from anhydrovinblastine, the

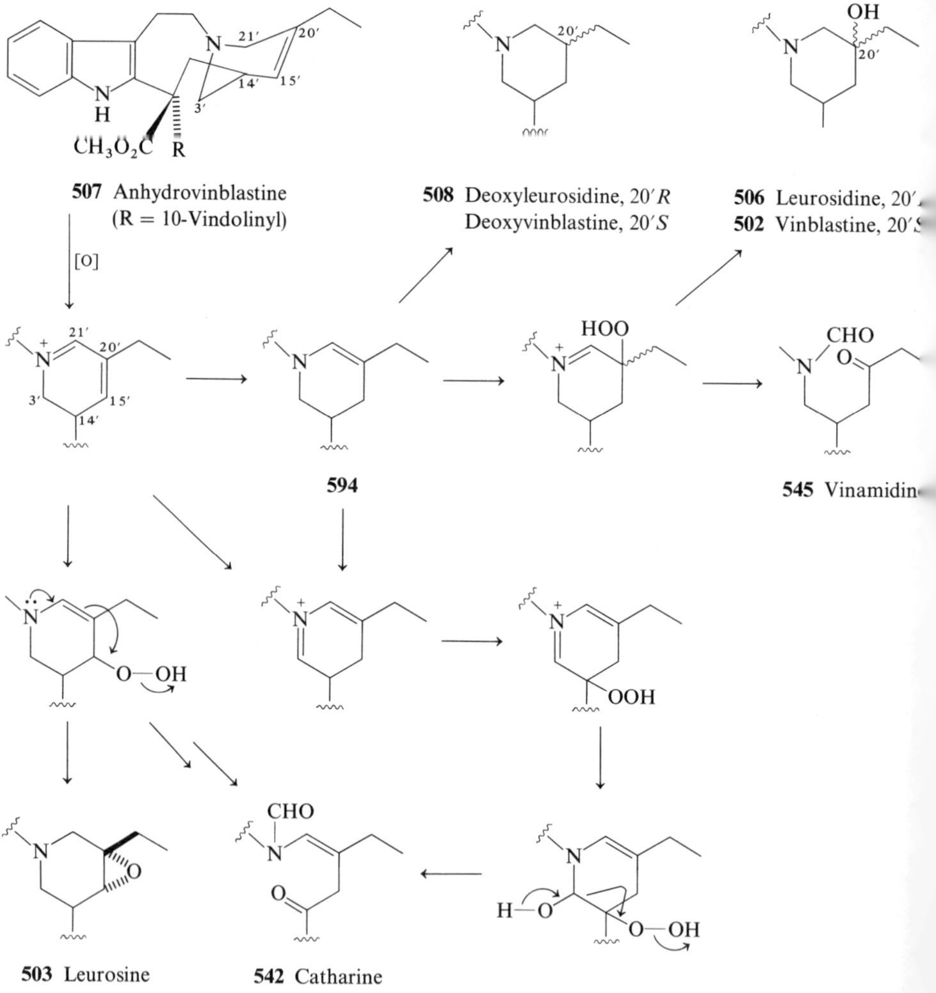

507 Anhydrovinblastine (R = 10-Vindolinyl)

508 Deoxyleurosidine, 20′*R*
Deoxyvinblastine, 20′*S*

506 Leurosidine, 20′
502 Vinblastine, 20′

594

545 Vinamidin

503 Leurosine

542 Catharine

SCHEME 27.

fact remains that aerial oxidation of anhydrovinblastine is a facile process that does not need to be enzyme mediated, and a recent re-examination of this reaction has revealed that all the alkaloids of the vinblastine group are produced (*479*). The oxidations were performed in acetonitrile solution, and in one experiment, conducted at 26° for 48 hr, the composition of the alkaloid mixture obtained was roughly similar to that observed in some extracts of *Catharanthus* species. In the oxidation the lone pair of electrons on N_b' are presumably involved, because anhydrovinblastine N_b'-oxide is inert toward air oxidation, and while the presence of moisture promotes the reaction, oxygen from the water is not incorporated into the oxidized alkaloids. On the basis of the available evidence, a series of mechanistic pathways for the oxidation of anhydrovinblastine was proposed (Scheme 27) (*479*).

The oxidation of anhydrovinblastine (**507**) by laboratory reagents has also been studied in an attempt to synthesize vinamidine (**545**). However,

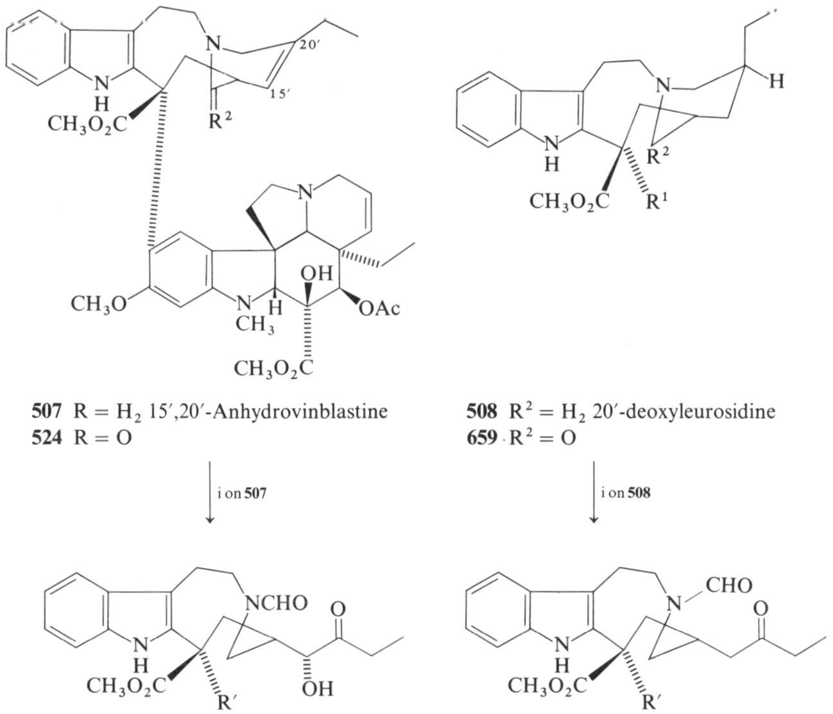

507 R = H_2 15′,20′-Anhydrovinblastine
524 R = O

508 $R^2 = H_2$ 20′-deoxyleurosidine
659 R^2 = O

i on **507**

i on **508**

658 15′-Hydroxyvinamidine

545 Vinamidine (catharinine)

R′ = 10-vindolinyl

Reagents: i, $KMnO_4$, acetone

oxidation of either anhydrovinblastine or leurosine (**503**) with potassium permanganate gave the related 3′-lactams **524** and **540**, together with a ketol (**658**) containing a formamide group, which proves to be 15′-hydroxyvinamidine. Fission of the α-ketol system in **658** by means of sodium periodate gave an aldehyde by loss of three carbon atoms, thus confirming that the oxidation of anhydrovinblastine had resulted in cleavage of the 20′,21′-bond with formation of a 15′-hydroxy-20′-oxo pattern of oxidation in the product. The 15′*R*-configuration postulated in **658** was assumed on the basis of retention of configuration at this position in the oxidation of leurosine (**503**). Because the 15′-hydroxyl group in **658** could not be removed, the oxidation of a substrate with a lower oxidation state was investigated. In fact, permanganate oxidation of 20′-deoxyleurosidine (**508**) gave some 3′-lactam **659**, together with 25% of a ring D′ cleavage product, identified as vinamidine (**545**) (*480, 481*).

The long series of experiments involving the synthesis of vinblastine analogs by the Polonovski coupling of catharanthine N_b-oxide derivatives with vindoline has finally culminated in the synthesis of vinblastine (**502**) itself (*482–484*). The rather unstable enamine (**594**), earlier prepared by the Polonovski reaction between 15,20*S*-dihydrocatharanthine N_b-oxide and vindoline (see below), can also be obtained by hydrogenation of anhydro-

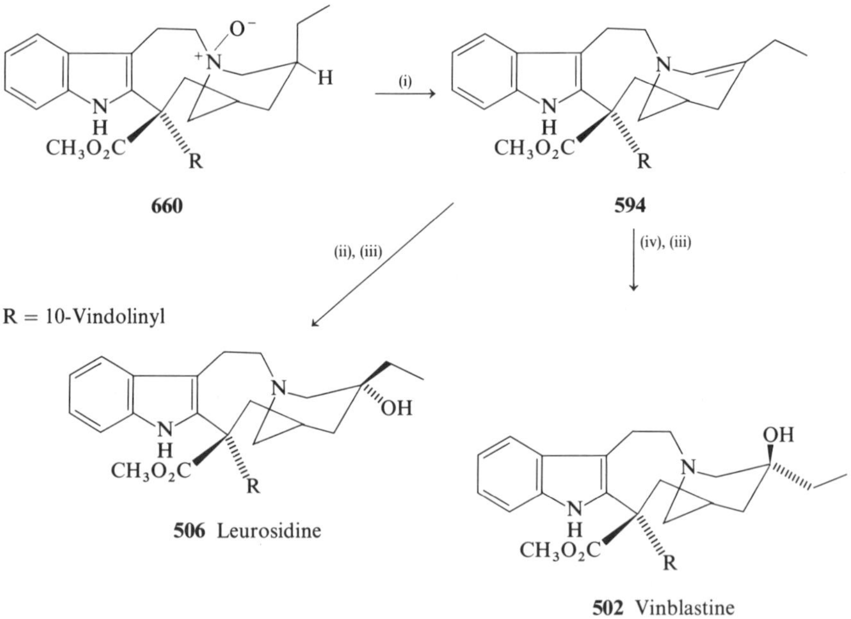

660

594

(i)

(ii), (iii)

(iv), (iii)

R = 10-Vindolinyl

506 Leurosidine

502 Vinblastine

Reagents: (i) TFAA, CH_2Cl_2; (ii) OsO_4; (iii) $NaBH_4$; (iv) $Tl(OAc)_3$.

vinblastine (**507**), which results in preferential addition of hydrogen at the α-face of the double bond, followed by oxidation (*m*-chloroperbenzoic acid) to the N_b-oxide **660**, and Polonovski reaction with trifluoroacetic anhydride. The effect of various oxidizing reagents on **594** was then examined. The earlier synthesis of leurosidine (**506**) following reaction with osmium tetroxide and sodium borohydride was confirmed, but a completely different result was observed when thallium triacetate was used as the oxidant. Attack now occurred at the β-face of the enamine double bond; consequently, reduction of the intermediate by means of sodium borohydride gave vinblastine (**502**) in 30% yield from the *N*-oxide **660**. The stereoselectivity of attack may be explained by postulating that osmylation and hydrogenation occur at the less hindered face of the double bond, but that reaction with thallium acetate is controlled by the orientation of the lone pair of electrons on N_b', which presumably coordinate with the reagent, and hence determine attack at the β-face of the enamine.

5′-Noranhydrovinblastine (**661**) was produced when the modified Polonovski reaction was applied to anhydrovinblastine N_b'-oxide (*485*, *486*),

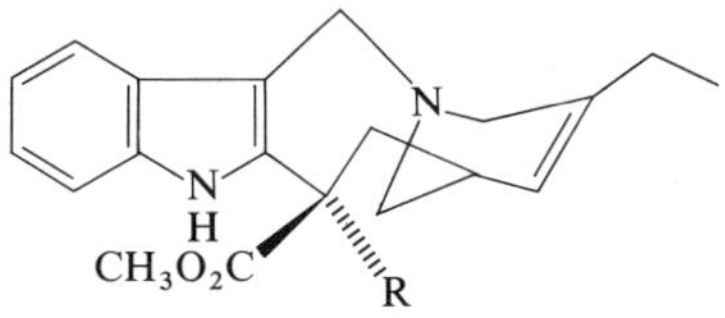

661 5′-Nor anhydrovinblastine (R = 10-vindolinyl)

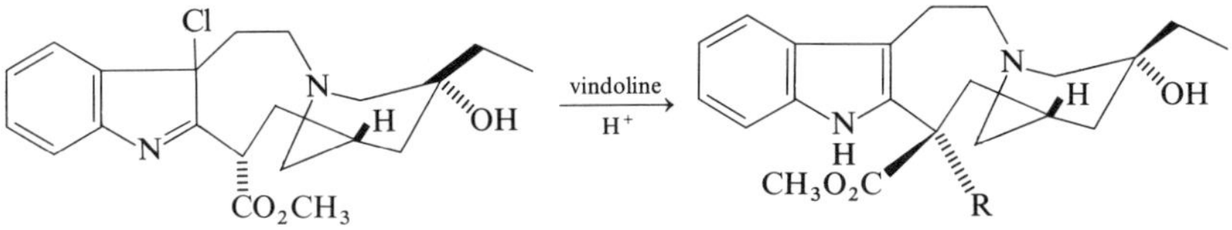

vindoline
H^+

662

663 R = 10-vindolinyl

and a variety of 5′,6′-seco products were formed when a range of nucleophiles was added (*486*).

Condensation of the chloroindolenine (**662**) of secopandoline with vindoline (**378**) under acidic conditions afforded 14′-epi-20-epivinblastine (**663**) (*487*), and a full paper on the partial synthesis of demethoxycarbonyldeoxyvinblastine (**568**) has appeared (*488*).

The Canadian group has reported (*489*) on the synthesis and conformational analysis of several isomers and analogs of vinblastine, and the French group has established (*491*) that 20′-deoxyleurosidine (**508**) is a precursor of vinblastine in *Catharanthus roseus* whole plants.

In further extractions of the leaves of *C. roseus* a new alkaloid has been isolated which was proved (*491*) to be 21′-oxoleurosine (**541**). This base also occurs, with 14 other bisindole alkaloids, in the aerial parts of *C. ovalis* (*492*). Aside from the vindoline 10,10′-dimer vindolicine, all these alkaloids are vinblastine relatives, namely, vinblastine (**502**), leurosine (**503**), vincristine (**501**), 21′-hydroxyleurosine, 20′-deoxyleurosidine (**508**), 20′-deoxyvinblastine, pleurosine (**505**), catharine (**542**), catharinine (**545**), vincathicine (**543**), vincovaline (**554**), vincovalinine, and vincovalicine. The structures proposed earlier (*279*) for vincovaline (**554**) and vincovalinine (16′-decarbomethoxyleurosine) have been confirmed and the structure of vincovalicine has been discussed. This last alkaloid appears to be composed of N_a-demethyl-N_a-formylvindoline attached at C-10 to an indolenine component, possibly a pseudoaspidospermane unit obtained by oxidation of a tetracyclic ibogane derivative with formation of a 3,7-bond. Structure **664**, provisionally proposed, is consistent with the available evidence; unfortunately, lack of material and the somewhat unstable character of vincovalicine have so far prevented a definitive structure determination.

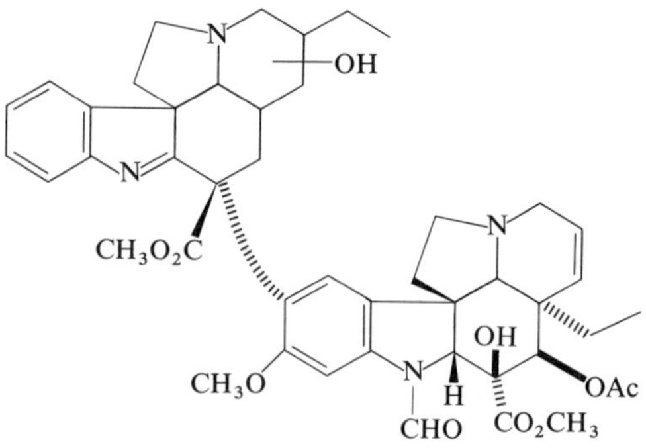

664 Vincovalicine (?)

XXVIII. *Aspidosperma*–Pseudoaspidosperma Type

Ervafoline (**665**) is the first member of a new group of alkaloids from *Ervatamia pandacaqui* (*493*) and *Stenosolen heterophyllus* (Vahl) Mgf. (Apocynaceae) (*494*), in which three bonds link the two monomer units. The structure was solved by X-ray analysis (*494*) and substantiated by an exhaustive study of the ^{1}H-NMR spectrum at 400 MHz (*495*).

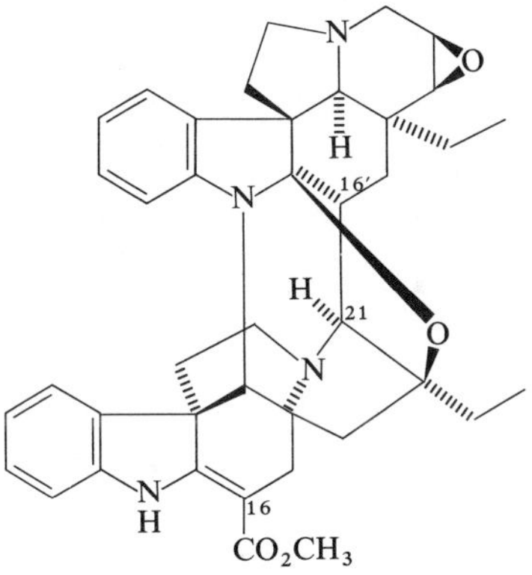

665 Ervafoline

XXIX. Bisindoline Alkaloids from *Petchia ceylanica*

Petchia ceylanica Wight has yielded three new alkaloids based on the vincorine system, namely, peceyline (**666**), peceylanine (**667**), and pelankine (**668**) (*496*). The structures were solved principally by interpretation of ^{13}C-NMR data, and the chemical shift of C-2 (97.5, 97.2) and C-2′ (96.3) in **666** and **667** should be noted.

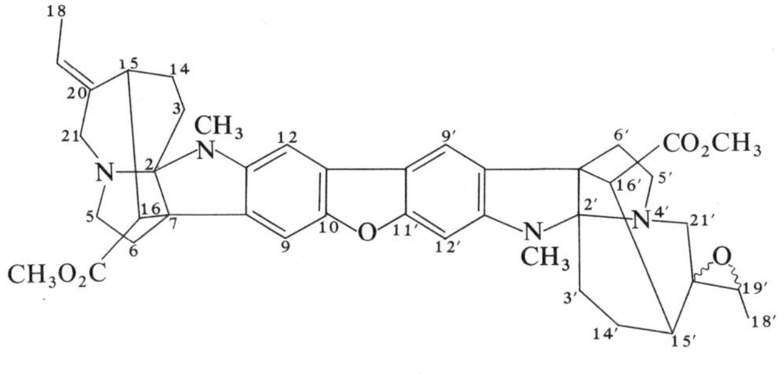

666 Peceyline

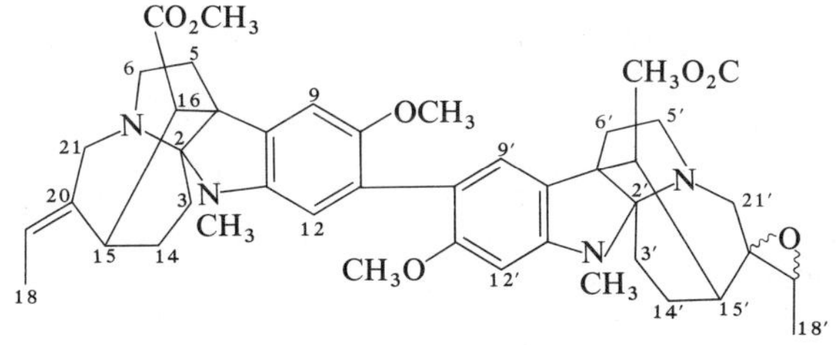

667 Peceylanine

668 Pelankine

REFERENCES TO THE APPENDIX

457. J. Garnier, M. Koch, and N. Kunesch, *Phytochemistry* **8**, 1241 (1969).
458. T. Hino, S. Kodato, K. Takahashi, H. Yamaguchi, and M. Nakagawa, *Tetrahedron Lett.* 4913 (1978).
459. K. P. Parry and G. F. Smith, *J. Chem. Soc. Perkin Trans. 1* 1671 (1978).
460. P. K. Battey, D. L. Crookes, and G. F. Smith, *Can. J. Chem.* **57**, 1694 (1979).
461. F. Tillequin and M. Koch, *Phytochemistry* **18**, 1559 (1979).
462. F. Tillequin, R. Rousselet, M. Koch, M. Bert, and T. Sevenet, *Ann. Pharm. Fr.* **37**, 543 (1979).
463. W. H. M. W. Herath, M. U. S. Sultanbawa, G. P. Wannigama, and A. Cavé, *Phytochemistry* **18**, 1385 (1979).
464. L. Angenot, C. Coune, and M. Tits, *J. Pharm. Belg.* **33**, 284 (1978).
465. E. Seguin and M. Koch, *J. Med. Plant Res.* **37**, 175 (1979).
466. J. Vercauteren, G. Massiot, T. Sevenet, J. Lévy, L. Le Men-Olivier and J. Le Men, *Phytochemistry* **18**, 1729 (1979).
467. M. J. G. Tits and L. Angenot, *Planta Med.* **34**, 57 (1978).
468. M. Tits, D. Tavernier, and L. Angenot, *Phytochemistry* **18**, 515 (1979).
469. J. Lamotte, L. Dupont, O. Dideberg, K. Kambu, and L. Angenot, *Tetrahedron Lett.* 4227 (1979).
470. R. I. Lores, *Rev. CENIC, Cienc. Fis.* **8**, 53 (1977).
471. B. Danieli, G. Palmisano, B. Gabeta, and E. M. Martinelli, *J. Chem. Soc. Perkin Trans. 1* 601 (1980).
472. H.-P. Husson, I. Chardon-Loriaux, M. Andriantsiferana, and P. Potier, *J. Indian Chem. Soc.* **55**, 1099 (1978).
473. M. Andriantsiferana, F. Pico, P. Boiteau, and H.-P. Husson, *Phytochemistry* **18**, 911 (1979).
474. A. Chatterjee, *J. Indian Chem. Soc.* **56**, 225 (1979).
475. B. Danieli, G. Lesma, G. Palmisano, and B. Gabetta, *Heterocycles* **14**, 201 (1980).
476. M. V. Lakshmikantham, M. J. Mitchell, and M. P. Cava, *Heterocycles* **9**, 1009 (1978).
477. A. I. Scott, F. Gueritte, and S. L. Lee, *J. Am. Chem. Soc.* **100**, 6253 (1978).
478. K. L. Stuart, J. P. Kutney, T. Honda, and B. R. Worth, *Heterocycles* **9**, 1391 (1978).
479. N. Langlois and P. Potier, *J. Chem. Soc., Chem. Commun.* 582 (1979).
480. J. P. Kutney, J. Balsevich, and B. R. Worth, *Heterocycles* **11**, 69 (1978).
481. J. P. Kutney, J. Balsevich, and B. R. Worth, *Can. J. Chem.* **57**, 1682 (1979).

482. P. Mangeney, R. Z. Andriamialisoa, N. Langlois, Y. Langlois, and P. Potier, *C. R. Acad. Sci., Ser. C* **288**, 129 (1979).
483. P. Mangeney, R. Z. Andriamialisoa, N. Langlois, Y. Langlois, and P. Potier, *J. Am. Chem. Soc.* **101**, 2243 (1979).
484. P. Potier, *J. Nat. Prod.* **43**, 72 (1980).
485. P. Mangeney, R. Z. Andriamialisoa, J.-Y. Lallemand, N. Langlois, Y. Langlois, and P. Potier, *Tetrahedron* **35**, 2175 (1979).
486. P. Mangeney, R. Z. Andriamialisoa, N. Langlois, Y. Langlois, and P. Potier, *J. Org. Chem.* **44**, 3765 (1979).
487. N. Kunesch, P.-L. Vaucamps, A. Cavé, J. Poisson, and E. Wenkert, *Tetrahedron Lett.* 5073 (1979).
488. J. Harley-Mason and Atta-ur-Rahman, *Tetrahedron* **36**, 1057 (1980).
489. J. P. Kutney, T. Honda, P. M. Kazmaier, N. J. Lewis, and B. R. Worth, *Helv. Chim. Acta* **63**, 366 (1980).
490. F. Guéritte, N. Y. Bac, Y. Langlois, and P. Potier, *J. Chem. Soc., Chem. Commun.* 452 (1980).
491. A. El-Sayed, G. A. Handy, and G. A. Cordell, *J. Nat. Prod.* **43**, 157 (1980).
492. N. Langlois, R. Z. Andriamialisoa, and N. Neuss, *Helv. Chim. Acta* **63**, 793 (1980).
493. P. Lathuilliere, L. Olivier, J. Lévy, and J. Le Men, *Ann. Pharm. Fr.* **28**, 57 (1970).
494. A. Henriques, C. Kan-Fan, A. Ahond, C. Riche, and H.-P. Husson, *Tetrahedron Lett.* 3707 (1978).
495. A. Henriques, S.-K. Kan, and M. Lounasmaa, *Acta Chem. Scand.* **33B**, 775 (1979).
496. N. Kunesch, A. Cavé, E. W. Hagaman, and E. Wenkert, *Tetrahedron Lett.* 1727 (1980).

——CHAPTER 2——

THE EBURNAMINE–VINCAMINE ALKALOIDS

WERNER DÖPKE

Sektion Chemie, Humboldt Universitat, Berlin, German Democratic Republic

I. Introduction

Since the eburnamine–vincamine alkaloids were last reviewed in Volume VIII of this series, many important developments have taken place in the chemistry of this group of alkaloids. Isolation studies have resulted in the addition of 16 new alkaloids whose structures are known with confidence. Many successful syntheses of this group of alkaloids have been recorded, including a large number that have emphasized the use of new approaches.

The absolute configurations of alkaloids belonging to this series have been established and progress has also been made in clarifying the relative and absolute stereochemistry of several. NMR spectroscopy and mass spectrometry have proved useful tools in the structure analysis of certain eburnamine–vincamine alkaloids, although, as yet, neither has been utilized extensively.

The alkaloids eburnamine from *Hunteria eburnea* Pichon and vincamine from *Vinca minor* are the prototypes of the indole alkaloid class eburnamine–vincamine. The alkaloids of this class are derivatives of ring skeletons of the

THE ALKALOIDS, VOL. XX

ISBN 0-12-469520-5

same constitution, namely of the 13a-ethyl-2,3,5,6,12,13,13a,13b-1-*H*-indolo[3,2,1,-d,e]pyrido[3,2,1,-i,j]-1,5-naphthyridins, sometimes also called "Vincan" (**1**) or "Eburnan" (**2**). Both skeletons occurring in diverse natural substances, however, are not identical, but enantiomeric. (*1*). This enantio-

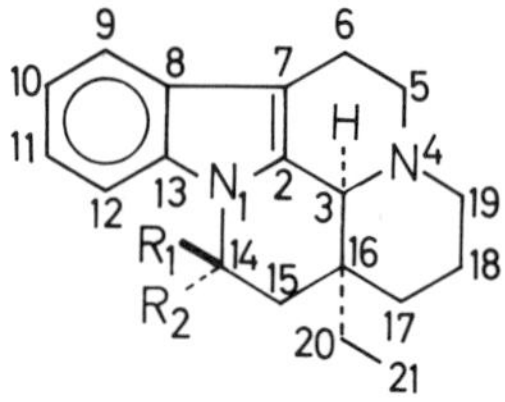

Vincan 1

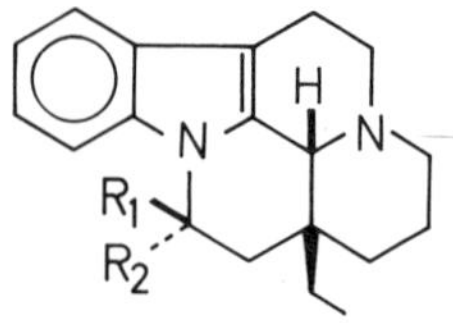

Eburnan 2

merism becomes evident when the acidic dehydration of eburnamine gives the unsaturated product (+)-eburnameinene in position 14,15, whereas in the case of vincamine, the α,β unsaturated ester is produced. The numbering of the vincan skeleton is shown above. (*2–5*) The IUPAC numbering for the ring system **3** of vincamine is shown above. The numbering corresponds to that proposed for vincamine by Trojanek and Blaha (*3*) and actually used by *Chemical Abstracts*.

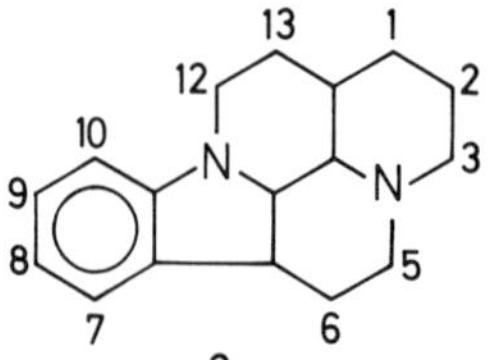

3

Although the genus *Vinca* has only a small number of species, it has become the object of intensive research activities. The concentrated effort of several groups of chemists, botanists, and pharmacologists has afforded comprehensive experimental material which has allowed the addition of the genus *Vinca* to the many other well-explored genera of Apocynaceae.

The genus *Vinca* is classified into three species with five varieties: i.e., *Vinca minor* L.; *Vinca major* L., with varieties *difformis* (Pourr.) Pich. and *major* (L.) Pich.; and *Vinca herbacea* Waldst. et Kit., with var. *herbacea* Pich., var. *libanotica* (Zucc.) Pich. and var. *sessilifolia* (A. DC.) Pich. Thus far, the following species have been chemically analyzed: *Vinca minor* L., *Vinca pubescens* Urv. [according to the above nomenclature, this species should be *Vinca major* L. var. *major* Pich.], *Vinca difformia* Pourr. [*Vinca major* L. var. *difformis* (Pourr. Pich.)], *Vinca major* L. *Vinca herbacea* Waldst. et Kit., and *Vinca erecta* Rgl. et Schmalh. [*Vinca herbacea* Waldst. et. Kit. var. *libanotica* (Zucc.)]. Of note is the stereochemical diversity in the group of eburnamine, aspidospermine, and quebrachamine alkaloids of *Vinca minor*. In addition to the occurrence of C-16 epimers, vincamine–epivincamine and eburnamine–isoeburnamine, some racemates have also been isolated, along with their (−) forms, i.e., (±)-eburnamonine. It has been shown independently in three different laboratories that the racemization had not taken place during the isolation procedure.

The occurrence of racemates and/or enantiomers in this class of natural products has been demonstrated in other plants, e.g., the isolation of (+)- and (−)-pyrifolidine, as well as (+)- and (−)-vincadifformine (*6,7*), and by the isolation of (+)-vincamine and (+)-minovincine from *Tabernaemontana rigida* (*7*). The genus *Tabernaemontana* of the family Apocynacee has been the subject of continued controversy with regard to problems of botanical classification. Therefore, the apparently exclusive production of alkaloids having the eburnamine type of skeleton of *T. rigida* (*7*) is quite interesting from the chemotaxonomic point of view, since no alkaloids of this structural type have been found in any species of *Tabernaemontana* previously examined. In a broader context, the alkaloids of *Vinca* species exhibit many features in common with the alkaloids from the closely related genera of the Apocynaceae (sub-family Plumieroideae) e.g., genera *Alstonia*, *Aspidosperma*, *Amsonia*, *Rhazya*, *Geissospermum*, *Tabernaemontana*, *Kopsia*, and *Catharanthus*.

In the past 15 years, at least 86 different alkaloids have been isolated from this genus of plants. Structure elucidation has been accomplished for all but 18 of these bases. The various *Vinca* species have each yielded the following number of alkaloids: *V. minor* (*37*), *V. major* (*15*), *V. erecta* (*23*), *V. herbacea* (*13*), *V. difformis* (*9*), and *V. pubescens* (*3*). Although our knowledge of the alkaloids from this genus is now rather complete, some work has been accomplished with regard to the stereochemistry and sterically nonselective as well as stereocontrolled synthesis.

II. The Alkaloids and Their Occurrence

Alkaloid (formula)	Structure number	Melting point (°C)	$[\alpha]_D$	Occurrence	Ref.
(−)-Eburnamonine $C_{19}H_{22}N_2O$	**27**	173–174	−85°	*Vinca minor*	*13*
11-Methoxyeburnamonine $C_{20}H_{24}N_2O_2$	**80**	169–170	−107°	*Vinca minor*	*8*
O-Methyleburnamine $C_{20}H_{26}N_2O$		181	—	*Haplophyton*	*14*

(*continued*)

II. The Alkaloids and Their Occurrence (*Continued*)

Alkaloid (formula)	Structure number	Melting point (°C)	$[\alpha]_D$	Occurrence	Ref.
17,18-Dehydrovincamine $C_{21}H_{24}N_2O_3$		218	+116°	*Crioceres longiflorus*	*12*
17,18-Dehydro-14-epivincamine $C_{21}H_{22}N_2O$		185	+30°	*Crioceres longiflorus*	*12*
(±)-Vincamine $C_{21}H_{26}N_2O_3$	**96**	235–236	—	*Tabernaemontana na rigida*	*16*
Base TR 2 (8-Epivincamine) $C_{21}H_{26}N_2O_3$		209–211	+1.8°	*Tabernaemontana rigida*	*16*
(±)-14-Epivincamine $C_{21}H_{26}N_2O_3$		—	—	*Tabernaemontana rigida*	*16*
Craspidospermine $C_{22}H_{23}N_2O_4$	**146**	—	−59°	*Craspidospermum verticillatum*	*17*
18-Dehydrovincine $C_{22}H_{26}N_2O_4$		211	+70°	*Craspidospernum verticillatum*	*11*
Cuanzine $C_{22}H_{26}N_2O_5$	**136**	196	−11°	*Voacanga chalotiana*	*18*
Vincarodine $C_{22}H_{26}N_2O_5$	**137**	235	−139°	*Vinca rosea*	*10*
Umbellamine $C_{41}H_{48}N_4O_4$	**142**	250	−217°	*Hunteria umbellata*	*15*
18-Dehydro-14-epivincine		—	—	*Craspidospernum verticillatum*	*11*
(+)-16-Methoxycarbonyl-20α,19,21α-eburn-16-ene		—	—	*Tabernaemontana rigida*	*16*
Dimethoxyeburnamonine $C_{21}H_{25}N_2O_3$	**79**	220	—	*Vinca minor*	*9*

III. Mass Spectrometry of Eburnamine-Type Alkaloids

1. (−)- and (+)-Eburnamonine (**4**)

The mechanism of fragmentation of all isomers is one in which the first step, the retro-Diels–Alder reaction in ring C, destroys the asymmetric center at C-3 and consequently makes the ions arising from the different

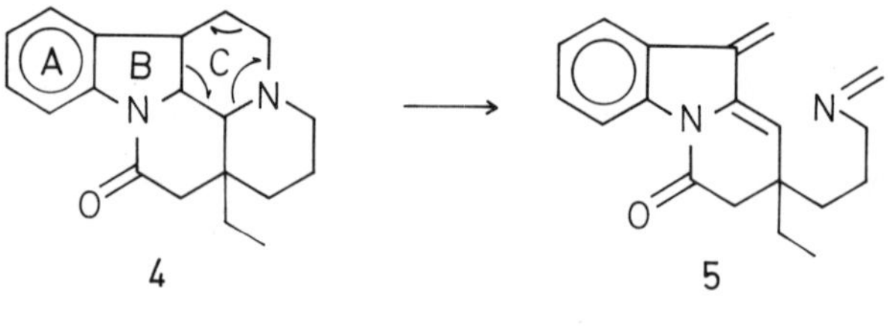

A B C N N O 4 → N N O 5

stereoisomers behave similarly. Thus mass spectrometry could not be employed to differentiate between the steric arrangement at C-3 and C-16 of similar compounds. (*19*)

2. Apovincamine (6) and Eburnamenine (7)

In the mass spectra of apovincamine and eburnamenine there were metastable peaks corresponding to the disintegrations $M^+ \rightarrow$ (M-29) and $M^+ \rightarrow$ (M-70). This is in agreement with the proposed loss of the corresponding radicals as a consequence of a strong allyl activation of linkages C-16—C-20 and C-16—C-17.

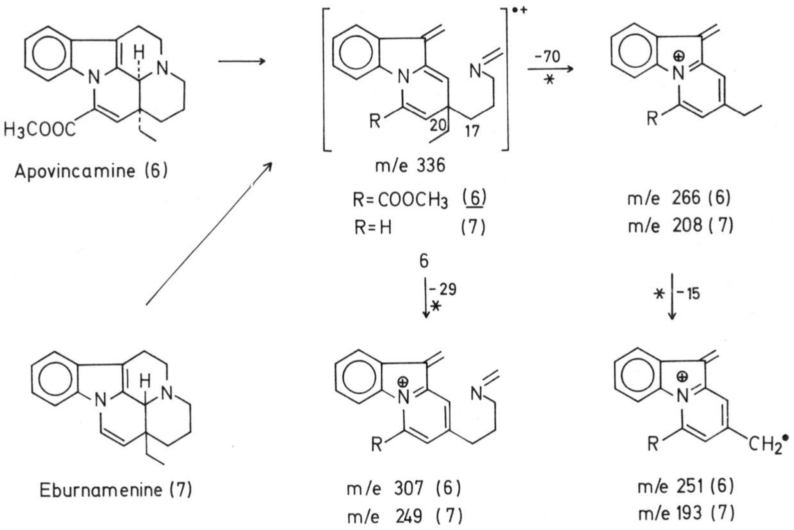
H3COOC
Apovincamine (6)
20
17
m/e 336
R=COOCH3 (6)
R=H (7)
-70
m/e 266 (6)
m/e 208 (7)
6
-29
-15
CH2•
Eburnamenine (7)
m/e 307 (6)
m/e 249 (7)
m/e 251 (6)
m/e 193 (7)

3. Eburnamine (8) and Isoeburnamine (9)

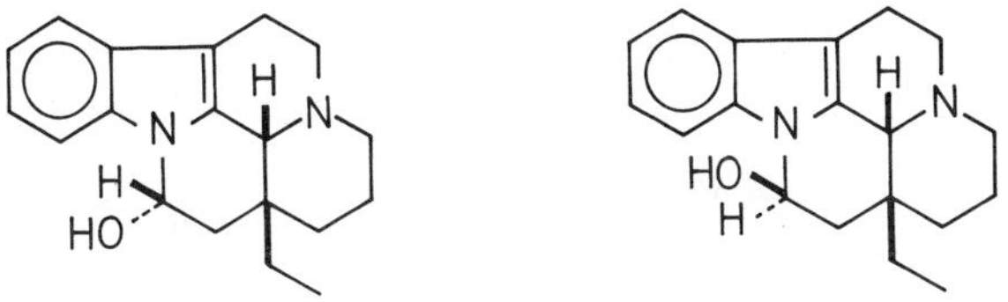

8 9

The mass spectra of these alkaloids were qualitatively similar; however, the main difference lies in the intensity of the (M-18) peak, relative to the molecular ion. This was 65% of M^+ in the spectrum of eburnamine, whereas in the isoeburnamine spectrum, it was 154%. This shows that under the same conditions, dehydration of eburnamine proceeds 2–5 times slower than that of the isoeburnamine molecular ions possessing a hydroxyl group in the less favored axial position.

4. Vincaminol (**10**) and Epivincaminol (**11**)

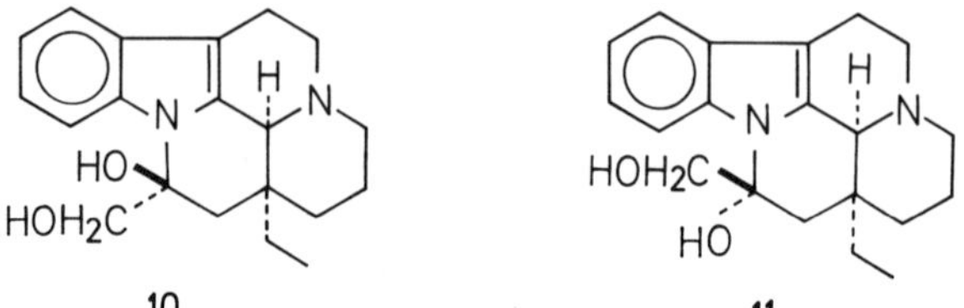

The fragmentation of both epimers proceeds essentially by two routes. Dehydration at the tertiary hydroxyl group gives rise to the ions (M-18) at *m/e* 308, which in turn disproportionately produces ions (M-18–29) and (M-18–70) at *m/e* 279 and 238 respectively. The second series is presumably started after retro-Diels–Alder collapse of ring C, by the loss of a radical CH_2OH to form the ions (M-31) at *m/e* 295. From the oxo form of these ions, a molecule of carbon monoxide is lost creating the ions (M-31–28) at *m/e* 267.

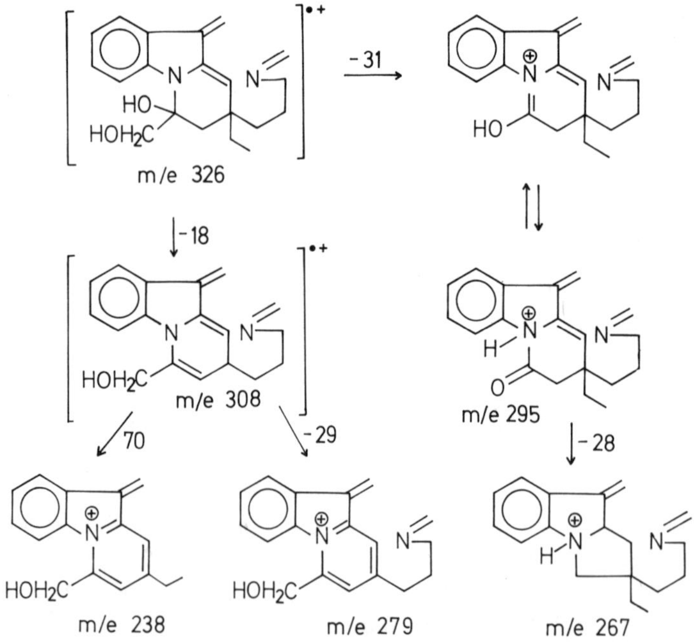

5. Vincamine and Epivincamine

The most intensive and characteristic fragmentation involves the elimination of a neutral component of 102 mass units and elemental composition $C_4H_8O_3$ corresponding to pyruvate methyl ester. This gives rise to the ion at *m/e* 252.

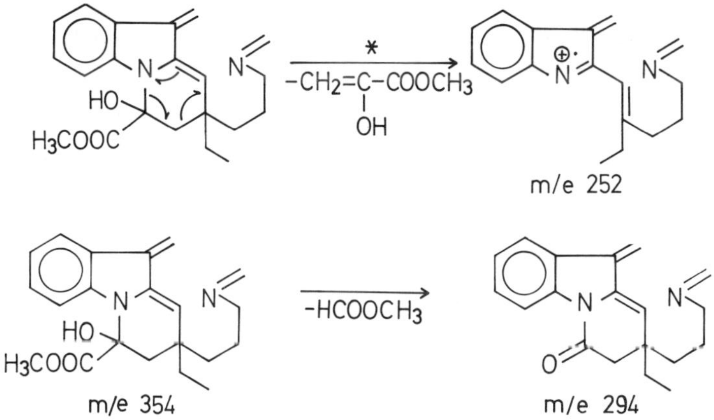

Another pathway involves the splitting of a methoxycarbonyl radical producing the ion (M-59) at *m/e* 295. This can again expel a molecule of carbon monoxide (M-59–28) to form ion *m/e* 267. The hydroxyl proton is not retained in the ion at *m/e* 294 as shown by the spectrum of the deutero analog. This ion could be formed by loss of methyl formate from the molecular ion. The subsequent disintegration of the ion (M-60) is similar to that of the eburnamonine molecular ion and leads to the formation of ions at *m/e* 293, 265, 237, 224, and 195. The base peaks in the mass spectra of vincamine and epivincamine are of different intensities.

IV. Eburnamine–Eburnamonine Alkaloids

A. Synthesis

Attention has been focused recently on the pentacyclic indole alkaloids which occur naturally in various plants, because vincamine and eburnamonine possess potent antihypertensive and sedative effects. Many conceptually different routes to eburnamonine have been employed with considerable success. The following synthesis (*20*) is a new approach in which the intramolecular cyclization of β-carboline derivatives of 3,4-dihydro-β-carboline and an α,β-unsaturated ketone play a central role.

Treatment (*20*) of the sodium salt of ethyl malonate **12** with iodo ester **13** gave the triester **14**. Partial hydrolysis of the ester, followed by decarboxylation of the resulting dicarboxyclic acid **15**, afforded the *tert*-butyl ester **16** which was converted to the diester **17**. Cleavage of the *tert*-butyl ester group gave rise to the acid **18**, which was converted to the acid chloride. The reaction of the latter with ethylene in the presence of aluminium chloride in ethylene chloride yielded the enone **20**. The reaction of 3,4-dihydro-β-carbolinium salt **21** with the enone **20** afforded the β-carbolinium salt **22** which was treated with triethylamine when two tetracyclic compounds **23** and **24** were isolated by alumina column chromatography in about equal quantity. The assignment of configurations at C-1 and C-12b of **23** and **24** were based on IR and NMR studies.

$COOC_2H_5$–HC(Na)–$COOC_2H_5$ + CH_3–CH_2–CH(I)–COOtBu → $(ROOC)_2$CH–CH(C_2H_5)–COOtBu

12 13 14 R=C_2H_5 15 R=H

CH_3OOC–CH_2–CH(C_2H_5)–COX ← ROOC–CH_2–CH(C_2H_5)–COOtBu

18 X=OH, X=Cl 16 R=H, 17 R=CH_3

CH_3OOC–CH_2–CH(C_2H_5)–C(=O)–CH=CH_2 + 21 ($ClO_4^{\ominus}$)

20 21

24 + 23 ← 22 (H_3COOC, H_5C_2, $ClO_4^{\ominus}$)

25 → 26 (N–NH–Ts) → 27

The tetracyclic ester **24** was cyclized at room temperature to form 17-oxoeburnamonine **25**. Reaction of **25** with *p*-toluensulfonylhydrazide gave the corresponding tosylhydrazone **26** which was then subjected to reaction with sodium cyanoborhydride/dimethylformamide to give (±)-eburnamonine **27**. Optical resolution of **27** was achieved by means of salt formation with (−)-*O*,*O*-dibenzoyltartaric acid in methanol, when the salt corresponding to (−) eburnamonine separated in crystalline form. When the salt formation was carried out with (+)-*O*,*O*-dibenzoyltartaric acid, the salt of (+) eburnamonine crystallized. In a similar manner, treatment of the other isomer **23** with sodium *tert*-butoxide afforded (−)-oxoisoeburnamonine **28** which was directly converted via the tosylhydrazone to (±)-3-isoeburnamonine **19**.

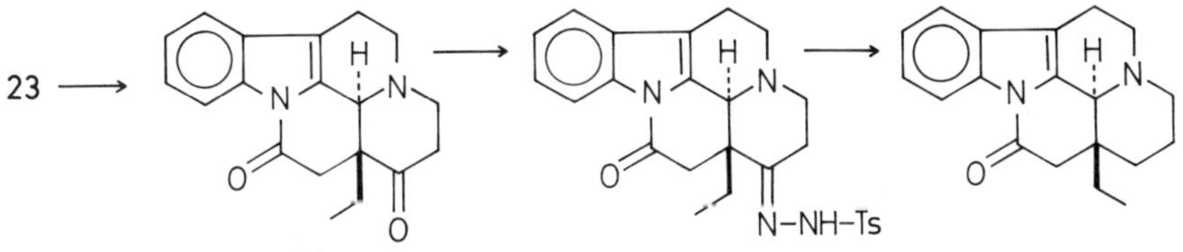

23 → 28 → 29 → 19

Another route is regiospecific alkylation (*21*) of the tricyclic lactam **30** which allows the construction of (±)-eburnamonine and (±)-eburnamine in overall yields of 67% and 54% respectively. Addition of **30** at −78° to a solution of lithium diisopropylamide gives rise to the dianion **31**. Alkylation of the dianion with methyl bromacetate afforded the lactam ester **32** in 95%

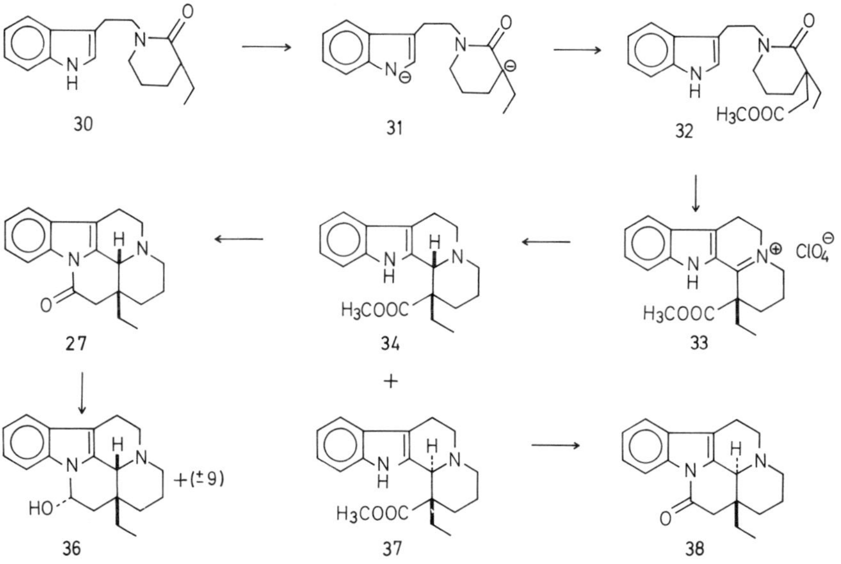

30 → 31 → 32 → 33 → 34 + 37; 34 → 27 → 36 +(±9); 37 → 38

yield. Cyclization of **32** was carried out in acetonitrile containing phosphorus oxychloride. The crude reaction mixture was treated with lithium perchlorate to give the immonium perchlorate **33**. Hydrogenation of **33** with 10% palladium charcoal gave a mixture of methyl eburnamoninate **34** and methyl epieburnamoniate **37**. This mixture of esters was cyclized, and the resulting reaction mixture was filtered through silica gel to give (±)-3-epieburnamonine **38** and (±)-eburnamonine **27**. Lithium aluminium hydride (LAH) reduction of synthetic eburnamonine in THF solution gave a 1:1 mixture of (±)-eburnamine **36** and (±)-isoeburnamine **9** in quantitative yield. The condensation (*22*) of enamine **39** and $CH_2{=}CClCN$ gave **40** which was reduced using Zn/HCl giving a mixture of **41** and **42**. This was followed by oxidation of the derived anion to yield (±)-eburnamonine **27** and (±)-3-epieburnamonine **38**. Oxidation of (−)-tetrahydroalstonine **35** with mercuric acetate lead to enamine **43**. This compound reacted with methyl chloracetate to give **44**, an analog of eburnamonine (*23*). A stereoselective synthesis (*24*)

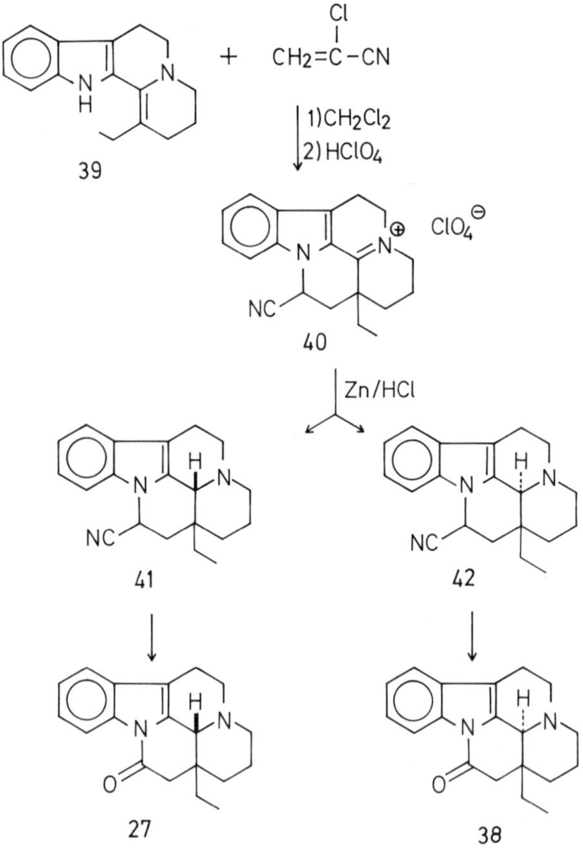

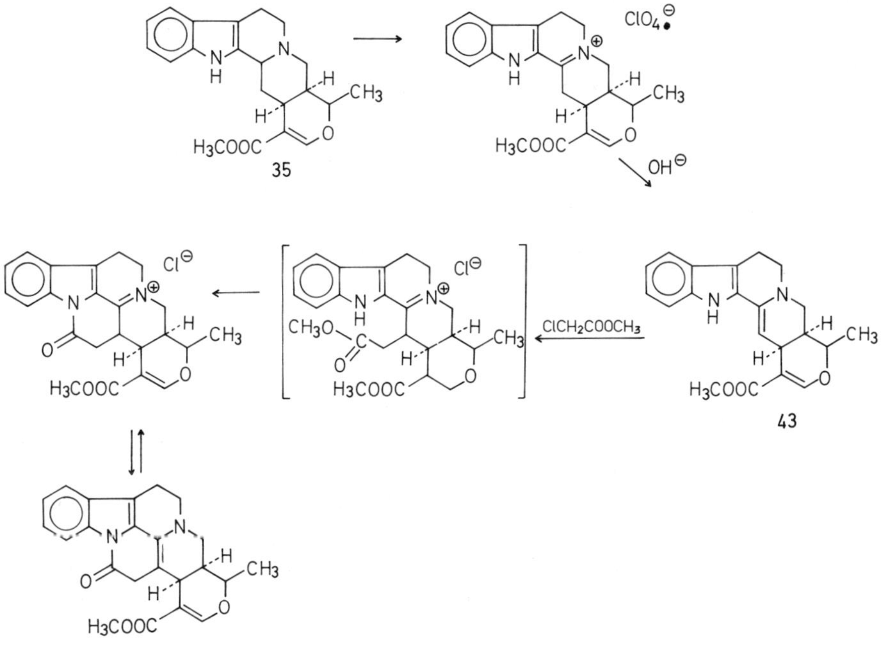

35

$ClO_4^{\ominus}$

$OH^{\ominus}$

$Cl^{\ominus}$

$ClCH_2COOCH_3$

43

44

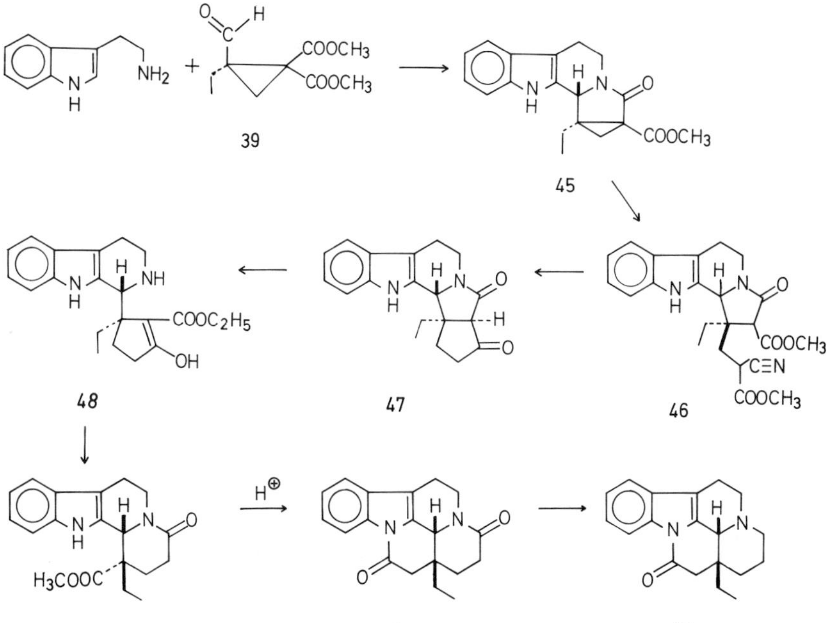

39

45

48

47

46

$H^{\oplus}$

49

50

27

of eburnamonine makes use of the regioselective ring opening of a cyclopropane derivative of defined configuration. In the condensation of cyclopropane **39** with tryptamine, **45** is obtained exclusively as the kinetic product. Ring-opening reactions occur with extraordinary regioselectivity, and with the lithium salt of the cyanoacetic acid, methyl ester **46** is obtained as the only reaction product. The lactam ketone **47** treated with triethyloxonium hexafluorophosphate followed by hydrolysis gives β-ketoester **48** that is transformed into **49** by treatment with methoxide and reesterification. Eburnamonine **27** is then produced via the dilactam **50**.

B. Rearrangement Reactions

Some rearrangements (*25*) in the aspidospermidine series are very interesting. Aspidospermidine oxide prepared from 1,2-dehydroaspidospermidine by per-acid oxidation, rearranged in acetic acid/triphenyl phosphine in the cold to 42% (+)-eburnamine and 5% (−)-eburnamenine. 14-Oxoaspidospermidine rearranged in acetic acid at 90° to yield 40% (−)-eburnamenine. On the other hand the oxime of the oxoaspidospermidine **40**, rearranged in $POCl_3$ at ambient temperature to the nitriles **41A** and **41B**. For this rearrangement, the following mechanism can be proposed. Hydrolysis and cyclization of **41B** and **41A** gave (−)-vincamone and (+)-epi-3-vincamone respectively.

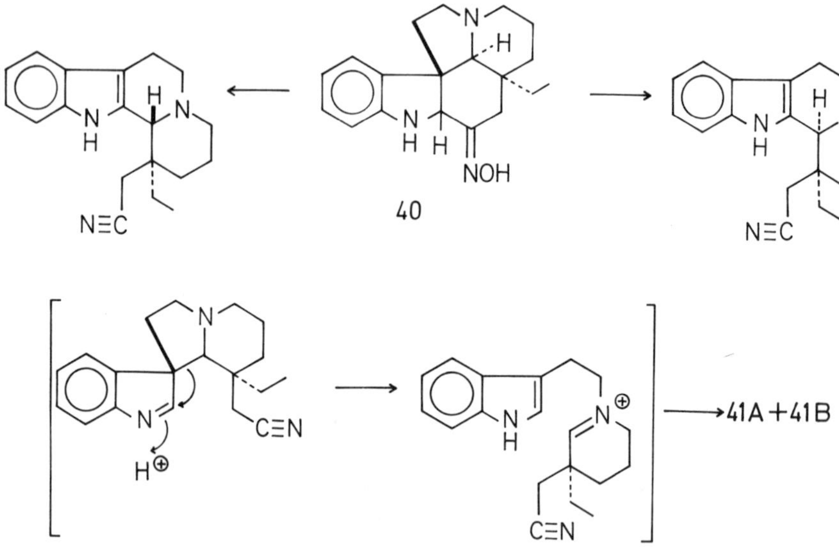

40

→41A+41B

Total synthesis (*26*) of (±)-eburnamine, together with the first total syntheses of (±)-homoeburnamenine and (±)-3-epihomoeburnamenine, was achieved. 2-Ethyoxycarbonyl-2-ethylcyclopentanone was reduced to the corresponding hydroxy ester, which was then dehydrated to the unsaturated ester **51**. Reduction of **51** to the primary alcohol **52** followed by a modified Oppenauer oxidation afforded the aldehyde **53**, that was converted by Wittig condensation with ethoxycarbonylmethylenetriphenylphosphorane into the unsaturated ester **54**. Hydroxylation of **54** by means of osmium tetroxide followed by acid hydrolysis gave a mixture of two stereoisomeric *cis*-diols **55** that was hydrogenated to a mixture of the corresponding saturated esters **56**. Condensation of **56** with tryptamine afforded the tryptamide **57**, which upon oxidation with sodium metaperiodate, gave the carbinolamide lactol **58** as two separable isomers. Cyclization of either isomer of **58** gave a mixture of 19-oxohomoeburnamenine **59** and 3-epi-19-oxohomoeburnamenine **60**, that when reduced afforded (±)-homoeburnamenine **61** and 3-epi-homoeburnamenine **62**. Hydroxylation of **59** gave a mixture of *cis*-diols **63**. Oxidation of **63** followed by removal of the *N*-formyl group gave (±)-19-oxoeburnamine **64**. Finally reduction of **64** gave (±)-eburnamine **8**.

51 R=COOEt
52 R=CH_2OH
53 R=CHO
54 R=CH=CH–COOEt

55

56 saturated ester

57

58

59 R=O
60 R=O, αH at C_3
61 R=H
62 R=H, αH at C_3

63

64 R=O
8 R=H

A completely different approach to total synthesis (*27*) embodied a rearrangement reaction. Such a reaction was developed in which *o*-halobenzylidene derivatives of 3-quinuclidinone **65** were converted to tetrahydropyrido[1,2-a]indoles. These products possess three of the rings, the

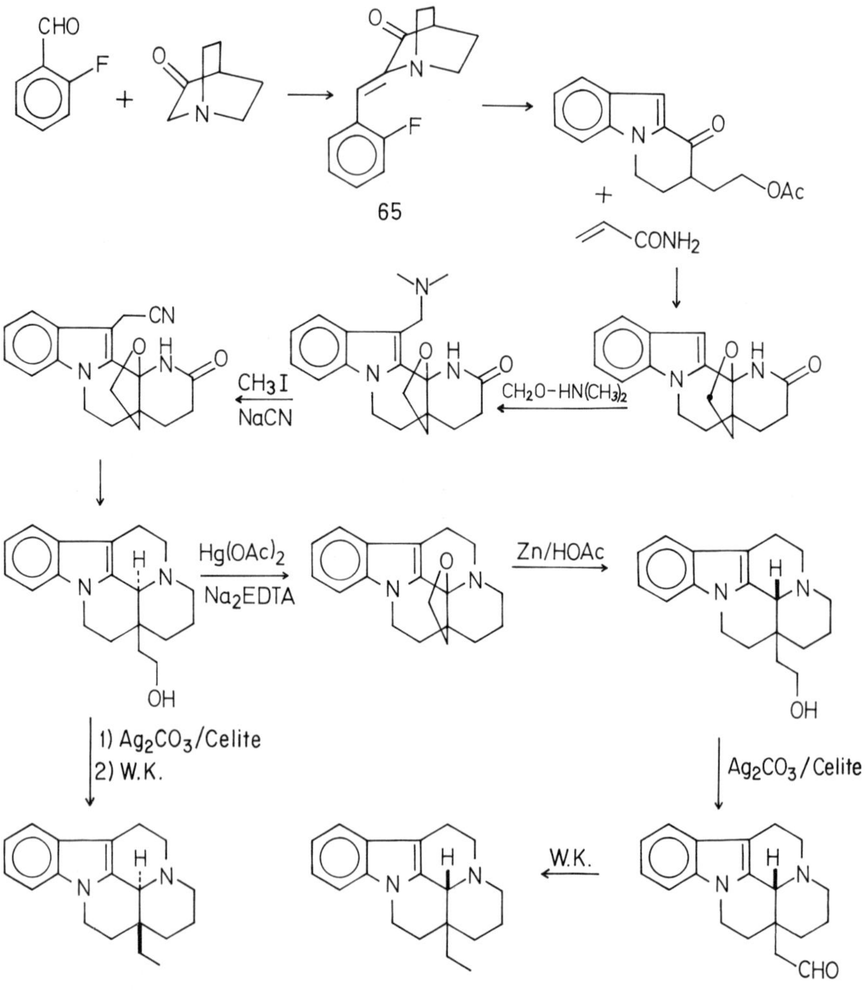

(±) epidihydroeburnamenine (±) dihydroeburnamenine

two-carbon side chain, and a strategically located ketone function which make them readily amenable to elaboration into the pentacyclic ring system of the eburnamine-type alkaloids. Another starting material (*28*) is the tetracyclic lactam **66** which forms two pairs of diol lactams **67** and **68** epimeric at the secondary hydroxy group. Periodate cleavage of either of the first pair gave eburnamine lactam **69** (X=O) reduced by LAH to eburnamine **8** (X=2H) whereas similar treatment of the second pair gave 3-isoeburnamine **70**. A synthesis of eburnane and homoeburnane derivatives

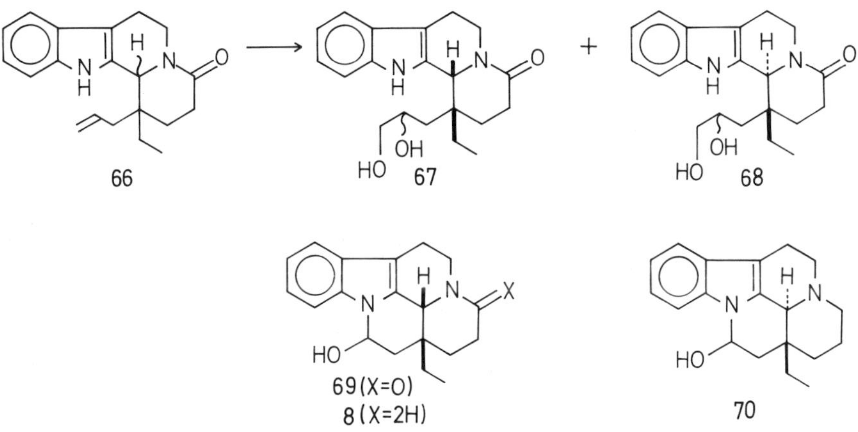

66 67 68

69 (X=O)
8 (X=2H)

70

oxygenated at C-15 was carried out as follows (*29–31*): Dihydro-β-carboline **71** condensed with 1-chloro-2-oxo-5-carbomethoxypentane to give the pyrido[3,4-b]indolium chloride **72**, which cyclized upon refluxing with $CH_3OH(C_2H_5)_3N$ to give the indolopyridonaphthylidinedione **73**. Analogous condensation (*31*) of **71** with 1-chloro-3-oxo-6-carbomethoxyhexane and cyclization yielded the indolopyridohomonaphthridine **74**. A stereo-

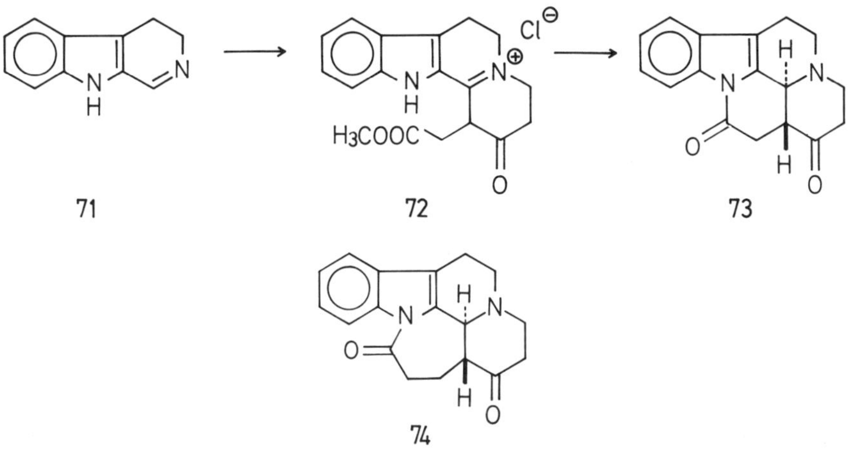

71 72 73

74

specific synthesis of (±)-desethyleburnamonine and of (±)-3-epi-desethyleburnamonine from a common intermediate was accomplished via the indoloquinolizidine **75**. Thus **75** alkylated with ethyl iodoacetate gave indoloquinolizidineum salt **76** that was treated with Zn/HCl to give (±)-3-epidesethyleburnamonine **77**. When **76** was treated with $NaBH_4$, it gave

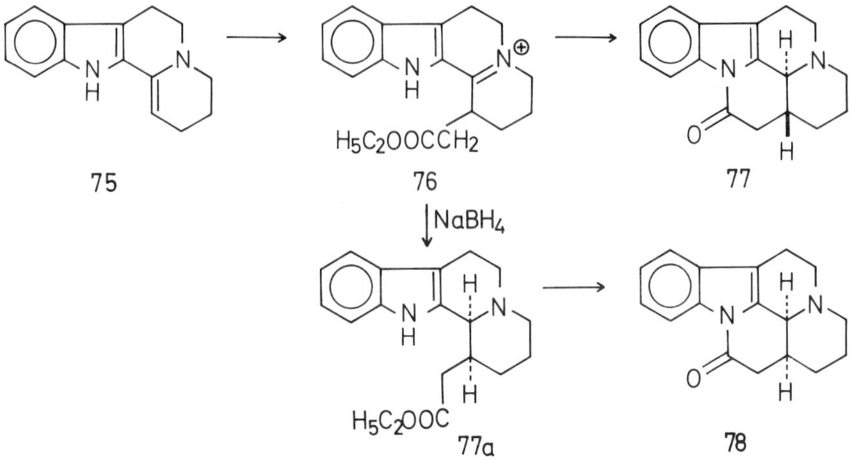

75 76 77

77a 78

indoloquinolizidine **77a** which was cyclized by C_2H_5ONa to (±)-desethyleburnamonine **78**. 3-Epidesethyleburnamonine was synthesized also by hydrogenation and cyclization of the indolylethylpyridinium salt **70** in an acidic medium.

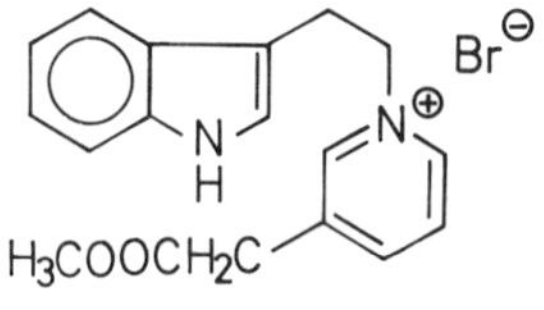

70

C. New Eburnamine Alkaloids

1. Dimethoxyeburnamonine

Dimethoxyeburnamonine, isolated from *Vinca minor* is an alkaloid (*9*) of the gross composition of $C_{21}H_{26}N_2O_3$, showing an eburnamonine spectrum that has been shifted towards higher masses by 61 units of mass and was assigned structure **79** after NMR investigations.

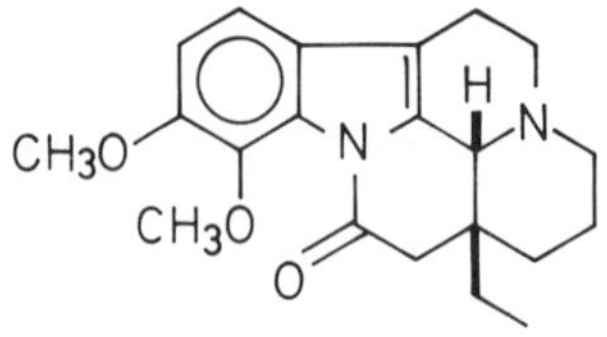

79

2. 11-Methoxyeburnamonine

Another alkaloid (*8*) from the leaves of *Vinca minor* shows in its mass spectrum the fragmentation behavior that is typical of the eburnamonine alkaloids. Its structure (**80**) was accordingly confirmed by isolation of the base after preparative thin layer chromatography from LAH reduction products of vincine.

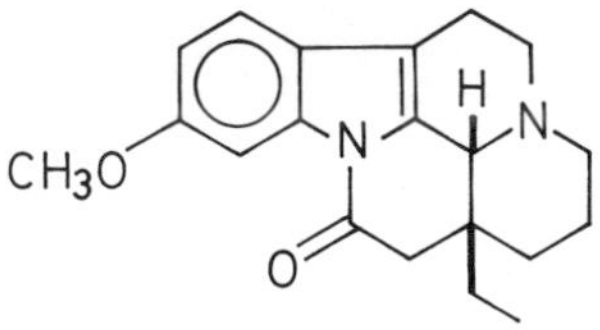

CH_3O H N N O

80

V. Vincamine Alkaloids

A. Synthesis

Over the past few years vincamine **96** has been found potentially useful in the medical treatment of cerebral insufficiency in man (*32*). Consequently, several sterically nonselective as well as stereocontrolled syntheses of racemic vincamine have been developed. Two nonstereoselective syntheses of the racemic form of vincamine are described in the literature (*33*). One of these has the following pathway (*34*).

Homoeburnamenine **61**, prepared as described in Section IV,B was oxidized to dihydroxyhomoeburnamenine **81** which was then subjected to oxidation under a variety of conditions. Unfortunately, the desired dioxo compound could not be obtained. Oxidation of **81** by means of Me_2SO: NEt_3: pyridine afforded two diastereomeric lactams **82**. Alkaline hydrolysis of the mixture of lactams gave the corresponding hydroxy acids, that were esterified to the methyl esters **83**. Finally, oxidation of the methyl esters by means of the Me_2SO:NEt_3: pyridine–$SO_3 \cdot H_2O$ reagent afforded (±)-vincamine **96**.

The first stereoselective total synthesis of (±)-vincamine was accomplished as follows (*35*). The reaction of the enamine **84** described by Wenkert and Wickberg (*36*), with methyl acrylate gave an adduct in excellent yield. However, either catalytic or chemical reduction of the adduct yielded the two possible stereoisomers, prepared earlier by Kuehne (*33*), in a ratio of almost 1:1. If the addition reaction was carried out with methyl α-acetoxy-acrylate, and the product reduced catalytically, only one stereoisomer **85** was obtained in good yield. With $NaBH_4$ as the reducing agent, **85** was

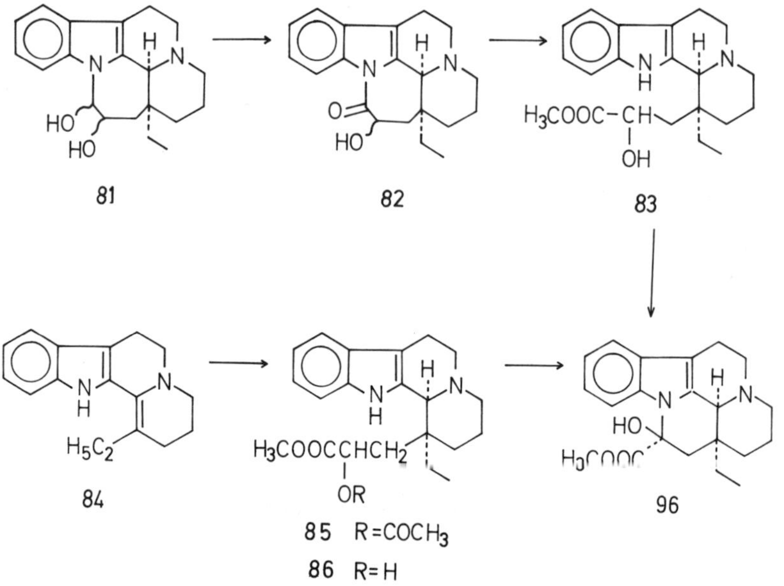

81 82 83

84 85 $R = COCH_3$ 86 $R = H$ 96

accompanied by a small amount of another epimer. Deacetylation of **85** furnished the alcohol **86** that was readily oxidized with Ag_2CO_3/celite to a mixture of vincamine and 14-epivincamine. When 14-epivincamine was boiled in xylene in the presence of Fetizon reagent, it readily epimerized to a mixture containing about 80% vincamine and 20% 14-epivincamine. Resolution of racemic vincamine was effected with dibenzoyl tartaric acid in methanol.

A high-yield stereospecific total synthesis of vincamine has also been reported (*37*). The intermediate lactam **89** was prepared from the lithium enolate of *tert*-butylbutyrate by alkylation with 1,3-dibromopropane to give the bromo ester **87** in 90% yield. The latter, **87**, was converted into the bromoamide **88** in 80% overall yield by the following reaction sequence. Treatment of **87** with *p*-toluenesulfonic acid in refluxing benzene gave the corresponding bromo acid that without purification was converted into its acid chloride. The crude acid chloride was then treated with a mixture of tryptamine hydrochloride and LAH to give the amide **88**. Reaction of **88** with potassium hydride gave the lactam **89**. Addition of **89** to a solution of lithium diisopropylamide produced the dianion **90**, which upon treatment with methyl-2-thiomethylacrylate affords the lactam ester **91** cyclized with $POCl_3$ in refluxing acetonitrile. The crude reaction mixture was treated with lithium perchlorate to give the immonium perchlorate **92**. Stereospecific reduction of **92** to the tetracyclic amine ester **93** could be realized in 98%

yield. The final ring of vincamine was introduced by oxidation of **93** to the sulfoxide **94**. Treatment of **94** with sodium hydride gave the lactam sulfoxide **95**. Direct conversion of **95** into vincamine **96** was accomplished by reaction of the lactam sulfoxide with acetyl chloride.

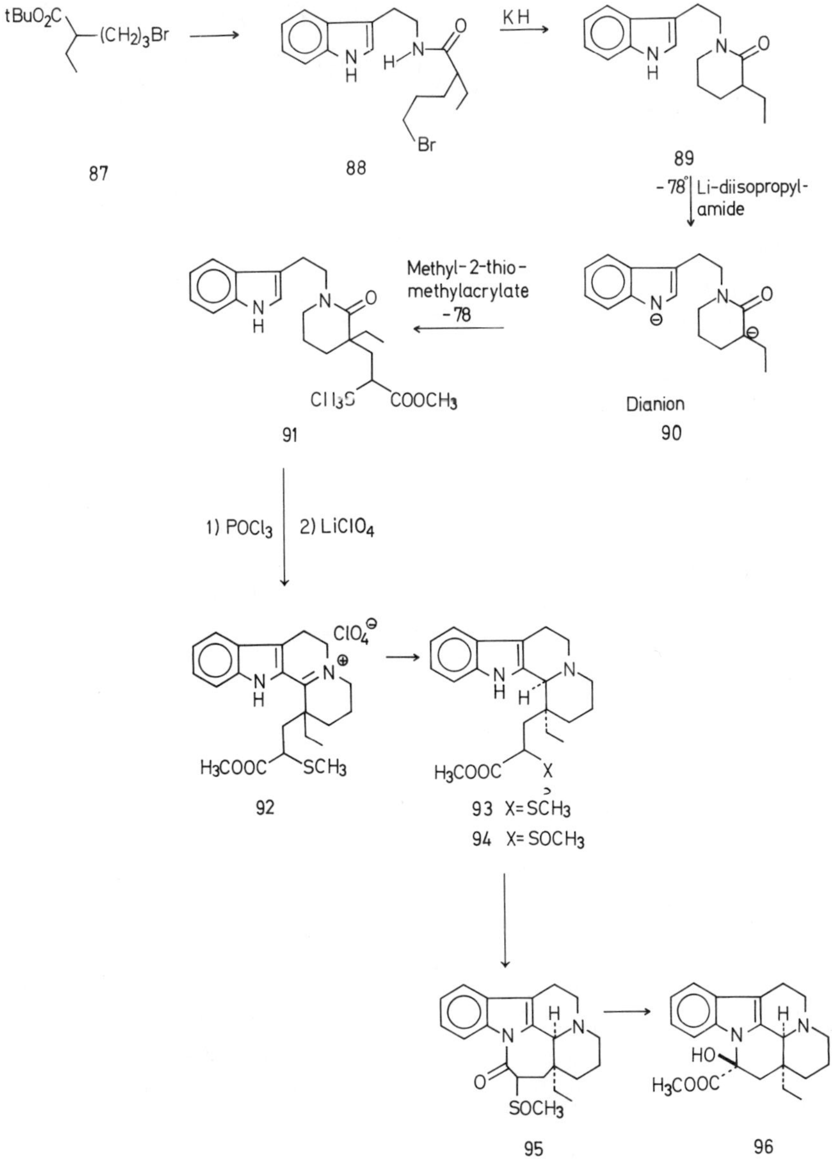

tBuO$_2$C–(CH$_2$)$_3$Br → 88 —KH→ 89 —−78° Li-diisopropylamide→ Dianion 90 —Methyl-2-thiomethylacrylate, −78→ 91 —1) POCl$_3$ 2) LiClO$_4$→ 92 → 93 X=SCH$_3$, 94 X=SOCH$_3$ → 95 → 96

87 88 89 90 91 92 93 94 95 96

The first total synthesis of optically pure natural vincamine which involves an early key intermediate containing the quarternary center C-16 with the natural chirality is shown in the following scheme (*38*):

COOC$_2$H$_5$ H C O → COOC$_2$H$_5$ O C H → COOC$_2$H$_5$ (C$_2$H$_5$O)$_2$HC → COOH (C$_2$H$_5$O)$_2$CH

COOC$_2$H$_5$ (C$_2$H$_5$O)$_2$CH OR

R=H

R=Tos

H N HN H$_5$C$_2$OOC (C$_2$H$_5$O)$_2$HC

N H O N (C$_2$H$_5$O)$_2$CH

80% HOAc

N H O N OHC

CH$_3$O C=HC H$_3$COOC

E-Isomer 60%

Z-Isomer 40%

ClO$_4^{\ominus}$ N$^{\oplus}$ 16 CH$_3$O C=HC H$_3$COOC

E-Isomer 37%

Z-Isomer 63%

H$_2$

CH$_3$O C=HC H$_3$COOC H N

H N 3 16 H$_3$COOC

Apovincamine

H N 3 16 HO 14 H$_3$COOC + H$_3$COOC OH H N

Vincamine (96)

During the reduction of the immonium salt, center C-16 selectively induces the stereochemistry at center C-3 which, at a later stage, together with center C-16 controls the epimerizable center C-14.

Natural vincamine was synthesized in an enantioselective manner (*39*) starting from the 2-ethyl-4-pentenal. Protection of the carbonyl group, hydroboration of the olefinic bond, bromination of the intermediate organoboron, and hydrolysis of the acetal furnished the bromaldehyde **97**. After conversion of the bromaldehyde to its silyl enol ether the latter was reacted with dihydro-β-carboline to obtain directly a 1:1 mixture of the tetracyclic aldehydes in 74% yield.

X=O
X=$(OCH_3)_2$
97 X=$(OCH_3)_2$
X=O
$(CH_3)_3SiOCH$
$(CH_3)_3SiO-HC$
cis/trans 1:1

Both isomers exhibit well-defined Bohlmann bands at 2750 and 2800 cm^{-1} in the IR as well as NMR signals of H—C_{12b} at $\delta = 3.70$ and 3.57 ppm, indicating their *trans*-quinolozidine structures. In the NMR spectra, the CH_3—C signal of both isomers appears at $\delta = 0.78$ and 1.02 ppm, respectively, showing an axial ethyl group for one of the isomers and thus the desired cis configuration. The undesired stereoisomers were recycled to the desired one using the related reversible Mannich reaction. This was followed by crystallization of the salt of the isomer with natural configuration and (+) malic acid. In the last step, this isomer was converted to natural vincamine in several steps including some known transformations. (+)-Vincamine and (−)-14-epivincamine are also prepared by the oxidation of (−)-vincadifformine **98** with $Pb(OAc)_4$ or a per-acid followed by a rearrangement of the indolenines obtained. Vincadifformaine reacts with lead tetracetate to give the indolenine **99** which rearranges in acetic acid containing sodium acetate

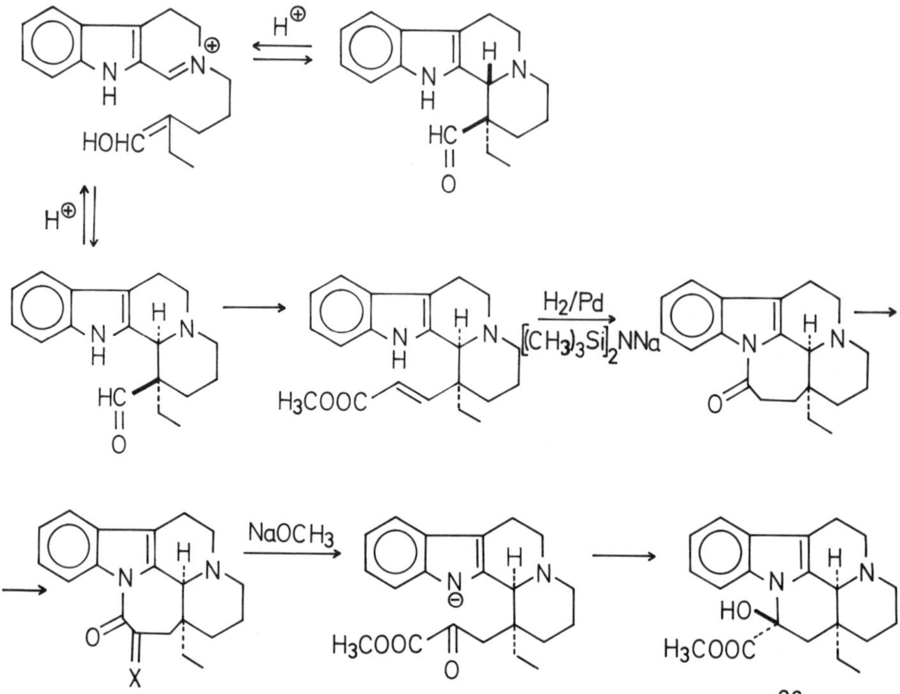

$H^{\oplus}$

$H^{\oplus}$

H_2/Pd

$[(CH_3)_3Si]_2NNa$

$NaOCH_3$

X = NOH

X = O

96

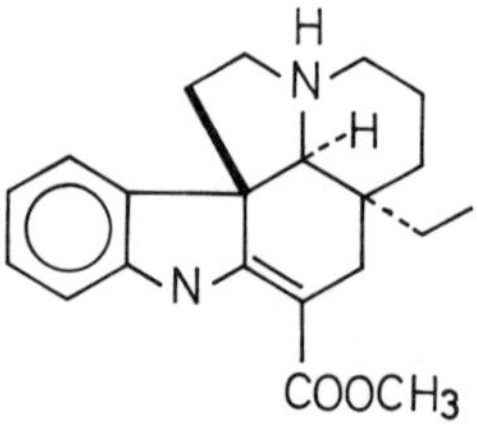

98

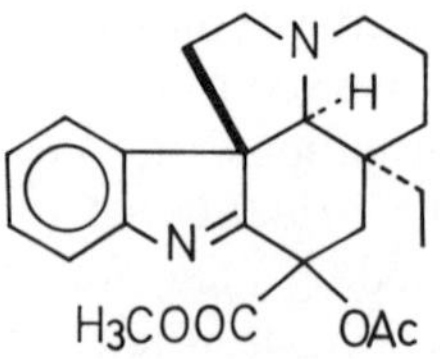

99

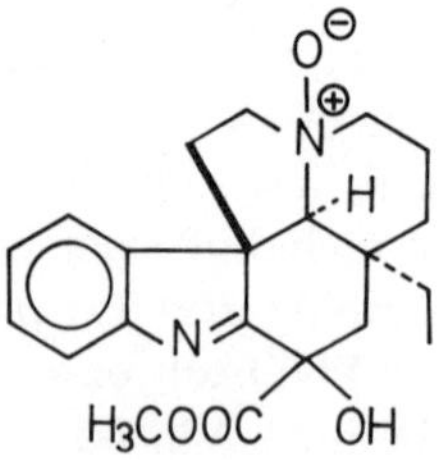

100

to give (+)-vincamine, (−)-14-epivincamine, and apovincamine (*40*). The indolenine-*N*-oxide **100** is obtained by the treatment of vincadifformine with 2 equivalents of *para*-nitroperbenzoic acid, and this *N*-oxide rearranged in acetic acid containing triphenyl phosphine to vincamine and 14-epi-vincamine.

Addition of α-acetoxyacrylicacid-(−)-menthylester to the enamine followed by catalytic hydrogenation gave the (+) and (−)-indoloquinolizines **101** (*41*) which were resolved with D-dibenzoyl tartaric acid and (−)-**101** was

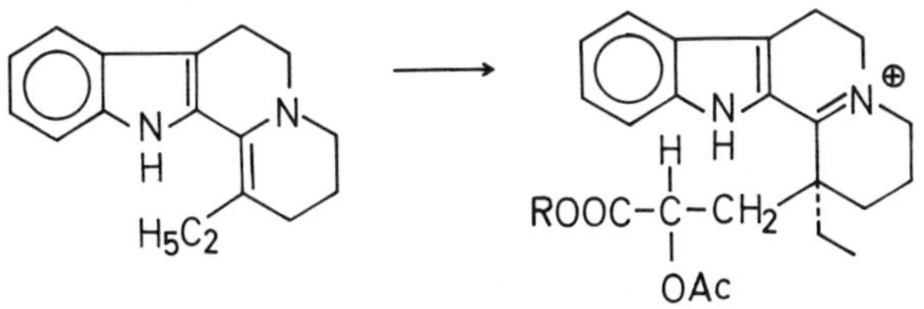

H_5C_2 ROOC–C–CH2 OAc 101

R=(−)-Menthyl

obtained in a combined optical yield of 41%. Sequential treatment of (−) **101** with Fetizon reagent and sodium methoxide gave a 65% yield of (+)-vincamine. A similar synthesis of (±)-desethylvincamine and vincamine (*42*) is illustrated below.

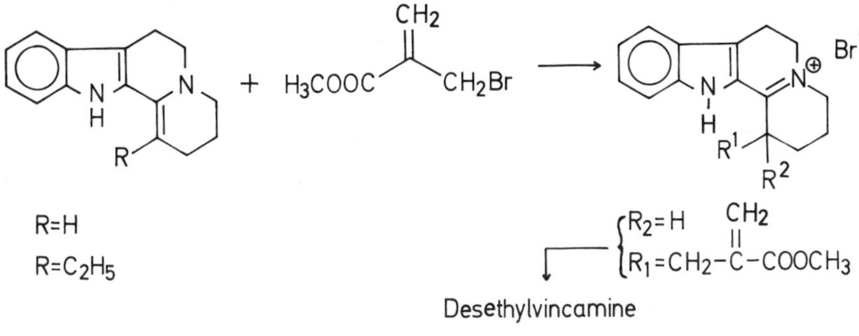

CH_2 + H_3COOC CH_2Br $Br^{\ominus}$ R^1 R^2

R=H

R=C_2H_5

R_2=H, R_1=CH_2–C(=CH_2)–$COOCH_3$

Desethylvincamine

B. Vincamine Reduction Products

When vincaminol **102** was heated at 100° in acetic acid, a mixture of deoxyvincaminal **103** and 14-epideoxyvincaminal was obtained which, upon immediate reduction with LAH and separation of the products by chromatography, gave deoxyvincaminol **104** mp 205°, $[\alpha]_D^{24}$ − 62° (c = 1 $CHCl_3$). 14-Epideoxyvincaminol **105** mp 170°, $[\alpha]_D^{24}$ − 80° (c = 1 $CHCl_3$), was prepared

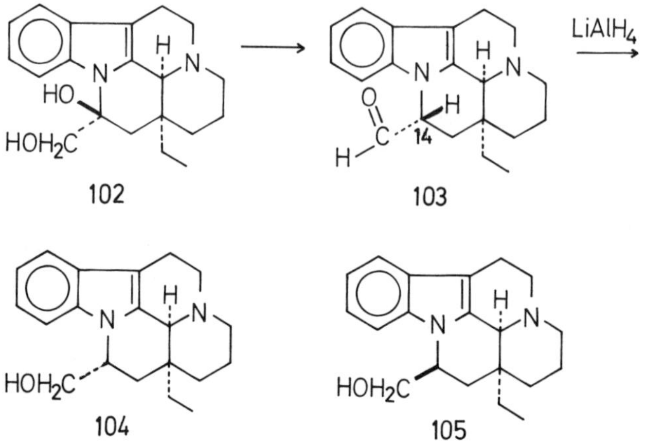

102 103 104 105

similarly from apovincaminol with the vincaminals as intermediates (*43*). 18*S*-ψ-Vincamine, 18*S*-14-epi-ψ-vincamine and 18*S*-ψ-apo-vincamine have been prepared from catharanthine by partial synthesis indicated below (*44*).

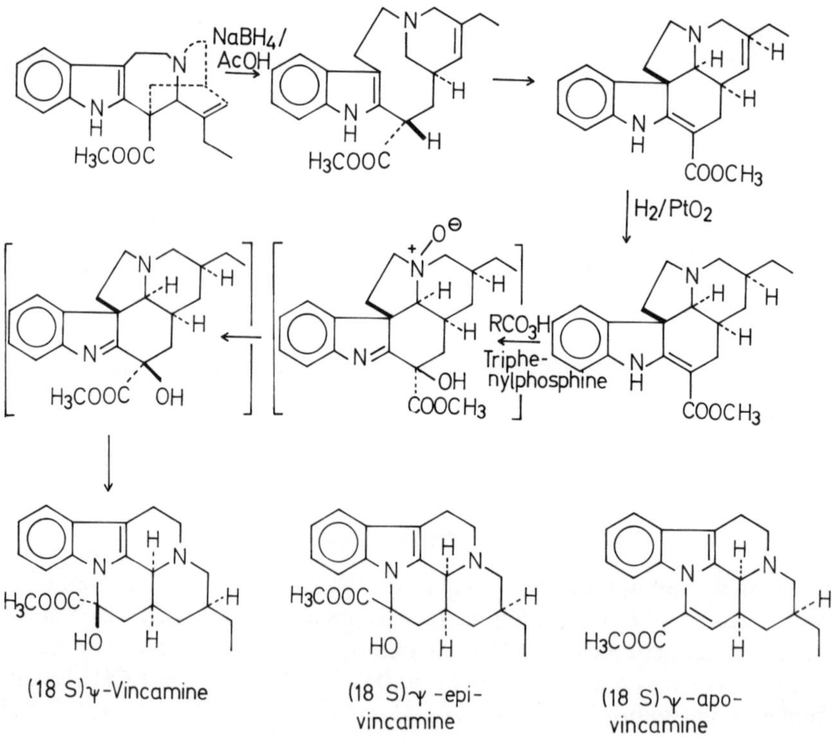

(18 S)ψ-Vincamine (18 S)ψ-epi-vincamine (18 S)ψ-apo-vincamine

C. VINCINE

Vinca minor possesses vincamine, its methoxy-containing analog vincine **125** (*45*), and other alkaloids. Because vincine has interesting pharmacological properties, a stereoselective synthesis of isovincine was accomplished. A solution of the lactone **106** and 5-methoxytryptamine in chlorobenzene was boiled for 8 hr. The crude amide **107** was cyclized with $POCl_3$ yielding **108**. The liberated base was reacted with methyl-α-acetoxyacrylate and the product was isolated in the form of its perchlorate. Catalytic reduction of the salt **109** proceeded with high stereoselectivity. The acetoxy group of ester **110** was hydrolyzed easily by the use of methanolic sodium methoxide. Reaction of **111** with Fetizon reagent yielded two products: the thermodynamically less stable 14-epi-isovincine **112** was obtained in 46% yield, and the more stable isovincine **113** in 18% yield. Epimerization of **112** to **113** was effected by methanolic NaOMe.

106 107 108

109

110 R=COCH3
111 R=H

112 113

VI. Absolute Configuration of Vincamine and Some Related Alkaloids

The absolute configuration of vincamine and some related alkaloids was established in two steps (*46,47,48*). The major evidence was comparison of the optical rotatory dispersion (ORD) curves of the (−)-1,1-diethyl-1,2,3,4, 6,7,12,12b-octahydroindolo(2,3-a)quinolizine **118** obtained by the degradation of (+)-vincamine **96** via vincaminic acid **115** with (−)-eburnamonine **27** and (+)-1,2,3,4-tetrahydroharman **119**, the absolute configuration of which was proven unequivocally by its ozonolysis (*1,2*) to (−)-*N*-carboxyethyl-D-alanine **120**. The second step involved the determination of the absolute configuration of the base **118** by comparison of its ORD curve with those of the yohimbane alkaloids of known configuration, i.e., (−)-corynanthine, (+)-pseudoyohimbine, (−)-rauwolscine, (−)-*β*-yohimbine, (+)-yohimbine, (−)-herbaeine. The results obtained prove the original assignment of the absolute configuration (*1,2*) of the alkaloids (+)-vincamine **96**, ( )-14-epi-vincamine **114**, and a series of intercorrelated enantiomeric alkaloids isolated from *Hunteria eburnea* Pichon and *Vinca minor*, (−)-eburnamonine **27**, (+)-eburnamonine **121**, (−)-eburnamine **122**, (+)-isoeburnamine **123**, and (+)-eburnamenine **124**. From this the absolute configuration of the eburnamine part of the dimeric alkaloid pleiomutine isolated from *Pleiocarpa*

96 R^1=$COOCH_3$, R^2=OH
115 R^1=COOH, R^2=OH
27 R^1R^2=O
114 R^1=OH, R^2=$COOCH_3$

118 R=αH

119 R^1=H, R^2=βH

120

121 R^1R^2=O
122 R^1=OH, R^2=H
123 R^1=H, R^2=OH
124 R^1R^2=H 14-15 double bond

mutica is also evident. On the other hand, the question of the absolute configuration of the bases found in *Aspidosperma quebracho blanco* (eburnamenine), *Pleiocarpa mutica* (eburnamenine and eburnamine), and *Rhazya stricta* (eburnamenine, eburnamonine, eburnamine) remains open, as their optical rotations are not published.

As a continuation of these studies a report about investigations on the optical rotatory dispersion of (+)-vincamine and the whole group of naturally occurring bases of the eburnane structure (at least 12 bases have been isolated), has been published (*49*). Considerable attention was paid to the stereochemistry of this group of compounds at asymmetric centers C-14, C-16, and C-3, but the results were hardly unequivocal. More interest has been devoted to the readily accessible alkaloids, namely (+)-vincamine **96**, (−)-14-epivincamine **114**, (−)-eburnamine **122**, (+)isoeburnamine **123**, and (+)eburnamonine **121**. In many instances the results were contradictory and the interpretation was a subject of criticism. The absolute configuration of these compounds has not been studied so far except for investigations of (+)-vincamine and (+)-vincine. Now there are new efforts to determine the absolute configuration of the eburnane alkaloids and to draw useful conclusions about the relative configuration at the corresponding asymmetric centers by comparison of ORD curves of these alkaloids with the dispersion curve of (+)-vincamine. The analysis of ORD curves of alkaloids of the eburnane type and comparison with those of yohimbane alkaloids made it possible to identify the particular Cotton effects and to determine the relationship between their sign and the absolute configuration of the corresponding chirality centers. In this way, the absolute configuration of (−)-11-methoxy-16α,3α-eburnan-14-one **126**, (+)-vincine **125**, (+)-vincaminine **127**, and (+)-vincinine **128** at carbon atoms C-16 and C-3 was determined (16 : *S*) and (3 : *S*).

Knowledge of the absolute configuration of alkaloids of the eburnane group opens new perspectives for the study of naturally occurring bases of the plant *Vinca minor* L. It is of interest that the representatives carrying a methoxycarbonyl function at C-14, namely, (+)-vincamine **96**, (−)-14-epivincamine **114**, (+)-vincaminine **127**, and (+)-vincinine **128** belong to the same stereochemical series characterized by an α-oriented ethyl group. On the other hand, the alkaloids lacking the methoxycarbonyl function, namely (−)-eburnamine **122**, (+)-isoeburnamine **123**, and (+)-eburnamenine **124**, are members of an enantiomeric series with the β-orientation of the ethyl group. The alkaloids of the aspidospermane (**130**) type display a like resemblance; the derivatives carrying an esterified carboxylic function at C-16 possess the same absolute configuration at C-20 as the carboxylated bases of eburnane series whereas the decarboxylated bases belong to the enantiomeric series (*49*).

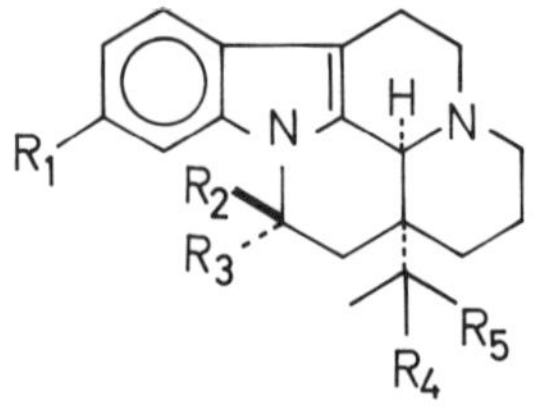

	R_1	R_2	R_3	R_4	R_5
125	OCH_3	OH	$COOCH_3$	H	H
126	OCH_3		=O	H	H
127	H	OH	$COOCH_3$	=O	
128	OCH_3	OH	$COOCH_3$	=O	

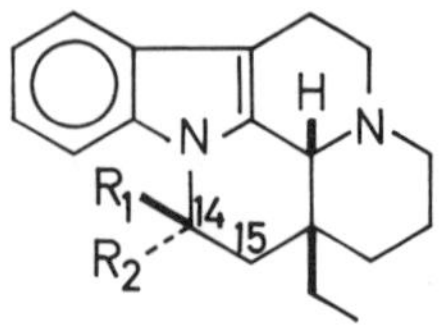

129 $R_1=R_2=H$
14,15 dehydro

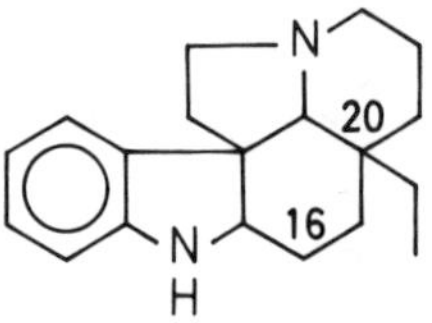

130

The ORD curve of **126** when compared with those of the other alkaloids of the same type bearing substituents other than a carbonyl function on C-14 was characterized by surprisingly high rotation values even in the comparatively long wavelength range of the spectrum. One reason for this anomalous behavior of **126** might be an inherent chirality effect of an indole ring conjugated with a carbonyl group. A similar situation might be expected also with compounds where the carbon–carbon double bond is conjugated with an indole nucleus (e.g., **131**) which would be an analogous to styrene derivatives.

The configuration of (+)-eburnamenine **124** at both chiral centers C-16 and C-3 is evident from known chemical correlations with alkaloids examined earlier. Because **124** is enantiomeric to (+)-14-methoxycarbonyl-16α,3α-eburn-14-ene **132** and (+)-11-methoxy-14-methoxycarbonyl-16α,3α-eburn-14-ene **133**, the shape of the respective dispersion curves is also virtually enantiomeric (*50*).

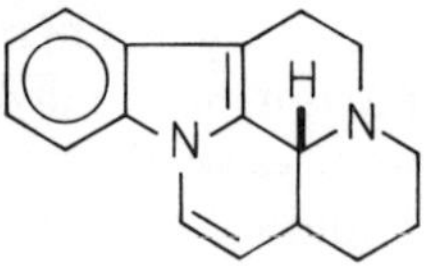

131

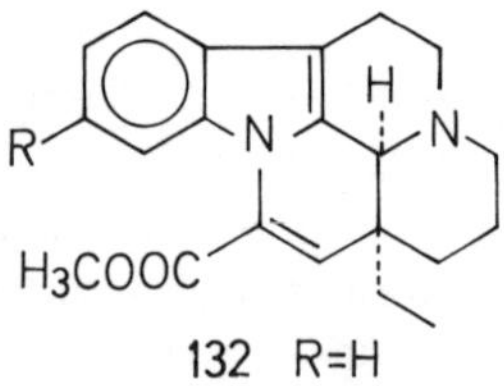

132 R=H
133 $R=OCH_3$

The structure and absolute configuration of vincamine has been determined by X-ray (*51*) analysis of the natural product and its hydrobromide methanolate. A comparison of both molecular structures shows that protonation and different crystal packing have no measurable effect on bond lengths and bond angles, but noticeable differences in torsion angles and consequently, in the finer details of the conformation, are observed.

The crystal data of vincamine are: space group $p2_12_12_1$, $z = 4$, $a = 7.708$, $b = 14.527$, $c = 16.776$ Å; vincamine hydrobromide methanolate: space group $p2_1$, $z = 2$, $a = 7.390$, $b = 17.843$, $c = 9.437$ Å, $p2$ 115.69°. The structures were solved from diffractometer data by the multi-solution and heavy-atom methods, and refined by least-squares techniques to $R = 0.042$ (1341 reflections) and 0.045 (2033 reflections). The results of the structural investigation confirm the earlier suggested configuration.

A. Cuanzine

The benzene extract of the root bark of *Voacanga chalotiana* collected in Angola (*52*) has been reported to contain a complex mixture of alkaloids. One of these is designated as cuanzine. The molecular ion peak at *m/e* 398.1842 proves cuanzine to possess the molecular formula $C_{22}H_{26}N_2O_5$ while the UV absorption at 226, 270, and 292 nm shows that this alkaloid contains a methoxyindole chromophore. The ^{1}H-NMR spectrum furnishes corroborative evidence for the nature of the substituent in the indole moiety by revealing a 1,2,3 hydrogen substitution pattern at $\delta = 6.44$, 7.19, and a singlet at $\delta = 3.40$ for a methoxy group. The presence of a carbomethoxy and of a hydroxyl group is secured by IR absorptions at 1746 and 3530 cm^{-1}, ^{1}H-NMR signals at $\delta = 3.33$ and 4.88, the last signal disappearing upon deuteration, and by the MS peaks at *m/e* 339 and 380 due respectively to the loss of $COOCH_3$ and H_2O. No acetyl derivative is obtained from cuanzine even under forcing conditions. The spectroscopic properties of the two products obtained by $NaBH_4$ reduction of cuanzine in THF suggested that the nitrogen of the indole moiety is substituted as indicated in the partial formula **134**. The product **135** derives from the reduction of the C—N linkage shown in part in structure **134**. Further reduction of **135** transforms the carbomethoxy group into a primary alcoholic function. The chemical

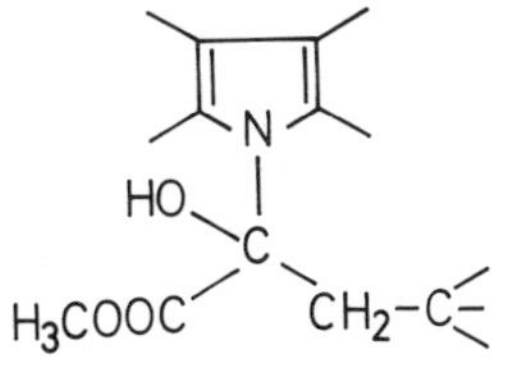

134

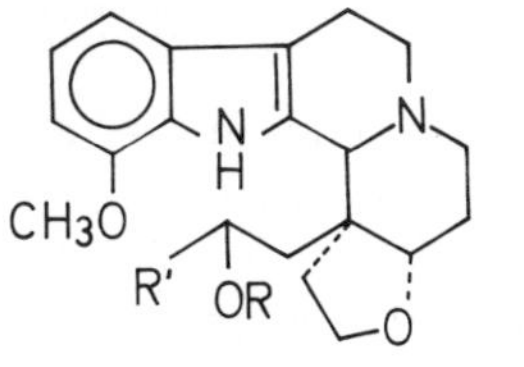

135

and spectroscopic evidence for the presence of **134** in cuanzine allows one to deduce that the new base possesses a vincamine-like structure, the remaining oxygen atom of cuanzine engaged in the formation of an ethereal linkage between C-18 and the D ring of the vincamine skeleton.

The mass spectrum of cuanzine is also compatible with the structure of **136** for this alkaloid. The most abundant fragment ions, in addition to M-18 and M-$COOCH_3$ already mentioned, correspond to the loss of 87 and 102 units of mass as in vincamine. In comparison with the fragmentation pattern exhibited by vincamine, the fragments derived from the elimination of the ethyl group at C-20 are obviously missing. However, cuanzine shows the presence of an intense ion at *m/e* 254, which is formed presumably through

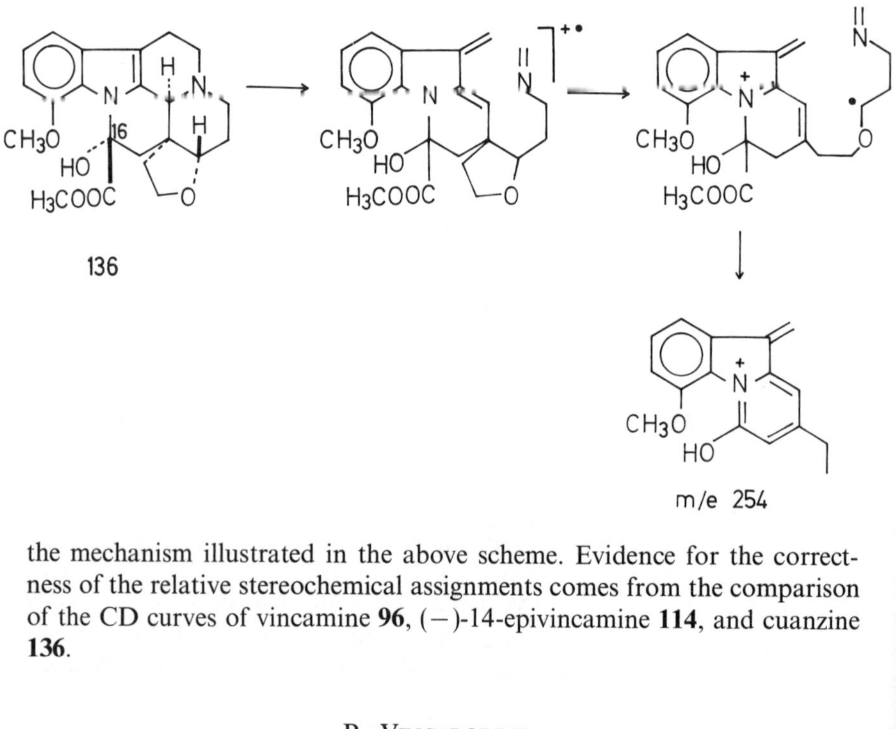

136

m/e 254

the mechanism illustrated in the above scheme. Evidence for the correctness of the relative stereochemical assignments comes from the comparison of the CD curves of vincamine **96**, (−)-14-epivincamine **114**, and cuanzine **136**.

B. VINCARODINE

Vincarodine (*53*) is one of the minor alkaloids in the leaves of *Vinca resea* L. (*Catharanthus roseus* G. Don). While the base, [mp 235° − 238° (dec.), $\alpha_D^{24} = -139°$ ($CHCl_3$), pK_a 5.9 (66% DMF)], was believed to be a dimer (*4*), examination of the spectral properties of the reisolated alkaloid and its derivatives indicate it to be a monomeric ($C_{22}H_{26}N_2O_5$) compound related to alkaloids of the vincamine type.

Acetylation of the alkaloid yielded a monoacetate **138** and LAH reduction a diol **139** which was acetylated to a diacetate **140**. These data, substantiated by elemental analyses, high resolution mass spectra, and IR, UV, and ^{1}H-NMR spectra demonstrated the presence of an aromatic methoxy group, a carbomethoxy unit, a secondary hydroxy function, and an ether methoxy group in vincarodine. The close relationship of the latter and vincine, apparent from the similarity of their IR and UV spectra, was of particular importance.

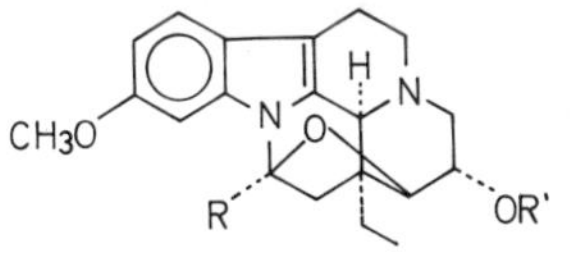

R=COOCH3, R'=H 137
R=COOCH3, R'=Ac 138
R=CH2OH, R'=H 139
R=CH2OAc, R'=Ac 140

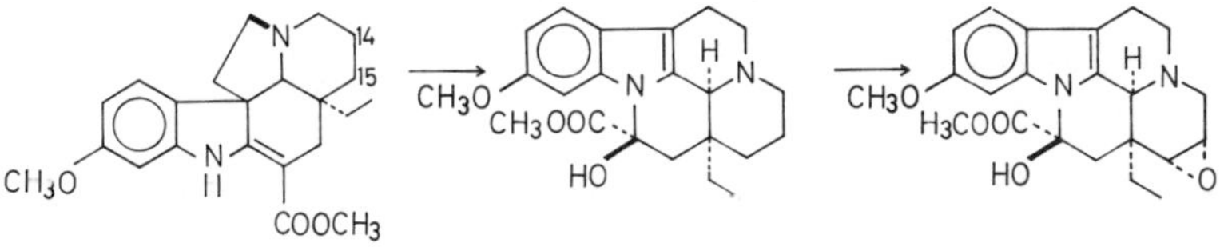

141

125

The location of the secondary hydroxy group and the nature of the oxide unit were deduced from analyses of the mass and NMR spectra of vincarodine and its derivatives. The high-resolution MS of the alkaloid exhibits several high intensity peaks, one of the most significant of them is the fragment *m/e* 296, reminiscent of the characteristic fragment from vincamine and vincine. Other peaks represent the loss of water and the carbomethoxy and ethyl groups as well as a more deep-seated fragmentation.

One of the most abundant peaks in the mass spectrum of vincarodinol **139** is ion **h**, the alcohol equivalent of fragment *m/e* 296, indicating that reduction of the ester function of the alkaloid did not modify a major fragmentation mode of the ring skeleton. In conjunction with considerations of the biogenesis of vincarodine, the mass spectral findings suggested formula **137** for the natural base. The 220 MHz–NMR spectra of vincarodine and its acetate **138** are consistent with the proposed structures in chemical shifts and geminal, vicinal and long-range coupling constants. The final solution to this problem came from X-ray analysis of vincarodine hydrobromide. Vincarodine hydrobromide ($C_{22}H_{26}N_2O_5 \cdot HBr$) crystallizes from isopropanol–ethanol, (mp 206°–209°) in the common orthorhombic system $p2_12_12_1$ with $a = 7.088$, $b = 11.89$, and $c = 31.02$ Å. The largest available single crystals were only 0.07 mm on an edge.

The structure was solved by standard heavy-atom methods and refined by full-matrix least-squares techniques. The final *R*-factor was 0.122 for the

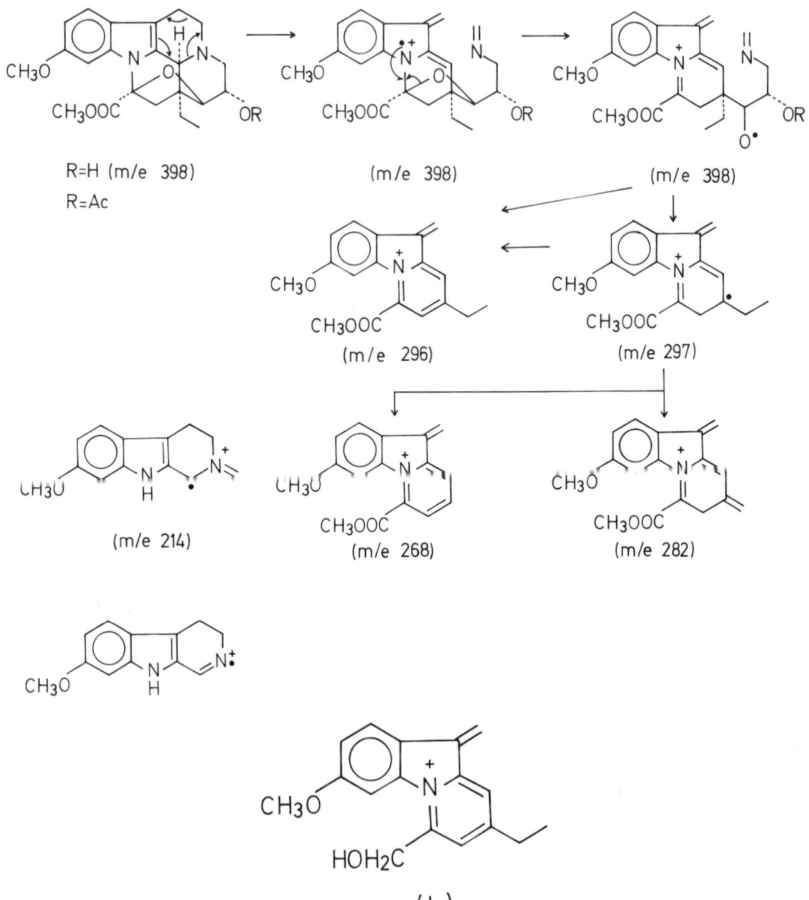

R=H (m/e 398)
R=Ac

(m/e 398)

(m/e 398)

(m/e 296)

(m/e 297)

(m/e 214)

(m/e 268)

(m/e 282)

(h)

observed reflections. The final structure and relative stereochemistry is shown in the following formula. Bond distances and angles, which were subject to the high standard deviations of 0.2 Å and 3 Å because of the limited data, generally agreed well with accepted values. No intermolecular distances less than 3.4 Å were found.

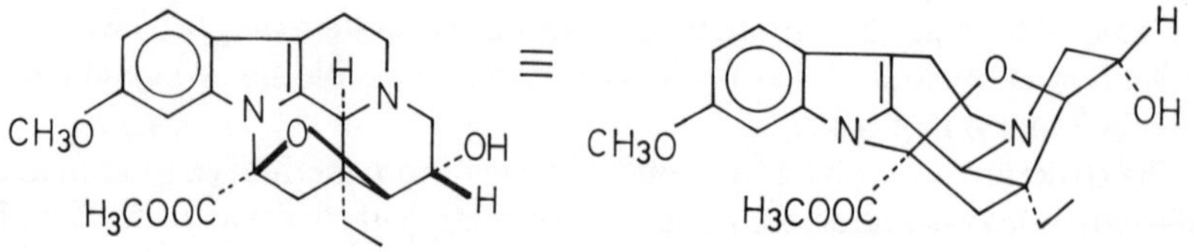

By analogy with the last stages of the biogenesis of vincine involving oxidative transformation of 11-methoxytabersonine **141**, the simplest explanation for the biogenetic origin of vincarodine requires the epoxidation of vincine followed by intramolecular displacement within the resultant 17,18-epoxyvincine.

C. UMBELLAMINE

Umbellamine, $C_{41}H_{48}N_4O_4$, a dimeric indole–indoline alkaloid **142** (*54*), was isolated from the root bark of *Hunteria umbellata* (k. Schum.) Hall. F.; it is probably identical with the alkaloid hunterine the structure of which is

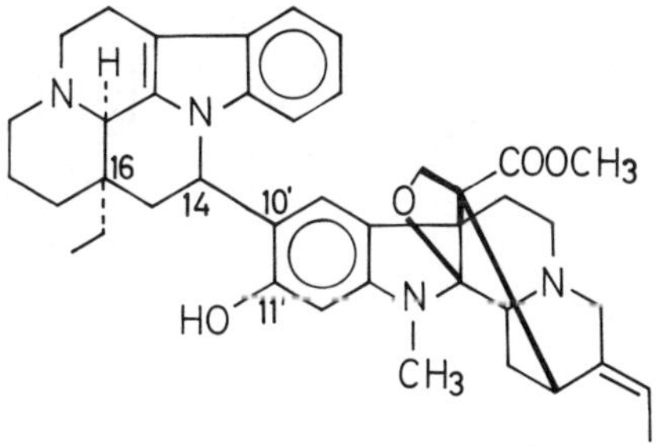

142

unknown. Thermolysis yielded (+)-eburnamenine, whereas a detailed mass spectrometric study of **142**, its *O*-methyl and *O*-acetyl derivatives, and the derived diol **143**, revealed the presence of a phenolic pseudoakuammingine group as a secondary part. The benzene nucleus of the latter is linked to the C-14 atom of a dihydroeburnamenine component.

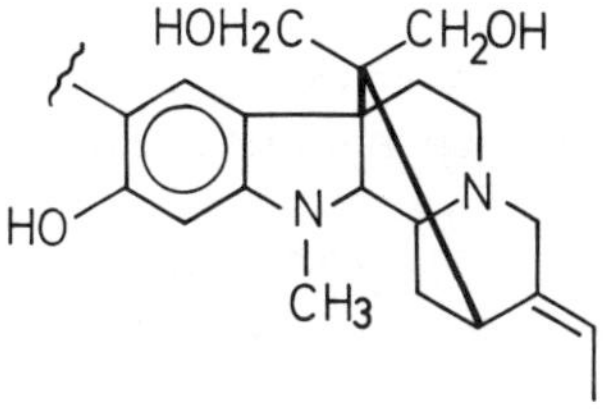

143

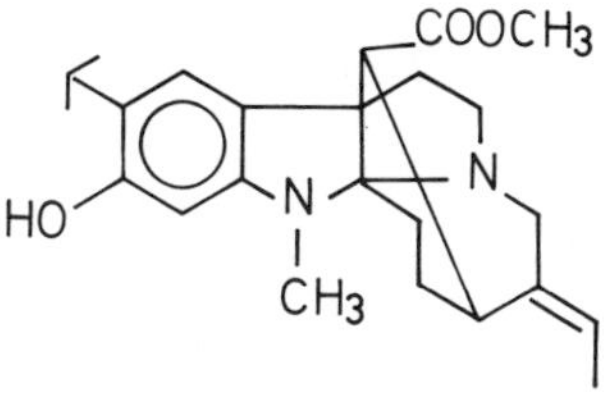

144

An NMR study of umbellamine (**142**) and its derivatives showed that the phenolic hydroxy group is either at C-10′ or C-11′ and that the dihydroeburnamenyl residue is connected to C-11′–C-10′ of the pseudoakuammingine residue. A choice between these alternatives was possible by investigating the hydroxylated base **144** arising from the pseudoakuammingine component, and obtained in small quantities by treatment of umbellamine with zinc or tin in acid. Its constitution, deduced by UV, NMR, and mass spectrometry, suggested that the hydroxy group is at C-11′. If this is correct, the dihydroeburnamenyl residue is linked to C-10′ of the pseudoakuammingine residue.

D. Craspidospermine

Craspidospermine **146**, $C_{22}H_{23}N_2O_4$, is an alkaloid of the same structural type. It occurs as an amorphous base in *Craspidospermum verticillatum* and its structure and stereochemistry was evident from spectroscopic investigations (*17*).

Partial synthesis of the alkaloid starts from 17,18-dehydrovincine **145** (*55*). Irradiation of **147** in the presence of oxygen and methylene blue for 6 hr also gave 35% craspidospermine (*56*) via a conjugated immonium compound. Vincamone **148** reacted similarly to give the corresponding 18,19-dehydrocompound **149**. Irradiation of vincamine **96** in acetone in the presence of oxygen for 18 hr gave 12% criocerine **149**.

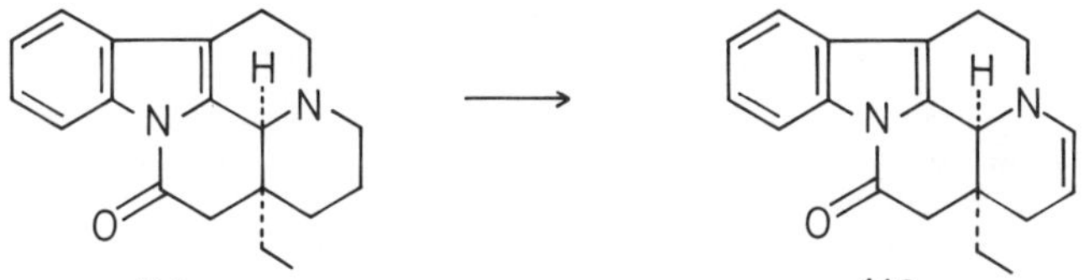

References

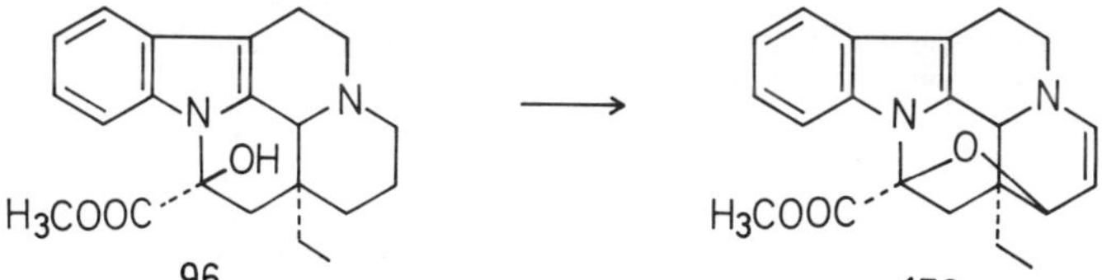

1. H. Pfäffli and H. Hauth, *Helv. Chim.Acta* **61**, 1682 (1978).
2. J. Le Men and W. I. Taylor, *Experientia* 508 (1965).
3. J. Trojanek and K. Blaha, *Lloydia* **29**, 149 (1966).
4. M. Hesse, "Indolalkaloide in Tabellen." Springer-Verlag, Berlin and New York, 1968.
5. W. Döpke, "Ergebnisse der Alkaloidchemie." Akademie-Verlag, Berlin, 1976.
6. M. Hesse, "Indolalkaloide in Tabellen." Springer-Verlag, Berlin and New York, 1964.
7. M. P. Cava, S. S. Tjoa, Q. A. Ahmed, and A. J. DaRoche, *J. Org. Chem.* **33**, 1055 (1968).
8. W. Döpke, H. Meisel, R. Gründemann, and G. Spiteller, *Tetrahedron Lett.* 1805 (1968).
9. W. Döpke, H. Meisel, and G. Spiteller, *Pharmazie* **23**, 99 (1968).
10. G. H. Svoboda, M. Gorman, A. I. Barnes, and A. I. Ulivier, *J. Pharm. Sci.* **51**, 518 (1962).
11. C. Kan-Fan, R. Besselievre, A. Cave, B. C. Das, and P. Potier, *C. R. Acad. Sci. Ser. C.* **272**, 1431 (1971).
12. A. Cave, A. Bonquet, and B. C. Das, *C. R. Acad. Sci. Ser. C.* **272**, 1367 (1971).
13. W. Döpke and H. Meisel, *Pharmazie* **21**, 444 (1966).
14. M. P. Cava, S. K. Talapatra, K. Nomuar, J. A. Weisbach, B. Douglas, and E. C. Shoop, *Chem. Ind.* 1242 (1963).
15. Y. Morita, M. Hesse, and H. Schmid, *Helv. Chim. Acta* **52**, 89 (1969).
16. M. P. Cava, S. S. Tjoa, Q. A. Ahmed, and A. J. Da Rocha, *J. Org. Chem.* **33**, 1055 (1968).
17. C. Kan-Fan, H. P. Husson, and P. Potier, *Bull. Soc. Chim. Fr.* 1227 (1976).
18. B. Gabetta, E. M. Martinelli, and G. Mustich, *Fitoterapia* **45**, 7 (1974).
19. V. Kovacik and I. Kompis, *Collect. Czech. Chem. Commun.* **34**, 2809 (1969).
20. L. Novak, J. Rohaly, and C. Szantay, *Heterocycles* **6**, 1149 (1977).
21. J. L. Herrmann, G. R. Kieczykowski, S. E. Normandin, and R. H. Schlesinger, *Tetrahedron Lett.* 1801 (1976).
22. A. Buzas, C. Herrison and G. Lavielle, *C. R. Acad. Sci. Ser. C* **283**, 763 (1976).
23. Y. Rolland, J. Garnier, J. Mahutean, and J. Poisson, *C. R. Acad. Sci. Ser. C* **285**, 583 (1977).
24. F. Klatte, U. Rosentreter, and E. Winterfeldt, *Angew. Chem.* **89**, 916 (1977).
25. G. Hugel, B. Gourdier, J. Levy, and J. Le Men, *Tetrahedron Lett.* 1597 (1974).
26. G. H. Gibson and J. E. Saxton, *Chem. Commun.* 799 (1969).
27. D. L. Coffen, D. A. Katonak, and F. Wong, *J. Am. Chem. Soc.* **96** 3966 (1974).
28. L. Castedo, J. Harley-Mason, and T. J. Leeny, *Chem. Commun.* 1186 (1968).

29. F. H. Imbert, C. Thal, H. P. Husson, and P. Potier, *Bull. Soc. Chim. Fr.* 2705 (1973).
30. F. H. Imbert, C. Thal, H. P. Husson, and P. Potier, *Bull. Soc. Chim. Fr.* 2013 (1973).
31. C. Thal, T. H. Imbert, H. P. Husson, and P. Potier, *Bull. Soc. Chim. Fr.* 2010 (1973).
32. L. Szpony and K. Szasz, *Arch. Exp. Pathol. Pharmakol.* **236**, 296 (1959); M. Aurrousseau, *Chim. Ther.* 221 (1971).
33. M. E. Kuehne, *J. Am. Chem. Soc.* **86** 2946 (1964).
34. K. H. Gibson and J. E. Saxton, *Chem. Commun.* 1490 (1969).
35. C. Szantay, L. Szabo, and G. Kalaus, *Tetrahedron Lett.* 191 (1973).
36. E. Wenkert and B. Wickberg, *J. Am. Chem. Soc.* **87**, 1580 (1965).
37. J. L. Herrmann, R. C. Cregge, J. E. Richman, C. L. Semmelhak, and R. H. Schlesinger, *J. Am. Chem. Soc.* **96**, 3702 (1974).
38. P. Pfäffli W. Oppolzer, R. Wenger, and H. Hauth, *Helv. Chim. Acta* **58**, 1131 (1975).
39. W. Oppolzer, H. Hauth, P. Pfäffli, and R. Wenger, *Helv. Chim. Acta* **60**, 1801 (1977).
40. G. Hugel, J. Levy, and J. Le Men, *C. R. Acad. Sci. Ser. C* **274**, 1350 (1972).
41. C. Szantay, L. Szabo, and G. Kalaus, *Tetrahedron* **33**, 1803 (1977).
42. C. Thal, Th. Severet, H. P. Husson, and P. Potier, *C. R. Acad. Sci. Ser. C* **275**, 1295 (1972).
43. L. Olivier, J. Levy, and J. Le Men, *C. R. Acad. Sci. Ser. C* **268**, 1443 (1969).
44. J. Le Men, C. Sigaut, G. Hugel, J. Le Men Olivier, and J. Levy, *Helv. Chim. Acta* **61**, 566 (1978).
45. G. Kalaus, L. Szabo, J. Horvath, and C. Szantay, *Heterocycles* **6**, 321 (1977).
46. J. Trojanek, Z. Koblicova, and K. Blaha, *Chem. Ind.* 1261 (1965).
47. J. Trojanek, Z. Koblicova, and K. Blaha, *Abh. Desch. Akad. Wiss. Berlin Kl. Chem. Geol. Biol.* **3**, 491 (1966).
48. J. Trojanek, Z. Koblicova, and K. Blaha, *Collect. Czech. Chem. Commun.* **33**, 2950 (1968).
49. K. Blaha, K. Kavkova, Z. Koblicova, and J. Trojanek, *Collect. Czech. Chem. Commun.* **33**, 3833 (1968).
50. K. Blaha, Z. Koblicova, and J. Trojanek, *Collect. Czech. Chem. Commun.* **34**, 690 (1969).
51. H. P. Weber and T. I. Petcher, *J. Chem. Soc. Perkin Trans. 2*, 2001 (1973).
52. E. Bombardelli, A. Bonati, B. Gabetta, E. M. Martinelli, G. Mustich, and B. Danieli, *Tetrahedron* **30**, 4141 (1974).
53. H. Neuss, E. Boaz, J. L. Occolowitz, E. Wenkert, F. M. Schell, P. Potier, C. Kan, M. M. Plat, and M. Plat, *Helv. Chim. Acta* **56**, 2660 (1973); J. P. Kutney, G. Cook, J. Cook, I. Itoh, J. Clardy, F. Fayos, P. Brown, and G. H. Swoboda, *Heterocycles* **2**, 73 (1974); N. Neuss, M. Gorman, N. J. Cone, and L. L. Huckstep, *Tetrahedron Lett.* 783 (1968).
54. Y. Morita, M. Hesse, and H. Schmid, *Helv. Chim. Acta* **52**, 89 (1969).
55. L. Chevolot, A. Husson, C. Kan-Fan, H. P. Husson, and P. Potier, *Bull. Soc. Chim. France* 1222 (1976).
56. R. Buehelnans, P. Herlem, H. P. Husson, F. Khuong-Huu, and M. T. Le Goff, *Tetrahedron Lett.* 435 (1976).

INDEX

B

C

D

E

P

Q

R

S

Y